Y0-BTB-652

NOUS·SOMMES·PRETS

SIMON FRASER UNIVERSITY
W.A.C. BENNETT LIBRARY

VM 365 T43 2003

TECHNOLOGY AND APPLICATIONS OF AUTONOMOUS UNDERWATER VEHICLES

OCEAN SCIENCE AND TECHNOLOGY
A series of books in oceanology
ISSN 1561-5928

Volume 1
CHEMICAL SENSORS IN OCEANOGRAPHY
Edited by Mark S. Varney
ISBN 90-5699-255-4

Volume 2
TECHNOLOGY AND APPLICATIONS OF AUTONOMOUS UNDERWATER VEHICLES
Edited by Gwyn Griffiths
ISBN 0-415-30154-8

Also available from Taylor & Francis
OCEANOGRAPHY AND MARINE BIOLOGY: AN ANNUAL REVIEW
Edited by R.N. Gibson, Margaret Barnes and R.J.A. Atkinson
Founder Editor - Harold Barnes
ISSN 0078-3218

TECHNOLOGY AND APPLICATIONS OF AUTONOMOUS UNDERWATER VEHICLES

Edited by
Gwyn Griffiths

Taylor & Francis
Taylor & Francis Group

LONDON AND NEW YORK

First published 2003
by Taylor & Francis
11 New Fetter Lane, London EC4P 4EE

Simultaneously published in the USA and Canada
by Taylor & Francis Inc,
29 West 35th Street, New York, NY 10001

Taylor & Francis is an imprint of the Taylor & Francis Group

© 2003 Taylor & Francis

Typeset in New Baskerville by
Newgen Imaging Systems (P) Ltd, Chennai, India
Printed and bound in Great Britain by
TJ International, Padstow, Cornwall

All rights reserved. No part of this book may be reprinted or
reproduced or utilized in any form or by any electronic,
mechanical, or other means, now known or hereafter
invented, including photocopying and recording, or in any
information storage or retrieval system, without permission in
writing from the publishers.

Every effort has been made to ensure that the advice and information in
this book is true and accurate at the time of going to press. However,
neither the publisher nor the authors can accept any legal responsibility
or liability for any errors or omissions that may be made. In the case of
drug administration, any medical procedure or the use of technical
equipment mentioned within this book, you are strongly advised to
consult the manufacturer's guidelines.

British Library Cataloguing in Publication Data
A catalogue record for this book is available from the British Library

Library of Congress Cataloging in Publication Data
A catalog record for this book has been requested

ISBN 0-415-30154-8
ISSN 1561-5928

CONTENTS

CONTRIBUTORS

Suleiman M. Abu Sharkh
School of Engineering Sciences,
University of Southampton, UK

Mikhail D. Ageev
Institute for Marine Technology
Problems, Russian Academy of Sciences,
Russia

Autosub Science and Engineering
teams, UK and USA

Anders Bjerrum
Maridan AS, Denmark

D. Richard Blidberg
Autonomous Undersea Systems
Institute, Durham, USA

Attilio Brighenti
SATE s.r.l, Italy

E.D. Brown
Cardiff University, UK

Peter H. Burkill
Southampton Oceanography Centre,
University of Southampton

Simon Corfield
QinetiQ, UK

Alex Cunningham
Department of Physics and Applied
Physics, University of Strathclyde, UK

Edwin F.S. Danson
Edwin Danson Associates, UK

Russ E. Davis
Scripps Institution of Oceanography,
USA

Charles C. Eriksen
School of Oceanography, University of
Washington, USA

Jonathan Evans
Ocean Systems Laboratory, Heriot-Watt
University, UK

James S. Ferguson
International Submarine Engineering
Ltd, Canada

Robin Galletti di Cadilhac
SATE s.r.l, Italy

Derek Graham
QinetiQ, South Arm, Rosyth Royal
Dockyard, UK

Gwyn Griffiths
Southampton Oceanography Centre,
University of Southampton, UK

Christopher Hillenbrand
Naval Undersea Warfare Center,
USA

Bjørn Jalving
Norwegian Defence Research Institute,
Norway

Clayton P. Jones
Webb Research Corporation, USA

David Lane
Ocean Systems Laboratory, Heriot-Watt
University, UK

Lawrence C. Langebrake
University of South Florida, USA

Nicholas W. Millard
Southampton Oceanography Centre,
University of Southampton, UK

Roland Rogers
QinetiQ, UK

Peter Stevenson
Southampton Oceanography Centre,
University of Southampton, UK

Nils Storkersen
Norwegian Defence Research Institute,
Norway

Karstein Vestgård
Kongsberg Simrad AS, Norway

Andrew T. Webb
Southampton Oceanography Centre,
University of Southampton, UK

John D. Woods
Imperial College London, UK

FOREWORD

This monograph documents the transition from AUV dream to reality. Today they are being used in scientific research, defence, surveying and industry. The dream began back in the early 1980s, when far-sighted engineers worked out how it would be possible to create unmanned vehicles that could perform many of the tasks then undertaken by manned ships and submersibles. The prospect of using mobile robotic systems in the deep ocean to complement satellites in space was immediately attractive. The potential benefits included cost reduction, and the ability to operate at sites that can never be reached by ship-based systems (e.g. under floating glaciers), or where ship operations are slow or dangerous, notably in sea ice and rough weather.

In 1985 Brian McCartney introduced me to the potential of AUVs for oceanographic research with the promise of making more effective use of costly research vessels. After detailed briefings on the engineering challenge and risks, I decided to advise the UK Natural Environment Research Council to back the *Autosub* project. From my perspective as Director of Marine and Atmospheric Sciences, *Autosub* would solve some major problems confronting oceanographers. First was the high cost of research vessels. Ships will always be needed to support scientific exploration of the ocean; so much remains to be discovered, and nothing can replace the scientist on board when plans need to be changed in the light of an unexpected discovery. But, at the end of the twentieth century, oceanography was also moving into operational monitoring, which could best be performed by AUVs.

With hindsight, it is remarkable that the greatest of all oceanographic projects, the World Ocean Circulation Experiment (WOCE 1990–97) was completed successfully with research ships stopping every 50 km along prescribed trans-ocean lines to lower instruments and sampling bottles to the ocean floor. Over 24,000 such hydrographic stations were made during WOCE at immense cost in ships and scientists' time at sea. If we were planning WOCE in the twenty-first century, we would use AUVs to perform the task, exploiting *Autosub* and the like to make hydrographic profiles to full ocean depth along inter-continental sections.

There will probably never be another scientific project like WOCE, but operational global models depend critically on the assimilation of routine observations, like those collected for global weather prediction. Various schemes are being assessed, including the ARGO array of drifting buoys. Such observations are inevitably sparse; the mean spacing between floats may be ten times the width of the Gulf Stream. The operational role of AUVs in ocean prediction will probably be for adaptive sampling. Real-time diagnosis of an ensemble forecast reveals where gaps in observations are limiting its skill. AUVs on standby in mid ocean, or at sensitive locations such as the continental slopes, or overflow regions, will be deployed to fill these gaps under instruction over satellite. Mobility under remote command will permit an affordable observing system for operational oceanography, providing as a by-product the raw data needed to improve our understanding of the ocean. A fleet of AUVs will provide the affordable scalability needed to cope with increasing demand for operational observations.

Many other, equally exciting applications are featured in this monograph. So too are the daunting engineering problems that have to be solved in making the dream become reality. The progress reported here is impressive, and augurs well for the future.

John Woods
Imperial College
London, 2002

INTRODUCTION

The technology of Autonomous Underwater Vehicles (AUVs) has formed the subject of very many conferences over the last few decades. Well established AUV conference series, such as those organised by the Autonomous Undersea Systems Institute and the Institute of Electrical and Electronic Engineering in the United States, have recently been joined by others such as the Unmanned Underwater Vehicle Showcase in the United Kingdom. Many of the breakthroughs in AUV technology have been reported at such conferences and papers from their Proceedings contribute to the bibliography of this book. In contrast, reports on the applications of AUV technology are distributed throughout the literature. Papers are emerging in the peer-reviewed scientific literature, in professional journals and on the web sites of companies and institutions.

This book brings together technology and applications in a series of 17 chapters written by leaders in their fields. Despite their scope, these chapters obviously cannot cover the entire breadth of recent developments in AUV design, construction and operation. However, they do provide either more detail or breadth that is usually possible in conference or journal papers. The emphasis is on civilian applications of AUVs. However, recognising that the driving force behind many developments lies with the military, one chapter reviews defence applications.

In Chapter 1, Evans and Lane describe the issues and applications of using hardware-in-the-loop simulation (HILS) when designing and integrating AUV sub-systems. The range and characteristics of sub-systems within complex ocean robots mean that not all can be simulated in real-time. This requires some form of synchronisation between those systems that can run in real-time, whether they are real or simulated, with those that cannot. The authors describe methods to achieve synchronisation that avoid the pitfalls of earlier strategies such as time-slicing. Five examples of how HILS was applied to different projects are discussed ranging from the simulation of a simple sensor to a simulated mission task involving concurrent mapping and localisation.

In many respects the current generation of AUVs are limited in their performance because of constraints on the energy they can carry. Abu Sharkh, Griffiths and Webb in Chapter 2 discuss power sources for AUVs. They review the performance of existing secondary batteries and point to some of the new types of cell that may prove attractive in the future. Packaging emerges as an important issue. The very high intrinsic energy densities of high temperature cells are difficult to achieve in practice. In contrast, Lithium–polymer cells look particularly attractive due to their high energy density, low density and tolerance of being operated at ambient pressures. While fuel cells may provide the energy needed for AUVs in the longer term, technical and safety issues continue to challenge designers. The authors review briefly electromechanical energy storage devices or flywheels. While these may have a role in providing short bursts of very high power, they do not look attractive for long endurance AUV missions.

The second part of the chapter concentrates on primary batteries, in particular the manganese alkaline cell that is used in a number of AUVs. A simple model is derived for the behaviour of the cell's terminal voltage and capacity under different loads and at different temperatures. This model is tested in the form of a fuel gauge algorithm against data from a manganese alkaline pack used on the Autosub AUV.

In Chapter 3, Davis, Eriksen and Jones focus on underwater gliders, otherwise known as buoyancy-driven vehicles, designed for ranges in excess of 3,000 km. Instead of using propellers, these vehicles achieve their forward velocity by converting the force generated by changing their volume and buoyancy into lift on wings. For three of the four gliders described the change in volume is achieved using a high-pressure pump. The fourth glider design uses the volume change associated with melting a material with a freezing point in the range of ocean temperatures. The authors discuss the advantages and disadvantages of different forms of buoyancy generation, the hydrodynamic performance of different hull shapes and the challenges of communication, especially maintaining the antennas clear of the surface. For these micro-power vehicles sensors pose two sets of problems: ensuring low power consumption and low impact on the hydrodynamic drag.

The authors describe and illustrate two classes of glider missions. The first class is the so-called virtual mooring, where the vehicle maintains station but spirals in the vertical to provide profiles of water column properties at a 'fixed' location. The second class has the vehicles gliding along a transect executing a sawtooth pattern in the vertical.

Another innovative approach to providing long endurance is the subject of Chapter 4 by Blidberg and Ageev. Using solar energy to recharge the AUV's batteries allows the range of a vehicle to be increased dramatically. To achieve this, the vehicle must surface daily for recharging and a significant part of the daytime must be spent on the surface. Factors that affect the conversion efficiency of solar radiation to electrical energy and to propulsion effort are discussed, as are the hydrostatics and hydrodynamics of a vehicle with a rather unusual shape. Typical mission scenarios are translated into operating schedules that meet the ocean sampling and the energy management and recharging requirements. Long-term biofouling of the photovoltaic arrays is an issue that is discussed, as are the effects of temperature and surface waves. The chapter concludes with a description of how the control system of the solar-powered AUV was designed using the CADCON tool, a flexible and open simulation environment.

Most AUVs feature advanced materials in their construction to deliver the required depth and range performance. In Chapter 5, Stevenson and Graham describe the influence of materials on the structural design of AUVs. In structural design, the goal is to combine high strength and stiffness with minimum weight at an affordable cost. While fibre composites perform well due to their low density, they have anisotropic properties. There are also difficulties in manufacture of thick section composites. The chapter explores in detail how these material properties affect the design of pressure vessels as unstiffened monocoques and as ring-stiffened cylinders. The use of ceramic pressure vessels is discussed, as is the provision of extra buoyancy in the form of closed cell and syntactic foam. Significant changes in material properties can take place over the range of pressure and temperature seen by a deep-diving AUV. These require to be established through measurement, as they exhibit dependence on shape and construction.

In Chapter 6, Galletti and Brighenti describe a general approach to designing docking systems for AUVs. Docking has emerged as a key supporting technology for many of the applications foreseen for AUVs and related vehicles in the offshore industry. Vehicles that use underwater docks to recharge their power sources, to upload data and download new mission plans become independent of surface support. This removes a major risk in AUV operations – recovery to a surface platform. The authors take a systems design approach to the problems of docking systems, beginning with analysing mission requirements, moving through concept generation and development of the most promising solution. The technical issues cluster around four topics: homing, docking, garaging and launch and recovery. The Chapter discusses the merits of several options for each of these topics, with examples of systems that have been used in the ocean.

The majority of AUVs use propulsion based on electric motors and screw propellers. In Chapter 7, Abu Sharkh reviews the theoretical background of propeller action relating motor torque and power requirements to its basic dimensions, output thrust and output efficiency. These mathematical models of propeller design are compared to test data to illustrate the effect of design parameters including blade length, blade thickness, hub diameter and rotation rate. As the author makes clear, the torque and power requirements of the propeller are in conflict with each other. Optimisation strategies are discussed that take into account the size, cost and performance of all of the components within the propulsion system. The Chapter concludes with a brief review of electric motor technology suitable for use in AUVs.

Danson, in Chapter 8, reviews the emerging roles of AUVs in the offshore industry and examines the economics of their use. With towed systems, especially when deep-towed, there are a number of limitations that lead to high cost. These include limited position accuracy when running lines, long and slow end of line turns, long run-ins and run-outs and slow speed. Danson shows that the AUV overcomes all of these limitations. He then goes on to compare the time taken on various survey designs for a conventional deep-towed system and for an AUV. The chapter provides an analysis of the features desired by offshore operators of AUVs against the specification of leading vehicles and looks to the development of combined remotely operated and autonomous vehicles capable of carrying out intervention tasks.

Chapter 9 describes a range of AUV applications in ocean science. It is based on the missions of the Autosub vehicle during the Natural Environment Research Council 'Autosub Science Missions' programme (1998–2001) and the work of eighteen scientists and nine engineers. The chapter opens with a description of the programme, how it was conceived and how it was executed. Experience has shown that this form of focussed programme was an effective mechanism for introducing a new technology to a broad marine science community involved in basic and applied research.

On a large AUV such as Autosub several different sensors can be deployed to study processes. In one example, sidescan sonar, turbulence probes and an acoustic air-bubble resonator were combined to provide simultaneous measurements of dissipation, breaking waves, bubble clouds and Langmuir circulation in the upper ocean. In another example, several different types of acoustic velocimeters were deployed in a terrain-following mission to study the spatial and temporal variations of flow and turbulence over sandbanks.

The AUV has proven to be an excellent platform for quantitative acoustic measurements. In an elegant experiment in the North Sea the hypothesis that fish do not avoid the fisheries research vessel *Scotia* was tested. The same instrumentation was used to determine if there was a difference in the population density of krill in open waters and beneath sea ice in Antarctic waters and to measure ice draft and hence ice thickness.

Chapter 10 is an overview of historical and current aspects of the naval use of autonomous underwater vehicles. Corfield and Hillenbrand put the need for military advantage from technology into a historical context, and describe the trend to take humans out of areas of danger. Their vision goes beyond taking man out of the water to taking man out of the mission execution loop. For this to happen, vehicles will need to have a far higher degree of sensing and autonomous decision making capacity than today's vehicles. The authors go on to explore the defensive and offensive roles for AUVs and conclude that the key sub-systems of fail-safe target identification, target motion analysis and weapon fire control will require significant long-term research and development. The chapter closes with an examination of the trends in AUV energy and propulsion systems, control systems, launch and recovery, navigation and communications from a military perspective.

In Chapter 11, Jalving, Vestgård and Stokersen discuss the application of AUVs to detailed seabed surveys. The enabling technology is described, based on the systems used within the HUGIN family of vehicles. The authors focus on ensuring that accurate digital terrain maps result from AUV surveys. A thorough and detailed analysis is provided of the factors that affect the accuracy of depths and position in the final maps for various combinations of sensors and systems. Following brief descriptions of the development and field experience with the HUGIN vehicles, several examples of bathymetric maps and computer visualisations of data are provided. These illustrate a number of applications: pipeline route survey, cold-water coral reef environmental survey, proof of concept for fisheries stock assessment, mine-hunting demonstration and oil and gas field bathymetric surveys.

Bjerrum, in Chapter 12, describes the MARTIN vehicle and how it has been used to survey complex environments. After reviewing the design of the MARTIN family of vehicles, the author discusses the attributes of the vehicle control system, in particular their choice of control system and computer architecture. In adopting the CAN bus, the designers adopted a decentralised computer network, which has proven reliable and flexible. Accurate navigation is a prime requirement for a vehicle operating in complex environments, and the MARPOS Doppler-aided Inertial Navigation System was developed specifically for this AUV. Combining a ring-laser gyrocompass with an inertial platform, an acoustic Doppler velocity log and DGPS satellite position data using a Kalman filter has provided a robust and accurate system. Application areas for the vehicle have included: archaeological surveys; mine reconnaissance; underwater crash site investigations and seabed surveys in underwater diamond provinces.

Using AUVs as cargo carriers is an application that shows great promise. In Chapter 13, Ferguson describes the *Theseus* AUV, which was specifically designed to lay optical fibre cable onto the seabed. With a cargo capacity of 660 kg the vehicle can lay a cable of up to 220 km in length. The design considerations for a cargo carrying AUV are discussed, including managing the change in buoyancy as the payload is dispensed. The second part of the Chapter describes trials of the vehicle and its systems and an expedition to the Canadian Arctic to lay a 175 km fibre optic cable beneath permanent pack ice.

Semi-submersible AUVs were first developed to improve the coverage efficiency of hydrographic surveys. They combine several of the desirable attributes of surface vessels with a number of features of the true autonomous underwater vehicle. Ferguson begins Chapter 14 with a history of a family of semi-submersible AUVs before describing the principle of operation, the motive power and the advantages that follow from having a surface piercing mast. The factors affecting static and dynamic stability are considered, as are buoyancy control and communication. With their substantial motive power, semi-submersibles are capable of towing subsidiary vehicles. However, the dynamics of towing affect substantially the stability of a nearly neutral buoyancy vehicle. The issues of towcable tension, downforce, torque and pitch control are discussed.

Chapter 15 reviews the provision of sensors for AUVs. Langebrake begins by classifying sensors into two groups: those providing situational information for the vehicle and those providing the data for the user. Concentrating on sensors for scientific research, the Chapter explores the many interrelated design variables for sensors to be used in low-power, autonomous applications. In many cases, commercially available sensors can be used directly, or by adapting them to an AUV. Nevertheless, the attributes of an AUV can call for sensors to be especially designed. Several examples of such sensors are given, spanning biological, chemical and physical applications. The Chapter concludes by looking to the introduction of sensors based on micro-electromechanical systems that combine low-power, small size and ruggedness with the potential to fabricate complex autonomous systems for biological and chemical analysis.

In Chapter 16, Griffiths, Millard and Rogers examine the procedures and risks that concern the operation of AUVs. Various modes of operation and levels of supervision are considered and suggestions for tackling key issues are given. The technical risk to an AUV is examined using reliability data from the Autosub vehicle. Experience with Autosub is also called upon to illustrate the risks to personnel and how they can be controlled. The chapter closes with a discussion on operational risk for different scenarios and a note on liability and insurance issues.

The final chapter, by Brown, provides a concise treaties on the public law aspects of the legal regime governing the operation of AUVs. After introducing the international legal framework and jurisdiction over the various maritime zones the consent process is discussed in detail. The important issue of whether an AUV is a ship is raised, which leads to a discussion of how clauses in the UNCLOS convention may apply to AUVs. As the law on AUVs remains uncertain and, as a consequence, disputes may occur, the rules on settling disputes are introduced. Brown documents the wide variation in national law and state practice, with a number of examples, before considering the special case of operating in waters subject to the Antarctic Treaty System.

It is inevitable that the technology and applications of AUVs will develop further over time. The examples in this book can only show the breadth of their utility in the opening years of this century, although the authors do provide glimpses into the future for vehicle technology and for their uses. Despite this excitement, pervasive use of AUVs will require significant advances in technology to deliver the combination of cost reduction and increase in performance that has led to such radical and disruptive change in other industries. There is tremendous scope for further innovation beyond the scope of the topics discussed here. It is to the next generation of engineers and scientists in academia, industry and government laboratories that we look to deliver those advances. This book will have succeeded if it contributed in some small part to the stimulation of new ideas on how to design, construct or use Autonomous Underwater Vehicles.

Gwyn Griffiths
Southampton Oceanography Centre
Southampton, 2002

ABBREVIATIONS AND ACRONYMS

ABE	Autonomous Benthic Explorer
ADCP	Acoustic Doppler Current Profiler
ADV	Acoustic Doppler Velocimeter
Ag–Zn	Silver Zinc (batteries)
AINS	Aided Inertial Navigation System
ALACE	Autonomous Lagrangian Circulation Explorers
amu	atomic mass units
ASW	Anti Submarine Warfare
AUSI	Autonomous Undersea Systems Institute
AUV	Autonomous Underwater Vehicle
CADCON	Co-operative AUV Development Concept
CASI	Compact Airborne Spectral Imager
CCAMLR	Convention for the Conservation of Antarctic Marine Living Resources
CCD	Charge Coupled Device
CEE	Comprehensive Environmental Evaluation
CFRP	Carbon Fibre Reinforced Plastic
CFS	Canadian Forces Station
CML	Concurrent Mapping and Localisation
COLREG	Convention on the International Regulations for Preventing Collisions at Sea
CORESIM	Core Simulation engine
COTS	Commercial off the shelf
CTD	Conductivity, Temperature Depth (instrument)
CW	Continuous Wave
DGPS	Differential Global Positioning System
DoD	US Department of Defence
DTM	Digital Terrain Model
DVL	Doppler Velocity Log
EEZ	Exclusive Economic Zone
ehp	effective horsepower
EM	Electro Magnetic
emf	electromotive force
EU	European Union
EV	Electric Vehicle

G-D States   Geographically Disadvantaged States
GFRP   Glass Fibre Reinforced Plastic
GPS   Global Positioning System
HF   High Frequency
HILS   Hardware-in-the-loop Simulation
HLA   High Level Architechture
hp   horsepower
IEE   Initial Environmental Evaluation
IMO   International Maritime Organisation
IMU   Inertial Measurement Unit
INU   Inertial Navigation Unit
INS   Inertial Navigation System
IOC   Intergovernmental Oceanographic Commission
kWh   kilowatt hour (equivalent to 3.6 MJ)
L/D   length to diameter ratio
LBL   Long Base Line (navigation)
LEOS   Low Earth Orbiting Satellites
L-L States   Land-locked States
MCM   Mine Counter Measures
MEMS   Micro Electro-mechanical systems
MFSK   Multiple Frequency Shift Keying
MIT   Massachusetts Institute of Technology
MSR   Marine Scientific Research
NERC   Natural Environment Research Council
n mile   nautical mile
NTN   Non-traditional Navigation
OAS   Obstacle Avoidance Sonar
ODAS   Ocean Data Acquisition System
ONR   US Office of Naval Research
PAR   Photosynthetically Available Radiation
PEM   Proton Exchange Membrane
php   propeller horsepower
POM   polymethylene oxide
ppb   parts per billion
PV   Photo Voltaic
RF   Radio Frequency
ROV   Remotely Operated Vehicle
SATCOM   Satellite Communications
SAUV   Solar Autonomous Underwater Vehicle
shp   shaft horsepower
SOLAS   Safety of Life at Sea (convention)
SSBL   Super-Short Baseline (navigation)
TCP/IP   Transmission Control Protocol/Internet Protocol
UNCLOS   United Nations Convention on the Law of the Sea
USB(L)   Ultra Short Baseline (navigation)
UTC   Universal Time Co-ordinated
UUV   Unmanned Underwater Vehicle
VR-Nav   Virtual Reality navigation

1. DISTRIBUTED HARDWARE-IN-THE-LOOP SIMULATION FOR UNMANNED UNDERWATER VEHICLE DEVELOPMENT – ISSUES AND APPLICATIONS

JONATHAN EVANS and DAVID LANE

Ocean Systems Laboratory, Heriot-Watt University, Edinburgh, Scotland, EH14 4AS, UK

1.1 INTRODUCTION

Development and integration of subsystems on advanced robots, such as unmanned underwater vehicles (UUV), can benefit from the availability of a hardware-in-the-loop simulation (HILS) facility. Although complete interoperability of simulated and real subsystems appears desirable, substantial additional complexity of data flows and hardware can be introduced. Where non-real time simulations are involved, methods of synchronising subsystems running at different speeds must be employed. These should take account of the realities of starting and stopping real subsystems. This chapter reviews some of these issues and presents CORESIM – a distributed HILS system based around High Level Architecture (HLA) (DoD, 1998). The system has been used to evaluate a time-slicing synchronisation approach, and to assist in the development of docking, servo control using visual feedback and concurrent mapping and localisation systems for autonomous underwater vehicles.

New developments in UUVs and subsea robots are leading to increasingly sophisticated systems. By necessity such complex systems must be broken into subsystems to make the engineering and supervision of the whole more manageable. These can include mechanical hardware, electronics and software for propulsion, manipulation, navigation, control, communications and the user interface (see Figure 1.1). The development cycle of each subsystem and the interdependency on other subsystems varies from component to component. The development process can be further complicated by subsystem reuse and geographical separation of development teams – perhaps even in separate organisations.

There are many advantages for design, integration and testing in having a HILS system to mix simulations and real subsystems and have them interact. However, without substantial computing resources, some simulations may not run in real-time. Furthermore, both virtual and real worlds will be involved for subsystems such as manipulation and propulsion (e.g. Jefferson, 1985; Frangos, 1990; Brutzman *et al.*, 1992; Alles *et al.*, 1994; Kuroda *et al.*, 1996; Lane *et al.*, 1998). Our motivation is to combine simulations and real hardware that operate together to:

- test and integrate remote subsystems;
- aid more rapid migration onto real hardware;
- evaluate work-tasks *a priori*;

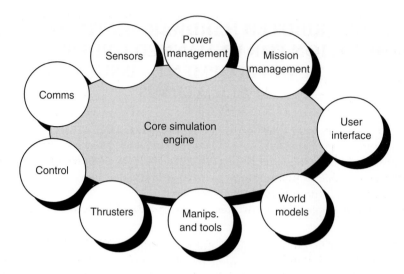

Figure 1.1   Typical unmanned underwater robot subsystems.

- replay missions *a posteriori*;
- enable remote presence;
- facilitate downstream simulation based design and acquisition.

This chapter examines issues of interoperability and synchronisation where real- and logical-time subsystems are concurrently mixed with real and virtual worlds. It discusses an architecture designed to investigate and develop solutions to these 'hybrid simulation' problems. The architecture, called CORESIM (core simulator), is an open framework for integrating vehicle subsystems and immersing them in a simulated environment.

The architecture has been applied to Autonomous Underwater Vehicle (AUV) and Remotely Operated Vehicle (ROV) applications, but the approach is equally applicable in the aerospace, manufacturing, nuclear, space, agricultural, food processing and construction industries.

1.2   CORESIM: AN ARCHITECTURE FOR INTEROPERABLE HILS

HILS is the operation of a hardware device or system with its inputs and outputs coupled to a computer simulation instead of its real working environment. The aerospace industry was among the first to develop the technique as a means of testing flight control systems, although applications have since diversified (e.g. Alles *et al.*, 1994; Maclay, 1997). In the subsea domain, HILS has been used to study control and visualisation (Brutzman *et al.*, 1992) and to augment available vehicle sensor data (Kuroda *et al.*, 1996) and for subsystem testing (Dunn *et al.*, 1994).

Although superficially similar to straightforward HILS, the ability to mix simulated and real subsystems introduces difficulties where sensors, actuators and a surrounding environment all interact. Specifically:

- Where real-time operation is precluded (by simulation complexity or by communication overheads due to the distribution of the simulations – perhaps at remote locations using the Internet) some means of synchronising real and non real-time (logical time) throughout the simulation is required.
- Maintaining interoperability as simulations and real subsystems are exchanged requires changes in data flows, which must be invisible from within subsystems.
- The interoperable-simulation must maintain a simulated state of the world, which is calibrated with the real world.
- Data must remain consistent throughout the distributed system during operation.

Figure 1.2 presents a logical view of the relationship between simulated and physical subsystems in a hybrid simulation. To the left are simulated sensors, actuators and an environment for initial testing and validation. In the centre are prototype physical sensors, actuators and an environment for initial HIL subsystem testing. To the right lie the real sensors and actuators, as deployed within the final system. CORESIM employs a common communications spine infrastructure. Initially, all simulations are employed. As real subsystems are integrated and tested, so simulations are removed, until the real robot is realised.

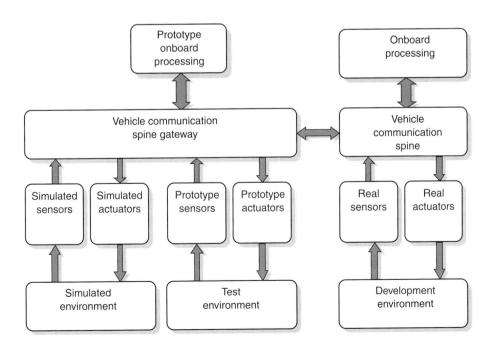

Figure 1.2 Hybrid HILS approach.

1.3   SEEKING INTEROPERABILITY

Ideally, we wish complete interoperability of simulations and real subsystems, in a true plug-and-play approach. However, not all combinations are possible or even useful (Lane *et al.*, 1998, 2000a). We examine these in further detail.

1.3.1   Real-sensor/Simulated Actuator

Consider the case where a selection of real sensors (and signal processing) have been integrated, but (some of) the robot actuators (and control systems) remain simulated. For proprioceptive sensors (e.g. compass, Doppler velocity log and inertial navigation system) to behave consistently with the simulated vehicle dynamics, they must be mounted on a motion platform and animated with the output of the vehicle model. This requires additional hardware (e.g. a Cartesian robot mounted over a test tank) with sufficient motion bandwidth to accommodate characteristic movements and software integrated within the communication spine.

The same applies to exterioceptive sensors (e.g. sonar, video). However, in addition, the sensed real world must remain consistent with the virtual world in which the dynamic simulation moves. This requires either complete *a priori* knowledge of the real world (i.e. *not* unstructured) or some additional sensing and processing to update the virtual world to remain consistent.

In both cases, the pure plug-and-play HIL architecture must be modified to include additional data flows between processes and hardware (see Figure 1.3). Some of these only exist to support the HILS aspects, and will not form part of the final in-water robot. Since the complexity of these systems can be equal to, or surpass, that of the final robot itself, care must be taken in specifying the accuracy requirements of the simulation, and careful consideration must be given to the practical (and useful) combinations of simulated and real subsystems.

1.3.2   Simulated Sensor/Real Actuator

Consider now the inverse case, where proprioceptive and exterioceptive sensors are simulated, but some of the real robot actuators have been integrated (e.g. connector insertion, real vehicle platform in water, thruster mounted on Cartesian robot etc.).

In the case of a real vehicle, with a simulated docking connector arm mounted on the front, reaction forces from the arm motion and contacts must be applied to the real vehicle to obtain realistic behaviour (Figure 1.4). The obvious way is to use the thrusters, but this then limits the available thrust vector for control, and no longer models the final in-water system. Alternatives include an appropriate mechanical mechanism, which does not limit the real vehicle's natural dynamic behaviour, or possibly a model-based control strategy, where the model being followed is a coupled vehicle and arm system.

For the case of a simulated vehicle and real arm, the base of the arm must be animated with the vehicle motion (as for the proprioceptive sensor above), using a Stewart platform or similar. However, in addition, the reaction forces at the base of the arm must be measured and applied to the simulated vehicle dynamics. Fortunately,

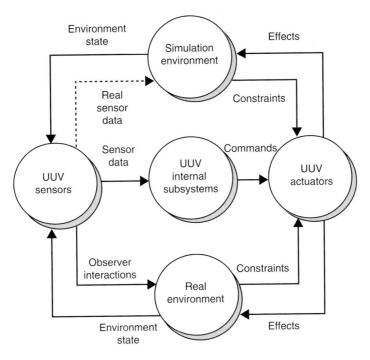

Figure 1.3 Hidden data flows for interoperable-HIL.

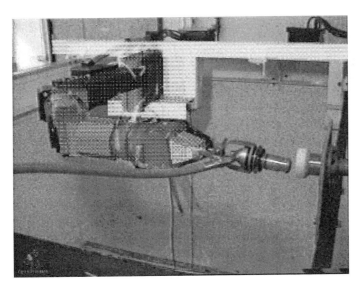

Figure 1.4 Simulated and real worlds during tele-assisted connector insertion (simulated manipulator over real video image for comparison).

typical UUVs move slowly, and so hydrodynamic drag and other inertial effects on the arm that are caused by the vehicle motion will be relatively small. Once again, additional data flows, hardware and complexity is involved. Although it is theoretically possible to simulate most (if not all) of the hybrid combinations, the complexity of implementing a completely interoperable HILS involving all UUV subsystems, may be more than practical considerations merit. However, there remain many interesting combinations, which, if simulations were available, would aid the development of more complex UUV components together with greater autonomy in the control system.

1.4   SYNCHRONISATION

Ideally, all simulations would run in real time, and a real-time clock would be used throughout the system to keep all the components (both real hardware and simulations) synchronised. In practice, however, some simulations require substantial computing resources to achieve real-time execution and remain faithful. This is also related to the level of detail or sophistication of the simulation involved. Sonar devices (sidescan, multi-beam etc.) are prime examples: simple models are available, but very sophisticated simulations involving complex physical models of sound propagation and reflection within the environment require significant computing resources.

If we allow non-real time simulations within the interoperable-HILS system, the issue of synchronising logical and real-time is raised. Two classical approaches (Jefferson, 1985; Frangos, 1990; Fujimot, 1990; Leivant and Watro, 1993; Alles *et al.*, 1994; Raynal and Singhal, 1996) are the following:

- To run everything in *lock-step*. The simulation supervisor issues 'tic' to commence the next logical time interval, and awaits 'toc' from all subsystems before repeating.
- A more opportunistic strategy using *time-slicing*.

Lock-step is intuitively the easiest, but slows the progress of logical time to that of the slowest simulation. Real subsystems must also be interruptible and freezable over very short time intervals (milliseconds). This is not helpful where velocity control methods are used in a real subsystem, for example, terrain-following motion control.

Opportunistic strategies using time-slicing are more helpful, in that longer time intervals are used for interrupting and freezing a real subsystem while simulations catch up. In the classical approach, subsystems proceed at their own pace using their own local copy of logical time. When new input data is required but unavailable, fast local proxy simulations are used as an approximation, and the subsystems proceed. If, when the correct data is available, there is a discrepancy with the proxy output, time is re-wound, anti-messages are sent out to negate interim outputs from the subsystem and the system continues.

The presence of anti-messages makes the above prone to *thrashing*. Anti-messages continually re-wind time in adjacent subsystems, so that no progress is made, or logical time continually rewinds to the beginning. For our activities we have therefore employed a variation on time-slicing, where fixed intervals are used to coarsely synchronise all subsystems.

Here, simulation subsystems are executed for a fixed period, while real subsystems remain frozen. Fast, local proxy simulations are used to predict the missing real outputs. After the fixed period, the real subsystems are executed, and their outputs compared with those of the proxies. If there are discrepancies above a given threshold, logical time is rewound (rollback) in the simulated subsystems, and processing repeated using the newly available data from the real subsystems.

The advantage of this approach is that real hardware can run for a reasonable length of time between frozen periods, and there should be no more that one logical time re-wind per subsystem in each fixed period (i.e. no thrashing). The disadvantage is that interaction between the real and simulated subsystems is not properly modelled during each fixed period. Determining the ideal period length is thus a balance of simulation accuracy (i.e. short periods) and real subsystem run length (i.e. long periods).

In practice, the proxy simulations can be used to approximate either the real subsystems, or the HIL simulations. To illustrate this method, here is a pseudo code fragment for using the proxy simulations to approximate the *real* subsystems:

```
while ()
   {
      pause_time = sim_time + T;
      while (sim_time < pause_time)
      {
         get sim_inputs;
         log sim_inputs;
         run sim;
         log sim_outputs;
         sim_time++;
      }
      while (real_time < sim_time)
      {
         get logged inputs;
         run hardware;
         if (logged outputs invalid)
            ROLLBACK;
         else
         real_time++;
      }
   }
```

and a pseudo code fragment for using the proxy simulations to approximate *simulated* subsystems:

```
Hardware process:
   while ()
   {
      get inputs from proxy;
      log inputs;
```

```
    run hardware;
    log output;
}
```

Simulation process:
```
while()
{
    get inputs;
    if (proxy inputs invalid)
        ROLLBACK;
    else
        get logged output;
}
```

Some observations made after practical experience of modified time-slicing can be summarised thus:

- rollbacks keep subsystems synchronised, but inevitably destroy realism;
- space scaling helps reduce incidence of rollback (e.g. Cartesian robot moves only one-sixth the distance of the UUV simulation);
- increased error tolerance reduces rollback incidence;
- rollback requires that real subsystems must be resetable – that is, they can be re-wound to an earlier (stored) state.

Rollbacks and noisy sensors lead to some indeterminacy in simulation progression.

1.5 CORESIM IMPLEMENTATION

Concepts from above have been implemented across a communications software spine within a distributed computing environment. UNIX boxes, VMEcrate with the VME-exec real-time operating system and PC/Windows have been concurrently supported, linked by TCP/IP. The prototype CORESIM system initially supports limited inter-operability of sensors, including animation of a Cartesian robot over a laboratory test tank (Figure 1.5). We have also implemented the basic mechanisms to evaluate our modified approach to time-slicing.

The simulation infrastructure is provided by the US Department of Defense HLA (DoD, 1998) with vehicle simulations expressed as Federates (e.g. the human interface, bottom station, propulsion, sensing, vehicle dynamics and simulation management). A drag and drop interface allows users to assign processes to particular machines (Hunter, 1999).

A subscription communication protocol has been implemented within the HLA run-time infrastructure to allow hot-swapping of real and simulated HLA components at run-time. The HILS runs on a virtual machine, and processes publish data and write when available. Processes also subscribe to data, and receive using call-back or polling mechanisms. The run-time interface is responsible for message routing, which is thus invisible to the programmer. Data can be annotated with meta-information defining priority, reliability and more.

Figure 1.5   Cartesian robot carrying sonar and video sensors during simulated docking manoeuvres.

1.6   EXAMPLE APPLICATIONS

This section outlines some real subsea research and development applications that have used this interoperable-HILS architecture. The aim is to illustrate the benefits of such an approach on a range of applications from single sensors to collections of subsystems:

- Simulation of an electromagnetic current meter (ECM) on a vehicle, as an example of a generic speed log.
- Visual feedback control: real-time video based motion control for vehicle stabilisation.
- VR-Nav: Improving pilot navigation and task visualisation on ROVs and on monitored AUV missions.
- Concurrent mapping and localisation for AUVs.
- SWIMMER: An autonomous docking and workclass ROV deployment vehicle.

1.6.1   Simulation of an ECM

The aim of this work was to investigate the use of the CORESIM architecture to interoperably swap between a simulated sensor and a real sensor device 'mounted' on a simulated vehicle.

The real sensor was mounted on a Cartesian robot mounted above a test tank. The robot was free to translate the current meter in two degrees-of-freedom simultaneously, and at various accelerations and velocities.

A software simulation of the instrument was also constructed. To evaluate the opportunistic time-slicing mechanisms, the speed of the simulation could be varied, together

JONATHAN EVANS AND DAVID LANE

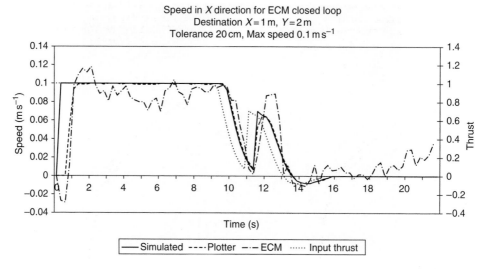

Figure 1.6 Speed in X direction during positioning – simulated, Cartesian robot, current meter and thrust. *See* Color Plate 1.

with the thresholds between the outputs of the real and simulated hardware that triggered a rollback event.

Figure 1.6 shows the results of a typical run, plotting speed in the X direction moving from the origin to position $X = 1\,\text{m}$, $Y = 2\,\text{m}$ at a demanded speed of $0.1\,\text{m s}^{-1}$. Four rollbacks took place over a 13 s period at 10.25, 10.5, 10.75 and 12.75 s.

Figure 1.7 shows plan view of the same run, performed twice at $0.5\,\text{m s}^{-1}$ with timeslice rollback error values of 5 and 20 cm. Plots show position values from the vehicle simulation, the Cartesian robot sensors and the current meter. Although the runs were similar, there was some non-determinisim caused by the noisy sensor.

The key parameters here are:

- the time-slice rollback threshold;
- the time-slice period;
- the run-up parameters (for instance, to allow the hardware to already be at the correct velocity at the synchronisation point).

The simulation environment can be programmed to systematically repeat the various paths with different threshold values and collect statistics on the number of rollbacks triggered, and the hardware 'run up' needed before the synchronisation point to continue succesfully the simulation from the point at which the simulation and hardware diverged. In this way, the system can help identify the necessary simulation, synchronisation and time-slice parameters for that particular sensor. These parameters can then used when the sensor simulation is integrated into a simulation of a larger system.

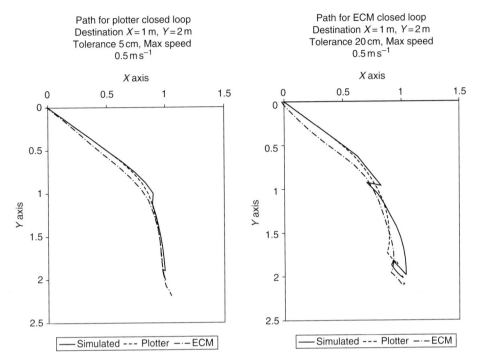

Figure 1.7 Varying UUV trajectory with varying time-slicing rollback error threshold. *See Color Plate 2.*

1.6.2 Visual Servoing: Real-time Video-based Motion Control for Vehicle Stabilisation

So far, in the field of underwater robotics, few attempts have been made to use vision sensors for control (Marks *et al.*, 1994; Rives and Borrelly, 1997; Negahdaripour *et al.*, 1999; Lane *et al.*, 2000b; Lotts *et al.*, 2000; Trucco *et al.*, 2000). However, vision shows some interesting features compared to classical positioning sensors when performing station-keeping tasks.

For example, magnetic compasses suffer from a low update rate and cannot be used in the vicinity of man-made ferromagnetic structures. More importantly, with the exception of depth sensors, which are both accurate and fast, on-board translation motion sensors (for surge and sway) are integrating sensors (e.g. accelerometers or Doppler velocity logs) hence they are subject to drift and therefore unsuitable for station keeping.

A standard video camera, however, is not subject to magnetic influence and has an update rate of 25 Hz. It can be used as a relative positioning sensor. In addition, if the absolute position of the target is already known, the vehicle's position in the world frame can also be calculated. Despite its short range (typically 3–10 m), and the need for substantial computing power, visual servoing (also called visual control) allows very diverse tasks such as station keeping (Marks *et al.*, 1994; Negahdaripour *et al.*, 1999; Lotts *et al.*, 2001) or pipe-following (Rives and Borrelly, 1997) to be carried out.

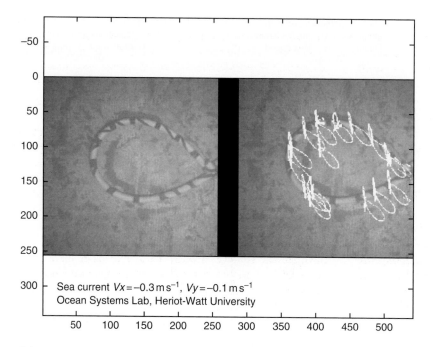

Figure 1.8   UUV visual stabilisation using CORESIM architecture to simulate the motion dynamics of the vehicle.

The CORESIM architecture was applied to this problem to allow the surge and sway motion of the vehicle to be simulated on a Cartesian robot mounted above a test tank. In this case, the computing resources available for the simulation were sufficient for real-time execution. The video processing and motion reconstruction was designed to be independent of the simulation architecture, enabling a later migration to real vehicle hardware. Figure 1.8 shows an example of the unstructured video image used for the servoing tasks. The second image shows the calculated motion of vehicle over time. The closed loops indicate the detected motion, and that returned the vehicle back to its original location.

The CORESIM architecture provides a clear interface between application and the vehicle controller – in this case a simulated vehicle. This emphasises the benefit of interoperability – the ability to integrate code developed from a simulated (or partly simulated platform) onto real hardware with the minimum of reconfiguration.

1.6.3   Concurrent Mapping and Localisation

Concurrent Mapping and Localisation (CML) is a technique for localising (positioning) an AUV within its environment. Objects within the environment are sensed (typically by sonar) and are used in conjunction with a vehicle model by the CML algorithm to build concurrently an absolute map of the environment and localise the vehicle in absolute co-ordinates.

Most AUVs are equipped with dead-reckoning sensors, such as a Doppler velocity log, inertial rate gyros, fluxgate or fibre-optic compasses. These types of sensors suffer from drift and offset errors and hence the error in the vehicle's position will grow without bounds. To fix the position of the vehicle in the world frame, absolute-positioning sensors can be used. Examples of underwater absolute-positioning sensors include acoustic ultra short, short and long baseline positioning systems. All of these systems require the vehicle to be within a volume of water covered by (typically) acoustic beacons, therefore restricting the vehicle's exploratory capabilities. This restriction has motivated research on CML. The basis of CML is to build a map and simultaneously localise the vehicle within the map as it is being built.

Our CML research has concentrated on the stochastic map approach (Smith *et al.*, 1990; Petillot *et al.*, 2001; Ruiz *et al.*, 2001). This is essentially an Extended Kalman Filter, combined with various techniques to contain the growth of the state vector that, if unchecked, would eventually prevent real-time solutions.

The eventual aim is to equip an AUV with the necessary sonar equipment and, in real-time, explore and navigate the vehicle through an unknown environment.

However, earlier in the development process, the simulation architecture was used to actuate the sonar within a test tank (using a Cartesian robot) to emulate the motion of the vehicle. This had the added advantage of providing ground truth data of the actual position of the (simulated) vehicle and the targets, allowing the errors within the CML algorithm to be assessed.

Figure 1.9 shows a typical simulated mission using the CML algorithm. In the first image, two objects are detected in the sonar image. As the vehicle moves forward (second image), the CML algorithm tracks the targets between successive frames, and using the range and bearing data, calculates the vehicle motion with respect to the targets. The simulated vehicle then moves right, and then returns to its original location (third frame). The sonar targets continue to be tracked, until the vehicle reaches it final location (fourth frame). The differences between the ground truth path (from the simulation) and the calculated path (from the CML algorithm) are small (as shown by the different line styles). Further extensions to the CML approach allow additional error bounding when revisiting and or redetecting previous objects.

Again, the interoperable simulation approach allowed valuable extra data to be obtained during the system development (for instance, the ground truth data) and then aided more rapid migration to the real hardware platform.

1.6.4 VR-NAV: Improving Pilot Navigation and Task Visualisation on ROVs

Where a communication link, which may use acoustic telemetry, optical-fibre or other technology, is possible between an AUV and its operator then it becomes feasible to use techniques developed for improving ROV navigation with AUVs. One such example is discussed below, where the additional feedback over a telemetry link, merging simulation and sensor data, would improve confidence that the autonomous mission was proceeding as planned.

One of the original applications developed with the help of the interoperable CORESIM architecture was VR-NAV (Lane *et al.*, 1998, 2000b) The aim of this project was to develop a Virtual Reality pilot's navigation aid that presents the ROV operator

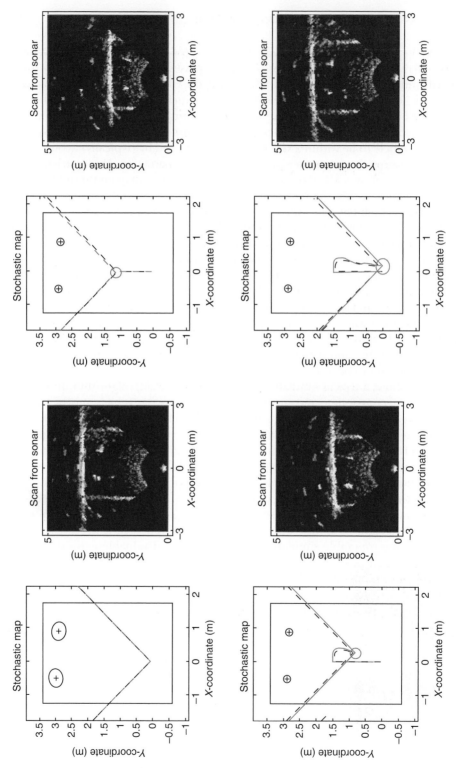

Figure 1.9   CML output and real sonar data showing the simulated AUV performing an 'out and back' mission.

with a 3D graphical rendering of the vehicle's current environment. Any number of arbitrary 'virtual camera' viewpoints can be selected and stored. The pilot is able to 'walkaround' the ROV, note obstructions, and better gauge routes to particular targets. It is not designed to replace traditional video and sonar displays, but instead augment the pilot's interface by fusing the information for a range of existing sensors, and present the information in a comprehensive form.

As a demonstration of the techniques, a prototype system was development and integrated onto a small inspection ROV (Hyball by Hydrovision, Aberdeen, Scotland).

The final system requires a real-time model (simulation) of the vehicle to be executed alongside the real vehicle. Various sensor information, including live sonar data, is used to keep the motion of the internally simulated vehicle to synchronise with the real vehicle. The output of the simulation is then used to generate a real-time 3D graphical output for the pilot (see Figure 1.10).

One of the key features in making the pilot's Virtual Reality display trustworthy is the synchronisation between VR-NAV's model of the vehicle's motion and reality. If the synchronisation drifts, then Virtual Reality display will quickly diverge from the real environment making it useless.

As part of the initial development process of VR-NAV, the distributed CORESIM architecture was used to simulate the vehicle, allowing the pilot to swim a simulated vehicle through a simulated environment using a Virtual Reality (hence simulated) display to observe.

This of course requires two completely separate simulations. The first is the VR-NAV 'on-board' simulation that attempts, through the use of sensor data, to maintain synchronisation with the external environment. However, during development, external sensors were not available hence simulated sensors were used, which sensed a simulated environment (maintained by CORESIM – as in Figure 1.2). Again the development process would have been considerably longer if a simulated vehicle and environment had not been available. The VR-NAV interfaces to the sensors were designed using interoperable interfaces, so when the design was implemented on the real vehicle, interface changes were minimised.

 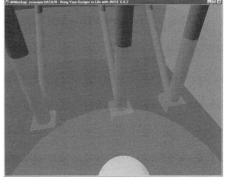

Figure 1.10 Left: A small inspection ROV (Hyball) approaching a submerged support structure. Right: VR-NAV output, showing the pilot's 3D display of the scene.

1.6.5   SWIMMER: An Autonomous Docking and Deployment Vehicle

As offshore exploration and production moves to increasingly greater water depths, the greater the reliance on remote intervention using UUVs. Any technology that can reduce the cost of deploying and using such vehicles could produce considerable savings (see Chapter 8).

The aim of the EU-Thermie funded SWIMMER project was to construct a prototype AUV capable of carrying a standard work-class ROV to a seabed installation, in effect an autonomous deployment platform (illustrated in Figure 1.11).

This would bring three immediate benefits:

- Remove the need for a permanent support vessel, with subsequent cost savings.
- Remove the long tether and umbilical with consequent improvement in vehicle dynamics and energy usage.
- Once docked, the work-class ROV can be operated completely conventionally via a subsea fibre-optic communications network and short (100 m) umbilical. This allows the ROV pilots to be stationed ashore.

The first fully autonomous sea trials of the prototype vehicle were in October 2001, in the Mediterranean Sea off St Tropez, France (Figure 1.11).

As part of the SWIMMER project a Short-Range Positioning System (Evans *et al.*, 2000, 2001a,b) was developed to allow the Swimmer AUV to position itself and dock onto the seabed installation. The system does not use deployed ultrasonic transponders, but instead uses profile data obtained from a high-speed, narrow-beam imaging sonar. The system is based on Active Sonar Object Prediction (Evans *et al.*, 2000). It uses computer-aided design information to derive a 3D model (or map) of the vehicle's environment, including a seabed docking station (Figure 1.12). This model is used to drive an active, that is, an attentive or focused, sonar object confirmation task.

As part of the initial development work of the short range positioning system, and well before the real vehicle had been assembled, the simulation facilities of CORESIM

Figure 1.11   Left: Illustration of the SWIMMER concept – an autonomous deployment vehicle for Workclass ROVs. Right: The SWIMMER AUV and payload ROV being recovered after autonomous docking on seabed docking station.

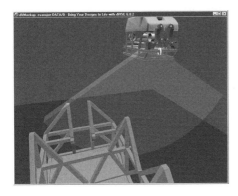 

Figure 1.12   Left: Graphical output of the simulation during a docking manoeuvre. Right: Real sonar data collected while simulating the docking manoeuvre.

were employed to emulate the motion of the vehicle, actuating the sonars as if the mounted on the real vehicle. A $\frac{1}{6}$ scale model of the docking station was placed in the centre of the test tank (pictured in Figure 1.5), and the positioning system's sonar heads were mounted on the robot's end-effector. A series of synchronised sonar and position data sets were collected, at varying speeds and approach angles (Figure 1.12).

Using the simulator it was possible to ground truth the sonar data, and provide a definite quality reference to which the quality of short-range positioning system's position output could be compared.

1.7   CONCLUSIONS

Although attractive, complete interoperability of simulated and real subsystems for subsea robot HILS is neither straightforward nor necessary in practice. However, there remain many interesting and practical combinations of simulated and real systems.

CORESIM has been a valuable tool in exploring some of the issues surrounding plug-and-play interoperable HIL. Its key successes are its flexible and extendable architecture; the interface design that allows easy migration from simulation to real hardware; its modified approach to time-slicing that allows the real hardware to be run in longer, more realistic sections of logical time; and its synchronisation strategy that helps reduce thrashing where the simulations continually rollback back, unable to make any progress.

Acknowledgements

The initial CORESIM research was funded by Shell UK Exploration and Production, Lockheed Martin, Halliburton Subsea Systems Ltd, Tritech International Ltd., Slingsby Engineering, Sea-Eye Marine and the Defence Evaluation and Research Agency (DERA) in conjunction with the UK Engineering and Physical Science Research Council (EPSRC) and the Centre for Marine and Petroleum Technology (CMPT) in the managed programme Technology for Unmanned Underwater Vehicles (TUUV), phase II – Core Simulation Engine 1997–1999. There have been many partners to many of the projects described, but we would like to give special mention to: Martin Colley, John Standeven and Vic Callagham (University of Essex); Neil Duffy and Jim Herd (Heriot-Watt University); Jeremy Smith, Jim Lucas and Kevin Keller (The University of Liverpool), Ioseba Tena Ruiz, Yvan Petillot and Jean Francois Lots (Heriot-Watt); and Ian Edwards (Halliburton Subsea Systems).

2. POWER SOURCES FOR UNMANNED UNDERWATER VEHICLES

SULEIMAN M. ABU SHARKH[a], GWYN GRIFFITHS[b]
and ANDREW T. WEBB[b]

[a] *School of Engineering Sciences, University of Southampton*
Southampton SO17 1BJ, UK and
[b] *Southampton Oceanography Centre, University of Southampton,*
Southampton SO14 3ZH, UK

2.1 INTRODUCTION

The power source of an underwater vehicle is the main component that determines its range of travel and the tasks that it can perform. Until recently, the choice of practical power sources for most applications was limited to lead–acid and silver–zinc batteries. Lead–acid batteries are a well-established technology and are available at low cost. However, they have a low energy density ($\sim$25 Wh kg^{-1}). Although silver–zinc batteries have a much higher energy density ($\sim$120 Wh kg^{-1}), they are expensive, and have a short life of 40–100 cycles compared to 1000 cycles for lead–acid (Carey *et al.*, 1992). There is, in practice, no all-purpose, ideal choice of power source for an autonomous underwater vehicle (AUV). Developers have sought to use a variety of battery types including nickel–cadmium, nickel–metal hydride, lithium ion and primary manganese alkaline and lithium cells, often using stacks of cells developed for the consumer or industrial markets.

In recent years considerable research and development work has been done on advanced power sources for road electric and hybrid vehicles. This on-going work is motivated by environmentally driven legislation aimed at reducing the harmful emissions of conventional internal combustion engine vehicles (e.g. IEA, 1993). Many in the AUV community have an expectation that breakthroughs and or cost reduction in electric power for road vehicles may deliver the ideal AUV power source. However, this potentially advantageous spin-off may prove difficult to achieve in practice. Electric cars need sources capable of delivering 5–50 kW at a capital cost per kW not too dissimilar from that of an internal combustion engine. In contrast, most AUVs have a power consumption an order of magnitude less, typically 0.2–2 kW. Furthermore, the AUV user would be willing to pay a higher price per kW than the car owner would. Much research and development effort is also directed at providing high energy density sources for low power consumption devices. For example, micro-fuel cells are becoming available for portable electronic equipment such as cellular telephones and laptop computers requiring 5–50 W. For these, the consumer is prepared to pay an even higher price per kW than the AUV user.

The ideal power source for an AUV should have a high energy density, high power density (higher charge and discharge rates), low cost, long life, low maintenance, high efficiency and wide operating temperature range. It should be safe and recyclable. In addition to these requirements there are additional considerations imposed by working in an underwater environment. Ideally, the power source should be non-gassing and in the case of batteries, the electrolyte should be spill-proof. The operation of the power source should also be independent of depth.

This chapter begins by reviewing the state of development of different types of rechargeable power sources that look promising for use in AUVs, and then briefly reviews the recent development of flywheel electromechanical batteries and supercapacitor energy storage systems. Although these latter technologies may not see service as the primary power source in AUVs, they may well have applications in particular circumstances. Section 2.3 discusses in more detail the electrochemistry, performance and application of primary manganese alkaline cells as used in the Autosub and other AUVs. The difficulties of estimating the energy remaining in the battery pack are explored and a simple fuel-gauge algorithm is derived based on electrochemical equations and laboratory performance data. The practicality of this fuel-gauge algorithm is demonstrated using data from a series of Autosub missions.

2.2 RECHARGEABLE ENERGY STORAGE

2.2.1 Secondary Electrochemical Batteries

Table 2.1 presents a summary of the performance of a number of the most promising types of secondary battery for use in an AUV. Primary batteries are considered in Section 2.3 but it is worth mentioning here that primary batteries may be attractive in some AUV applications due to their much higher energy density compared to the best of the secondary batteries.

Sealed lead–acid cells have been used as the power source in many AUVs, especially for engineering trials and short-duration missions. In the Autosub vehicle, for example, seven 12 V 90 Ah batteries in series, weighing 300 kg in total, provided ~7.5 kWh of energy for coastal trials. This was sufficient energy for up to 10 h operation giving a range of 70 km (Millard *et al.*, 1998).

Silver–zinc cells have been used extensively in AUVs. In the Odyssey IIb vehicle, for example, a 48 V 24 Ah pack weighing 10 kg provided power for over six hours of vehicle operation at a cost of under $200 per mission (Altshuler *et al.*, 1995). However, batteries formed from silver–zinc cells require very careful maintenance and record keeping on the state of charge to ensure safe operation. Standard silver--zinc cells also have a limited number of charge–discharge cycles, a limited life once the electrolyte has been added to the cell and a high capital cost. These limitations have led several AUV manufacturers to seek alternative rechargeable power sources. Nevertheless, significant research and development effort has been directed towards augmenting the silver–zinc chemistry in order to improve the energy density and the number of charge–discharge cycles. These efforts have involved modified anodes, for example adding bismuth, cadmium or lead oxide to the zinc oxide and using new electropermeable membranes to almost double the cycle life (Smith *et al.*, 1996).

Nickel–cadmium presents one commercially available proven alternative to lead–acid and silver–zinc. The higher energy density and longer life of nickel–cadmium cells, however, comes at a price that is ten times that of lead–acid for the same capacity. Careful thermal management is necessary when charging nickel–cadmium cells *in situ*.

Table 2.1 Performance of practical batteries.

Type	Wh kg⁻¹	Wh l⁻¹	W kg⁻¹	W l⁻¹	Temp Range (°C)	Cycle Life[a]	Energy Efficiency (%)	Status[b]
Sealed lead–acid	20–30	60–80	100	230	−40 to 55	700	68	A
Silver–zinc	100–120	180–200	400	660	−48 to 71	100	75	A
Nickel–cadmium	40–55	60–90	100	180	−40 to 60	1500	70	A
Nickel–metal hydride	50–70	100–150	145	330	ambient	1500	75	A
Lithium–ion	90–150	150–200	250	275	−20 to 45	600–1000	–	P
Lithium–solid polymer	130–190	170–240	–	–	ambient	300–3000		P
Lithium–solid polymer	>350	–	>560	–	ambient	>200		L
Sodium–sulphur	90–120	120–130	150	136	350	800	70	A
ZEBRA: Sodium–nickel chloride	110–120	110–120	120	160	300	700		P
Aluminium–oxygen	–	260	–	–	ambient	–	81	L/P
RAM: Secondary manganese	40–80	110–220	–	–	ambient	>25	–	A
Alkaline								

Sources: Based on information in Cornu (1989, 1998), Hawley and Reader (1992), Coates *et al.* (1993), Cockburn (1993), Dreher *et al.* (1993), Marcoux and Bruce (1993), Smith *et al.* (1993), Descroix and Chagnon (1994), Vincent and Scrosati (1997) and Huang *et al.* (2001).

Notes

a at maximum recommended depth of discharge (DOD), typically 80%.

b A = available commercially; P = prototype; L = laboratory; – data is not available.

During charging, the process is exothermic; in large packs the heat generated must be dissipated to avoid an excessive temperature rise that will reduce the charging efficiency. Repetitive shallow discharges of nickel–cadmium cells leads to a (reversible) decrease in capacity – the so-called 'memory effect'. The cell's full capacity can be restored through a deep-discharge followed by a normal recharge. Disposal of used nickel–cadmium cells requires special considerations due to the significant amount of toxic heavy metals present. Notwithstanding these complications, nickel–cadmium cells have been used in several AUVs, including those from Florida Atlantic University (Smith *et al.*, 1996) and Maridan (Madsen *et al.*, 1996).

Nickel–metal hydride batteries are under active development (e.g. Cornu, 1998). While these cells have a higher energy density than nickel–cadmium and a reduced, but not zero, 'memory effect', significant self-discharge may be an issue for use in some AUV applications. Vincent and Scrosati (1997) report capacity loss of 4–5% per day during storage due to hydrogen reacting with the positive electrode.

Lithium–ion secondary cells, with energy density of $\sim$150 Wh kg^{-1} are available and have been used successfully in a number of AUVs including the autonomous benthic explorer (ABE; Bradley *et al.*, 2000) and the Urashima (Aoki, 2001). In ABE the battery comprised 378 'D' size cells, each with a capacity of $\sim$4 Ah at an on load voltage from 4.2 to 3.0 V. Charging and monitoring circuits were built-in to the lithium–ion pack, operating on 14 modules in series of nine cells in parallel. This $\sim$5.4 kWh pack replaced an earlier 1.2 kWh gel lead–acid battery and gave the vehicle an endurance of nearly 30 h on the seafloor at a depth of 2200 m, some five times the endurance that was achieved with the lead–acid battery (Bradley *et al.*, 2000). In Urashima, the power source comprises three 130 V 100 Ah batteries in parallel, giving a total capacity of 39 kWh, sufficient for the vehicle to travel 100 km (Aoki, 2001). Improvements in the materials used within these cells, such as increasing the amount of lithium at the cathode, will lead to an increase in energy density but the cell chemistry limits will be reached at about 210 Wh kg^{-1} (Green and Wilson, 2001).

Lithium–solid polymer cells look promising for underwater vehicle applications because their solid-state construction allows their use at ambient pressure without the need for a pressure case (Huang *et al.*, 2001). Their solid-state construction also means that they are resistant to shock and vibration and, in principle, need not be restricted to cylindrical form factors. Experiments have been successfully carried out on small (ca. 10 Wh) lithium–polymer cells enclosed in oil-filled bags at pressures to 600 bar. As the technology matures, and higher capacity cells that operate at ambient temperature become available, lithium–polymer will become a feasible and cost-effective power source for AUVs.

Sodium–sulphur cells have high energy densities and long life at a reasonable cost, but they operate at 295–350 °C, which presents some handling and operational challenges. However, BAE Systems demonstrated that these could be overcome with proper engineering (Cockburn, 1993; Tonge and Cockburn, 1993). However, the support systems that needed to be installed in the vehicle to manage the cells led to a substantial decrease in the effective energy density. The battery itself comprised four ABB type B120 12 V 800 Ah batteries, each weighing 13 kg and rated at 9.6 kWh (738 Wh kg^{-1}), which is close to the theoretical energy density of 790 Wh kg^{-1} (Vincent and Scrosati, 1997). But with the battery management unit, the rigid metal framework, the electrical heating element, all inside a double-walled stainless steel outer container with fibreglass insulation *in vacuo* between the walls, the total system providing 38 kWh weighed 550 kg, equivalent to an energy density of 69 Wh kg^{-1} (Tonge and Cockburn, 1993). In-water tests were successful and the need to maintain the battery temperature above 295 °C was not a significant operational problem. Nevertheless, except in Japan, sodium–sulphur battery development has ceased (Dell, 2001).

Zebra batteries are a recent development – a class of cells that use liquid sodium metal as the anode and chlorides of various transition metals as the cathode. Cells using iron II, copper II, cobalt, nickel and chromium II chloride have been developed. The nickel chloride cell, for example, has an open circuit voltage of 2.58 V and a theoretical energy density of 750 Wh kg^{-1} (Vincent and Scrosati, 1997). Extensive trials of these batteries with capacities in the region 13–17 kWh have taken place in electric road vehicles. But, operating at 270–300 °C they share the insulation and preheating requirements of the sodium–sulphur battery described above. Nevertheless, energy densities of ~100 Wh kg^{-1} for complete systems have been achieved (Vincent and Scrosati, 1997).

Metal seawater secondary batteries based on magnesium or aluminium anodes and silver chloride, lead chloride or manganese dioxide can produce high energy densities, but at low power densities (Lee *et al.*, 1989; Rao *et al.*, 1989; Hasvold, 1993; Vincent and Scrosati, 1997). Open circuit voltages of 1.6–1.9 V are typical with energy densities of between 100 and 160 Wh kg^{-1}.

2.2.2 Fuel Cells and Semi-fuel Cells

In a true fuel cell the reactants are stored external to the cell and fed to the cell as required, while in a semi-fuel cell one reactant is stored externally and brought to the reaction cell. Alupower developed aluminium–oxygen semi-fuel cell prototypes for AUVs (Scamans *et al.*, 1994a,b). Trials included those on the ARCS AUV and a later version was tested by the Monterey Bay Aquarium Research Institute Dorado AUV in 2001. The Norwegian Defence Research Establishment developed an 18 kWh aluminium–hydrogen peroxide semi-fuel cell that has been successfully used on the Hugin II AUV for missions of up to 36 h (Vestgård *et al.*, 1998). This has been followed by the installation of a 35 kWh unit into the larger HUGIN 3000 vehicle as the standard power source (see Chapter 11), giving the vehicle an endurance of 40–50 h. Each charging cycle involves exchanging the hydrogen peroxide and each alternate charge requires the aluminium anode to be replaced. While these batteries have a high energy density, they are, to some degree, still under development and require very careful handling, servicing and maintenance.

The invention of proton exchange membranes (PEM) has been an important factor in the development of fuel cells capable of operating efficiently at low temperatures. Based on a PTFE backbone with sulphonic acid (SO$_3$H) groups attached, these membranes allow protons (H$^+$) to pass from anode to cathode, but pass little of the water generated at the cathode to the anode. Using suitable noble metal catalysts bonded to the membrane, areal current densities of 0.6–1.0 A cm^{-2} have been achieved at a cell potential of ~0.7 V (Hamnet, 1999). In principle, 30 such cells in series, each 10 cm by 10 cm (100 cm^2) could produce 60 A at 21 V; dc to dc converters would transform the supply to the voltages necessary for the AUV systems. Despite the very small volume of the fuel cell reactor, storage of the reactants remains a problem and it is not surprising that very few true fuel cells have been incorporated within an AUV to date. Meyer (1993) described the design and operation of a 15 kW PEM fuel cell for use in the Advanced Projects Research Agency (ARPA) 44″ AUV. This hydrogen–oxygen cell used cryogenic gas storage.

Cryogenic storage of hydrogen within an AUV is not an attractive option for most users. Fuel cells are under development that use room-temperature liquids such as methanol. The direct methanol fuel cell consumes a dilute solution of methanol in water and oxygen and generates carbon dioxide and water as the reaction products. It has the additional advantages of operating at low temperature (<150 °C) and reasonable pressures (<0.5 MPa), but the disadvantages that the conversion efficiency is low with a simple catalyst and that today's membranes are semi-permeable to methanol, which reduces the efficiency of the cell. Hamnet (1999) and others have shown that using promoters to aid the catalyst can increase efficiency such that current densities of $\sim 0.6\,\mathrm{A\,cm^{-2}}$ can be achieved at a cell potential of $\sim 0.4\,\mathrm{V}$. Using a methanol fuel cell within an AUV would pose engineering challenges in recycling part of the generated water and in disposing of the carbon dioxide.

At least one AUV developer is committed to producing a vehicle based on a fuel cell. After 2002 the power source for Urashima is intended to combine a 4 kW PEM fuel cell with a 130 V 30 Ah lithium–ion battery in a hybrid arrangement to give the vehicle a range of 300 km at $1.5\,\mathrm{m\,s^{-1}}$ (Aoki, 2001). The fuel cell uses compressed hydrogen and oxygen in gaseous form (at 50 MPa pressure) feeding a stack of 43 cells of 170 mm by 190 mm, weighing 2 kg (Aoki and Shimura, 1997). With the necessary support systems, including gas storage and the pre-heating battery, the total system weight is 70 kg, leading to an overall energy density of $103\,\mathrm{Wh\,kg^{-1}}$.

Clearly, the challenge for the future is to reduce the weight of the ancillary systems to achieve higher energy densities. A prototype fuel cell powered bicycle may show the way: the Hydrocycle™ uses a PEM fuel cell generating 670 W that weighs 0.78 kg with the hydrogen stored in a carbon fibre pressure vessel to achieve an energy density of $205\,\mathrm{Wh\,kg^{-1}}$ (Manhattan, 2001).

2.2.3 Flywheel Electromechanical Batteries

In a flywheel battery the energy is stored as kinetic energy in a spinning rotor made of strong but light material, supported by nearly frictionless magnetic bearings, and enclosed in a vacuumed housing to minimise windage loss. Energy is coupled into and out of the rotor via a highly efficient electric motor–generator and associated electronics.

While the concept of a flywheel is old, its realisation as a viable battery with an energy density that compares well with the best of secondary batteries has been possible only recently. The main factor in the rebirth of the flywheel as an electromechanical battery is the improvement of carbon fibre technology. The energy density of a flywheel is proportional to the ratio of the tensile strength to mass density of the rotor material. In other words, for high energy density, the rotor material should be strong and light, which are properties of carbon fibre.

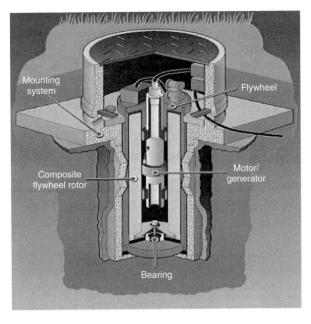

Figure 2.1   Schematic diagram of the Urenco Pirouette flywheel.

Currently, there are at least five development programmes of flywheel batteries for EV applications (Abacus Technology Corporation, 1993). The most notable of these is that by Trinity Flywheel Batteries Inc. of the USA, who is currently developing prototype flywheels in collaboration with the Lawrence Livermore National Laboratory and Westinghouse (Post *et al.*, 1993; Bowler, 1997). The Trinity development programme is expected to deliver a power flywheel having a rated capacity of 140 kW discharged over a period of 30 s. Flywheel Energy Systems Inc. is another major player in the field, with major expertise in winding carbon fibre rotors (Flanagen *et al.*, 1990).

In the United Kingdom Urenco is continuing the development of flywheel energy storage systems that was started by IES (Horner, 1996; Proud *et al.*, 1996). Prototype flywheels have been developed and the units are available for evaluation by potential customers. A schematic diagram of the IES flywheel is shown in Figure 2.1, and Table 2.2 summarises the performance of the Urenco Pirouette™ flywheel core unit. While firm efficiency figures for flywheels are not available, various developers have made claims of turnaround efficiency of about 90–92%.

The focus of most flywheel manufacturers including Trinity and Urenco is mainly on power quality and aerospace applications. The Urenco flywheel is designed for stationary applications and would not be suitable for use in a moving vehicle, nevertheless all the enabling technologies for the development of a flywheel for AUV applications are available, although considerable work is still needed.

Table 2.2   Performance envelope of the Urenco Pirouette™ flywheel.

Power	2 to 60 kW
Voltage	48 to 600 V dc
Stored energy	3–5 kWh (depending on configuration)
Charge–discharge cycles	$>10^6$
Response time	20–100 ms
Design lifetime	20 years
Dimensions	1.5 m × 0.5 m diameter

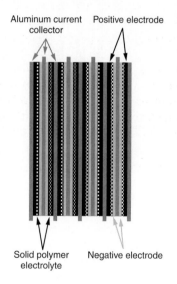

Figure 2.2   Schematic diagram of a supercapacitor.

2.2.4 Supercapacitors

Developments in new materials mean that 'super' or 'ultra' capacitors can rival batteries as means of short-term energy storage (Figure 2.2). Capacitors with power densities up to 4000 W kg^{-1} and energy densities of 15 Wh kg^{-1} are claimed to be possible (Eureka, 1998). Capacitors with power densities of 9 Wh kg^{-1} have been made, but the technology is still maturing and the costs remain high.

While the low energy density of supercapacitors may exclude them from being used in most AUVs, they could be attractive for some applications including untethered short-range UUVs or seabed vehicles operating from underwater rapid-charging stations. They may also be used in parallel with batteries to provide high power impulses.

2.3 PRIMARY ELECTROCHEMICAL CELLS

Despite needing to be replaced after a single use, primary batteries can be a cost-effective AUV power source when the cost of the battery is but a small part of the total cost of the project or task. This could easily be the case for AUVs engaged in scientific research or on missions for the military. The main advantage of primary over secondary cells is their high energy density. Primary lithium batteries have particularly high energy densities, for example, lithium sulphur dioxide at $\sim$300 Wh kg^{-1} and low rate lithium thionyl chloride cells have achieved 661 Wh kg^{-1} (Vincent and Scrosati, 1997). Primary lithium cells have been used successfully within AUVs, giving even small vehicles significant operating ranges. For example, the 30 kg REMUS AUV has a range of 88 km at a speed of 1.5 m s^{-1} with a primary lithium battery, compared to 27 km for the same configuration with a lead–acid battery (Remus, 2001).

At the other extreme, it has been proposed that the Theseus vehicle (see Chapter 13) could cross the Atlantic if its silver–zinc battery was replaced with a primary lithium pack. Unfortunately, the high cost of primary lithium cells works against their widespread use in AUVs.

In contrast, the manganese alkaline primary cell is ubiquitous. Billions of cells are manufactured each year for consumer and industrial applications. This brings such enormous economies of scale that they are a leading contender for powering medium endurance AUVs such as the Autosub (Griffiths et al., 1999) and the US Navy Oceanographic Office/Penn. State University AUV (Bradley, 2001), even though the cells can only be used once. However, in spite of their apparent simplicity, primary alkaline cells have a number of limitations that the AUV designer must consider when using them as a power source. These limitations include:

- the conversion efficiency from chemical to electrical energy is dependent on current drain and on temperature;
- the available electrical energy is dependent on temperature;
- the terminal voltage decreases with increasing current drain;
- significant self-heating of cells takes place on discharge;
- the terminal voltage decreases through the usable life of the cells.

Not all of these limitations are discussed in detail in manufacturers' literature. Commonly, performance data is only available in the form of graphs of capacity or terminal voltage at either constant current or when connected to a constant load resistance. In contrast, within AUVs using switched mode power supplies, the more common scenario is constant power drain, that is, current consumption increases as the battery terminal voltage decreases. Interpolating from constant current or constant load resistance to constant power is not straightforward. While detailed electrochemical models of manganese alkaline cells are available (Podlaha and Cheh, 1994a,b) they require such detailed knowledge of the physical–chemical parameters that they are impractical for use in an AUV fuel gauge. For example, Podlaha and Cheh (1994b) list 40 physical–chemical parameters that need to be determined for a complete model of an AA sized cell. In this section we derive a simplified model of the performance of manganese alkaline cells that is sufficiently robust for use in an AUV fuel gauge. The derivation of the model is described in some detail as the procedure can be used to derive fuel gauge algorithms for other primary cell chemistries.

Table 2.3 The major materials that make up a typical alkaline manganese cell with weights for a Duracell MN1300 cell, total weight of 139 g.

Component	Function	Weight (g)	Weight as %	Moles
Manganese dioxide	Cathode (with carbon)	52	37	0.60
Potassium hydroxide	Electrolyte	11	8	0.20
Zinc	Anode	21	15	0.32
Water	Ionic solvent	15	11	0.83
Carbon	Cathode	6	4	
Steel	Can and cathode current collector	25	18	
Brass	Anode current collector	3	2	
Plastic	Separator	6	4	
Paper		1	1	

Source: www.duracell.com

2.3.1 Electrochemistry of the Alkaline Manganese Cell

The alkaline manganese cell is a derivative of the Leclanché cell where that cell's electrolyte (usually a mixture of ammonium chloride and zinc chloride) has been replaced with a concentrated solution of potassium hydroxide. The complete cell reaction is complex, proceeding in stages involving the reduction of Mn^{+4} to Mn^{+2}. For practical purposes, however, only the first stage from Mn^{+4} to Mn^{+3} is of concern.

This one electron reduction of the manganese dioxide takes place over the voltage interval 1.5–0.9 V showing a characteristic plateau during mid-discharge. The second electron discharge from Mn^{+3} to Mn^{+2} is not well understood (Patrice *et al.*, 2001), but is occasionally seen as a plateau in the discharge voltage at ~0.9 V, but only at low discharge currents. The second electron discharge provides virtually no benefit in AUV applications given typical discharge currents of over 100 mA per cell.

Table 2.3 lists the components of a typical 'D' size manganese alkaline cell with the weight of the parts and moles of the active ingredients. The theoretical capacity of any cell is dependent on the moles of reactants available as discussed below.

2.3.2 Modelling the Cell's Capacity and Terminal Voltage under Load

The electromotive force (emf) of an electrochemical cell is given by the Nernst equation:

$$E_t = E_0 - \frac{RT}{nF} \ln \left(\frac{a_P^{v_P} a_Q^{v_Q}}{a_A^{v_A} a_B^{v_B} a_C^{v_C}} \right), \tag{2.1}$$

where R is the gas constant (8.3144 J kg^{-1} mol^{-1}), T is the absolute temperature (K), F is Faraday's constant (9.6485 $\times$ 10^4 C mol^{-1}), n is the number of electrons involved in the reduction process, a_i are the molal activities and v_i are the stoichiometric numbers of the reactants in the cell reaction:

$$v_A A + v_A B + v_C C \rightarrow v_P P + v_Q Q, \tag{2.2}$$

where A, B and C are reactants and P and Q are products, which for the first stage of the alkaline manganese cell reaction equates to:

$$Zn + 2MnO_2 + H_2O \rightarrow MnO \cdot OH + ZnO, \qquad (2.3)$$

that is, $v_A = 1$, $v_B = 2$, $v_C = 1$, $v_P = 2$ and $v_Q = 1$. During discharge, the reactants A, B and C are converted into products P and Q at a rate determined by the discharge of the cell. In practice, the molal activities can be approximated by the molar concentrations to give:

$$E_t = E_0 - \frac{RT}{nF} \ln \left(\frac{[P]^{v_P}[Q]^{v_Q}}{[A]^{v_A}[B]^{v_B}[C]^{v_C}} \right). \qquad (2.4)$$

Even with this approximation, the molar concentrations during discharge cannot be determined for a commercial cell. However, as the capacity to complete discharge (Q_T) of a cell is related to the moles of reactants $[M]$ through:

$$Q_T = [M]nF, \qquad (2.5)$$

and, following the same argument, the moles of products generated are related to the energy consumed, and the moles of reactants remaining is related to the energy remaining. Equation 2.4 can be written as:

$$V_t = V_h - \frac{\phi RT}{F} \ln \left(\frac{Q_l}{Q_r} \right) + \Delta\eta_c - \Delta\eta_a, \qquad (2.6)$$

where V_h is the cell voltage at half the useful life of the cell, ϕ is a constant, Q_l is the useful energy (J) extracted from the cell to time t, Q_r is the total useful energy remaining (i.e. $Q_T - Q_l$) and $\Delta\eta_a$ and $\Delta\eta_c$ are the anodic and cathodic overpotentials respectively. This is a simplification, as the theoretical capacity of a cell is rarely achieved. That is, the coulombic efficiency of a practical cell is less than 100%. In particular, the coulombic efficiency is reduced at high current drain.

As graphical summaries dominate manufacturer's data on capacity an experiment was conducted to test the viability of using Equation (2.6) to predict performance. The experiment was carried out with Rayovac industrial alkaline 'D' cells at three temperatures (20, 5 and $-10\,^{\circ}$C) and at nine power drains (0.1–0.3 W per cell in increments on 0.025 W). These conditions were chosen to be representative of those likely to be encountered within the Autosub AUV. A battery of five cells in series was used in the experiments, with the results normalised to apply to one cell. This averaging of five cells provides a more representative measure of performance, given the spread of characteristics of individual cells.

The heavy line in Figure 2.3 shows the measured cell voltage as a function of service hours for a power drain of 0.1 W at 20 °C. Automated measurements of voltage, current and temperature were taken every 20 min. The cell voltage at the end of its useful life was set at 1.0 V. The energy extracted per cell, Q_T, was calculated from the product of service hours (178.5) and power drain (0.1 W), converted to Joules, to give 64.26 kJ. The terminal voltage at half-life, V_h, was 1.28 V. Table 2.4 lists the values of Q_T and V_h for the nine power drains and three temperatures, which form the basis of the fuel gauge model.

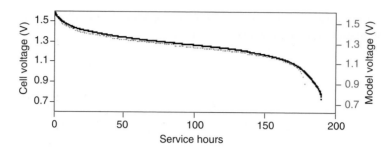

Figure 2.3 Voltage profile as measured for a Rayovac industrial 'D' cell at a constant power drain of 0.1 W and a temperature of 20 °C (heavy line) and as modelled using Equation (2.6) based on a usable capacity of 64.26 kJ and $\phi = 2.5$ (light line). The model voltage profile scale is offset by 0.02 V to separate the profiles.

Table 2.4 Initial energy capacity Q_T and half-life voltage V_h of Rayovac 'D' cells determined by experiment at 20, 5 and −10 °C and at power drains between 0.1 and 0.3 W per cell.

Power Drain (W)	20 °C		5 °C		−10 °C	
	Q_T (J)	V_h (V)	Q_T (J)	V_h (V)	Q_T (J)	V_h (V)
0.100	64,260	1.28	57,240	1.225	33,410	1.19
0.125	61,960	1.27	52,740	1.21	28,940	1.185
0.150	58,970	1.26	48,130	1.20	25,670	1.18
0.175	56,950	1.255	43,490	1.20	23,080	1.17
0.200	55,080	1.25	40,390	1.195	21,600	1.16
0.225	53,210	1.24	36,860	1.185	19,510	1.15
0.250	52,490	1.24	34,850	1.185	16,020	1.145
0.275	50,900	1.23	31,900	1.18	14,760	1.15
0.300	48,820	1.22	29,380	1.18	12,530	1.1

The ability of Equation (2.6) to predict the cell voltage based on the energy extracted was tested in a numerical experiment. The light line in Figure 2.3 shows the predicted cell voltage at one hour time-steps, with a constant power drain of 0.1 W, corresponding to an energy drain Q_t of 360 J h^{-1} based on Equation (2.6) and a total capacity Q_T of 64.26 kJ. The two lines – measured and predicted – have been separated by 0.02 V, as they overlie each other for much of the service life. At this power drain, Equation (2.6) can accurately model the cell discharge based on the initial capacity Q_T and V_h determined either through experiment or from manufacturer's literature at the temperature and current or power drain of interest. As the power drain increased, this simple model of cell voltage with service hours continued to perform well, Figure 2.4.

Having shown that the voltage discharge curve of a cell can be predicted based on the energy extracted, a simple quantity to calculate in real time within the AUV, and Q_T and V_h a fuel gauge for the AUV can be constructed if Q_T and V_h can be parameterised in terms of the current consumption and the cell temperature.

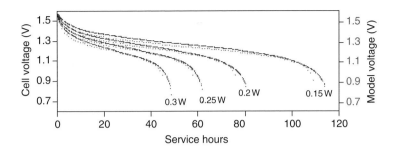

Figure 2.4 Voltage profiles as measured for a Rayovac industrial 'D' cell at a constant power drains of 0.15, 0.2, 0.25 and 0.3 W and a temperature of 20 °C (heavy line) and as modelled using Equation (2.6) based on the usable capacities shown in Table 2.4 and $\phi = 2.5$.

Vincent and Scrosati (1997) give a practical expression for the effect of temperature on the emf of an electrochemical cell based on thermodynamics:

$$E_T = E_{298} + \frac{(T - 298)\Delta S}{nF}. \tag{2.7}$$

where E_{298} is the emf at 298 K and ΔS the entropy change resulting from the cell reaction. However, these authors point out that the 'working voltage of a practical cell under load usually rises with temperature ... almost entirely due to a reduction in the internal resistance of the cell caused by an increase in the conductivity of the electrolytic phase and in the diffusion rates of the electroactive species'. Consequently, a general expression for the effect of temperature on the on-load voltage is not readily available. Instead, from the data in Table 2.4 a least squares linear model fit to V_h as a function of temperature and current, valid over 263–293 K, was calculated as:

$$V_h = 0.507 + 0.00266T - 0.258\,I. \tag{2.8}$$

with a root mean square error of 0.0094 V.

The initial capacity of a cell can be obtained from a least squares fit to the capacity data in Table 2.4 with a model based on current, temperature and the product of current and temperature:

$$Q_T = -251406 + 1106T - 101482I + 820((I - 0.1767) * (T - 278)). \tag{2.9}$$

Alternatively, the initial capacity can be expressed in terms of power drain, temperature and their product:

$$Q_T = -261678 + 1145T - 88716P + 820((P - 0.0.21) * (T - 278)). \tag{2.10}$$

The remaining capacity at time t and the on-load voltage of the battery can be determined from the following steps. First, calculate the energy extracted during the mth interval of Δt seconds:

$$Q_m = \left(\frac{V_m + V_{m-1}}{2}\right) \cdot \left(\frac{I_m + I_{m-1}}{2}\right) \cdot \Delta t. \tag{2.11}$$

The cumulative energy extracted to time t is:

$$Q_t = \sum_{m=1}^{m=t/\Delta t} Q_m. \tag{2.12}$$

The initial capacity estimated by Equation (2.9) is valid at a fixed temperature and a fixed current drain. In actual use, the current and temperature will vary during discharge. Hence the fuel gauge model must evaluate the effective capacity at each time step, with the current taken normalised by the number of parallel cells in the battery N_P, which for the current and temperature at time step m gives:

$$Q_{T_m} = -251406 + 1106 T_m - 101482 \cdot \left(\frac{I_m}{N_P}\right) + 820 \left(\left(\frac{I_m}{N_P} - 0.1767\right) \cdot (T_m - 278)\right). \tag{2.13}$$

The remaining capacity at the start of time step $m + 1$ is then:

$$Q_r = Q_{T_m} - Q_t. \tag{2.14}$$

The on-load voltage V_m at time step m can be calculated by substituting V_h from Equation (2.8) and Q_t and Q_r from Equations (2.13) and (2.14) into Equation (2.6) at each time step, with N_s being the number of cells in series within the battery:

$$V_m = \left(\left(\left(0.507 + 0.00266 T_m - 0.258 \left(\frac{I_m}{N_P}\right)\right) - \frac{\phi R T_m}{F}\right) \cdot \ln\left(\frac{Q_t}{Q_r}\right)\right). \tag{2.15}$$

These last two equations can be used as the basis of a simple fuel gauge for an AUV powered by Rayovac 'D' size manganese alkaline cells. For other cells, the same general procedure can be followed to derive the model parameters from experimental data.

2.3.3 Operational Experience with Alkaline Cells in the Autosub AUV

The Autosub-2 AUV uses manganese alkaline battery packs made up from 75 cells in series. While there is sufficient volume within the carbon-fibre pressure vessels for at least 70 such packs, the number installed depends on the weight in water of the scientific payload for the mission or campaign. That is, the available energy is limited by weight rather than volume. For Autosub's campaign in the Weddell Sea, Antarctica in early 2001 the battery consisted of 58 packs in parallel of 75 cells in series, distributed within four pressure tubes. Checks on the terminal voltage of the individual packs after removal showed that not all had been in-circuit during use, most probably due to problems with interconnections. Three sets of batteries were used during the campaign; the analysis below uses the data from the second set.

The Autosub data logger measures the overall terminal voltage of the battery, the current consumed from each tube and a spot temperature within each tube (the

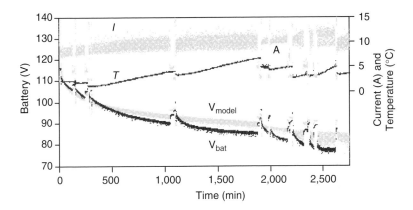

Figure 2.5 Comparison of the actual terminal voltage of the Autosub battery pack as measured during missions 250–259 (V_{bat}) compared to the model results (V_{model}) computed from Equation (2.15) using the measured average current consumption (I) and average battery pack temperature (T). During the mission marked 'A' the EK500 echo sounder was not drawing current.

tubes are 2.8 m long by 0.236 m diameter). Current consumption is far from steady, with variations of 3–5 A present mainly due to the current pulses drawn by the high-power acoustic instruments on-board. During the period marked 'A' on Figure 2.5 the current drawn was steadier, as the EK500 two-frequency scientific echo sounder was not working.

The second set of batteries provided power for missions 250–259, comprising a short test missions (1 km), several missions under icebergs (some 52 km in total), and two missions under sea-ice (60 and 63 km). Between some missions, the vehicle was recovered for maintenance checks. Figure 2.5 shows average battery temperature, current and terminal voltage spanning these ten missions. During use the battery pack temperature increased due to self-heating, dropping slightly during periods on deck. The current gradually increased as the terminal voltage fell, as most of the vehicle's systems operated on a constant power basis. A hindcast of the terminal voltage is also shown in Figure 2.5, based on the measured average temperature and current taking 56 of the 58 packs as being in-circuit. Over the first 75% of the time in service, the simulation was within 5 V of the measured terminal voltage. The increasing discrepancy with time is not surprising given the integrative nature of the problem. Ideally, temperature and current measurements should be available for each of the individual packs making up the battery to improve the reliability of the estimate. Also, it would be better to measure the true energy extracted, rather than from spot measurements of current and voltage at 15 s intervals.

Having established that a measurable quantity – terminal voltage – can be predicted with reasonable accuracy, Equations (2.12)–(2.14) can then be used to predict the energy remaining at each time interval. Hence, the remaining useful life can be predicted, taking into account the mission requirements in terms of current demand and operating temperature.

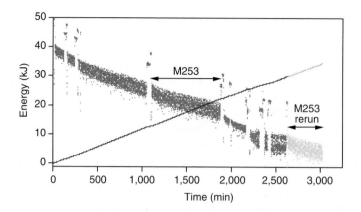

Figure 2.6   Cumulative energy extracted from each cell of the battery during missions 250–259 (thin black) followed by a simulated re-run of mission 253. The remaining energy at each time step (grey) is projected forward for the simulated re-run of mission 253 until it reaches zero.

Figure 2.6 shows the measured cumulative energy extracted from the Autosub battery pack during missions 250–259, together with an estimate of the energy remaining based on the fuel gauge model developed from Equations (2.12)–(2.14). The scatter in energy remaining results from the short-term fluctuations in current drain. Over the first 250 min the remaining energy decreases more rapidly than over the subsequent 1500 min. This is easily understood by noting that the temperature of the battery pack was at its lowest at the start of the period, see Figure 2.5. Again, from ~1800 min, where the temperature of the battery pack dropped, the energy remaining decreased at a higher rate. A re-run of mission 253 was added as a simulation after mission 259, using the fuel gauge equations to take the battery pack to depletion. This projection beyond actual use showed that some 400 min of operational use remained in the pack when it was removed from the vehicle at the end of mission 259, equal to some 16% of its capacity.

The Autosub team clearly changed the battery pack at the right time. At the start of the mission 253 re-run, as with all missions, the current drain was some 3 A higher than for the normal cruising speed for 2 min as the vehicle accelerated at full power on the surface prior to diving. This additional current drain, according to the model, reduced the battery pack terminal voltage below 1 V per cell and led to a near zero predicted energy remaining. After this short period on full power, as the vehicle dived and current consumption reduced, the estimate of remaining energy increased, and the terminal voltage rose above 1 V per cell. In practice, the motor controller would not have been able to keep a constant power into the motor and as a result the current consumption would have been lower than in the simulated re-run. Nevertheless, the result would have been a dive that would have been attempted at less than full power and the vehicle may not have been able to submerge. The fact that the vehicle had had sufficient power to complete mission 259, and had sufficient energy remaining for some 400 min of operation would not have been sufficient to guarantee that there was sufficient energy available on full power for the vehicle to dive. This information alone would seem to make the use of a fuel gauge algorithm worthwhile.

2.4 CONCLUSIONS

There is no universal power source for AUVs that combines all of the attributes required by vehicle designers and users. Rather, the choice of power source will depend on the particular requirements of the operational missions to be undertaken by the vehicle. Lead–acid cells remain a practical and low cost option for trials. Lithium–ion and lithium–polymer secondary batteries are attractive when there is a need to combine high energy density and frequent use that allows the high capital cost to be amortised over a large number of missions. Where there is a need for high energy density, but with less intensive use, primary manganese alkaline or lithium batteries may prove most economical.

Semi-fuel cells have established their operational reliability although at energy densities that still require significant volume for the energy source. Hydrogen–oxygen fuel cells and high temperature batteries remain the exotic choice for an AUV power source. The complexity, weight and volume of the ancillary systems that these power sources require mitigate against their widespread use, as the total energy density has yet to be sufficiently high to justify the risk. As additional research and development solve these issues with the ancillary systems surrounding fuel cells, the increase in effective energy density will begin to outweigh complexity and constructors and users are likely to adopt these novel sources.

Flywheels offer an exciting alternative that potentially have a high energy density, longer life and better efficiency than most available secondary batteries. However, the future of flywheel technology for vehicle applications (including AUVs) is uncertain. Supercapacitors have low energy densities but their very high charge/discharge rates may make them attractive for use in conjunction with low power density batteries to provide high power impulses for a short period of time.

Whatever the electrochemistry of the power source, the AUV user needs to be able to estimate the energy remaining during missions and prior to the next mission. Using as an example the manganese alkaline primary battery pack within the Autosub AUV we have shown how basic theory and data from experiments on capacity can be combined to derive a simple model for a fuel gauge algorithm for use in a vehicle.

Acknowledgements

S. M. Abu Sharkh's work was supported by the EPSRC and the CMPT and he is grateful to Mr John G. W. West for supplying copies of papers on batteries. Gwyn Griffiths and Andrew Webb thank all of the members of the Autosub team for the information that they have provided for this chapter.

3. AUTONOMOUS BUOYANCY-DRIVEN UNDERWATER GLIDERS

RUSS E. DAVIS[a], CHARLES C. ERIKSEN[b] and CLAYTON P. JONES[c]

[a] *Scripps Institution of Oceanography, La Jolla, CA 92093-0230, USA,*
[b] *School of Oceanography, University of Washington, Seattle, WA 98195-5351, USA and*
[c] *Webb Research Corporation, Falmouth, MA 02536, USA*

3.1 INTRODUCTION

Historically, the interior ocean has been mainly observed using instruments lowered from research ships or, later, suspended from moorings. Typical ship cruises last a month or two while moorings may last a year or two. The relatively high cost of these observation platforms has limited their number and, consequently, the spatial and temporal density at which the ocean has been observed. Initially this may not have seemed a serious hindrance because the ocean's circulation was thought to be largely steady with broad spatial scales outside a few concentrated boundary currents. Over the last 30 years, however, satellite remote sensing and intensive experimental ocean observations have belied this view and shown that the ocean is highly variable on time scales that are somewhat longer than those of the atmosphere and space scales of tens of kilometers, much smaller than those of the atmosphere. Even before the fullness of ocean variability was known, Stommel (1955) likened the oceanographic observational approach to meteorologists observing the atmosphere using "half a dozen automobiles and kites to which air sounding instruments were attached and doing all their work on dark moonless nights when they could not see what was happening in their medium."

The advent of satellite navigation and communication made possible a class of small, inexpensive instrument platforms that are changing the way the ocean is observed. Much as satellite remote sensing led to a quantum jump in our understanding of the ocean's surface, these new platforms provide a view of the interior ocean with much higher spatial and temporal resolution than is possible with conventional shipboard and moored instruments. Increased resolution is important in solving problems from the management of coastal resources to the prediction of climate change. Surface drifters are now mapping the changing surface circulation and reporting global sea surface temperatures as they vary on mesoscale and climatic time scales (Niiler, 2001). Autonomous profiling floats (Davis *et al.*, 1991) have already shown an ability to return routine real-time observations from all parts of the ice-free ocean at a low cost made possible by minimal dependence on research vessels. The international Argo program (Wilson, 2000) is now building up a global array of 3000 profiling floats, each of which will report an ocean profile every 10 days, providing a synoptic view of the ocean much as the World Weather Watch provides global weather information. A feature of both surface drifters and profiling floats is that they drift with the ocean currents, allowing the reliable measurement of these currents but making the location of these measurements uncontrollable and difficult to predict.

Even before the first autonomous floats were operating, Stommel (1989) envisioned a world ocean observing system based on "a fleet of small neutrally-buoyant floats called Slocums" that "migrate vertically through the ocean by changing ballast, and they can be steered horizontally by gliding on wings at about a 35 degrees angle ... During brief moments at the surface, they transmit their accumulated data and receive instructions ... Their speed is generally about 0.5 knot." This chapter describes a class of autonomous underwater vehicles (AUVs) that are realizing Stommel's vision. Because they can be constructed for the cost of a few days ship time, can be reused, are light enough to be handled from small boats, can operate for a year or more while covering thousands of kilometers, and report measurements almost immediately while being directed from shore, these vehicles can make subsurface ocean observations at a fraction of the costs of conventional instrument platforms. This cost reduction makes feasible a proliferation of instruments to substantially increase spatial and temporal density of ocean observations and, consequently, the range of scales that can be resolved.

The vehicles described here, autonomous underwater gliders, change their volume and buoyancy to cycle vertically in the ocean and use lift on wings to convert this vertical velocity into forward motion. Wing-lift drives forward motion both as the vehicles ascend and descend, so they follow sawtooth paths. The shallowest points on the sawtooth are at the surface where satellite navigation and communication are carried out, eliminating the need for *in situ* tracking networks. Four basic sampling modes for gliders have presented themselves. If forward motion is used to counter ambient currents and maintain position, gliders can sample virtually as a vertical array of moored instruments with a single sensor package. Moving from place to place yields a highly resolved section, although the slowness of advance mixes time and spatial variability. Gliders controlled remotely from a research vessel can form an array to describe the spatial and temporal context in which intensive shipboard measurements were embedded. Finally, the long operating lives and ability to sample densely suit gliders to missions where unusual events are sought and then studied intensely when found. With ranges of thousands of kilometers, durations of O (year), and control and global data relay through satellite, many new missions are anticipated.

This class of vehicles is distinguished by four inter-related operating characteristics: the use of buoyancy propulsion, a sawtooth operating pattern, long duration, and relatively slow operating speeds. At a fundamental level, generating forward motion from wings is similar to propulsion by propellor thrust. In gliders, electric or thermal energy is converted to pressure–volume work to change vehicle volume and generate relative motion that is converted to forward thrust by wing lift. In propellor vehicles, internal energy is converted to shaft rotation that provides the relative motion so that propellor blades can generate lift and vehicle propulsion thrust. Buoyancy propulsion is well suited to the performance objectives of this class of vehicle. It provides the vertical sampling needed in the stratified ocean where variability along a horizontal path often results mainly from vertical migration of patterns. Typical glide slopes, of the order 1:4, are much steeper than the slope of oceanic distributions, so each leg of a glider sawtooth produces the equivalent of an ocean profile. As Stommel envisioned, a primary objective of gliders is to observe ocean variability, which spans the energetic time scales from days to seasons that characterize meteorologically forced, mesoscale, and interannual variability. This demands operational lifetimes of months that require efficient conversion of energy to motion and minimization of hydrodynamic drag. Much of AUV drag is caused by forward motion of the hull and can be minimized only by streamlining and operating at slow speeds. The significant drag associated with lift can be reduced by using long, slender wings (or propellor blades) with high lift-to-drag ratios (see Chapter 7). While the high power-to-volume ratios needed for high speeds are easier to achieve with propellor systems, high lift-to-drag and high efficiency is more easily achieved in the simpler hydrodynamic environment of wings.

The underwater gliders discussed here were designed to fit into a particular sampling niche. They are small enough to be handled by a crew of 1–3 on small boats without the power assistance generally available only on research or survey vessels. They are inexpensive enough that individual projects might afford several – this translates to construction costs equal to that of a very few days of research ship time or a small mooring. They can sample frequently enough to resolve phenomena such as internal waves, fronts, the diurnal cycle, coastal variability and biological patchiness. Spanning depth ranges of O (1 km) in a few hours requires vertical speeds of O $(0.1\,\mathrm{m\,s^{-1}})$. Collecting long time series is made feasible by amortizing the costs of deployment/recovery over long operational durations of O (year). A high operating speed would, of course, be desirable, but this conflicts with the primary goals of low cost, small size and long duration. Yet, if gliders are to maintain station or occupy prescribed sections they must have operating speeds that are at least comparable to typical sustained large-scale currents averaged over the glider's operating depth. Localized currents may exceed $1\,\mathrm{m\,s^{-1}}$ while currents of O $(0.3\,\mathrm{m\,s^{-1}})$ are common. Depth-averaged currents are generally weaker than surface currents and gliders can operate in localized strong currents by drifting downstream as they cross and then make up lost ground in parallel regions of weak flow. Nevertheless, long periods operating at O $(0.3\,\mathrm{m\,s^{-1}})$ are needed and higher peak speeds would expand the operating area. Accurate on-surface navigation, the ability to accept simple commands from shore and to relay kilobytes of data to shore, reliable control of gliding performance and the ability to process data *in situ* were additional design requirements.

The functional design goals roughly translate to:

- endurance of O (year) at operating speeds of O $(0.3\,\mathrm{m\,s^{-1}})$ and vertical velocities of O $(0.1\,\mathrm{m\,s^{-1}})$;
- mass of O (50 kg), length of O (2 m), and maximum operating depths from O (100 m) to O (1,000 m);
- GPS navigation, an ability to receive simple commands and transmit kilobytes per day of data, and PC-level internal data processing; and
- construction cost of O ($50,000) and refueling cost of O ($3,000).

The technology developed to meet these objectives is very new. In Section 2, we describe four field-tested gliders, none of which was operational much before the year 2000. Each glider embodies a particular set of solutions to the major design challenges and evaluation of these solutions is not yet complete. It is likely that different combinations of features will appear in future vehicles. For this reason, the discussion in Section 3 focuses on specific design challenges and comparison of how they have been addressed. Section 4 describes observations from a few glider operations to give a flavor for what can be accomplished. This chapter concludes with some thoughts on suitable sensors and the future.

3.2   THE GLIDERS OF 2001

Twenty years after Stommel's futuristic article anticipating Slocums (named for Joshua Slocum, the first solo global circumnavigator), there exist three ocean-proven electric-powered gliders and a field-tested thermal-powered glider. The University of Washington (UW) "Seaglider" and Scripps Institution of Oceanography (SIO) "Spray" (Joshua Slocum's boat was named "Spray") are electric-powered vehicles optimized for use in the deep ocean where long duration is paramount. Webb Research Corp (WRC) optimized "Slocum Battery" for missions in shallow coastal environments and "Slocum Thermal" for long duration missions in waters with a well-developed thermocline.

3.2.1   Spray and Slocum Battery

Slocum Battery (Webb *et al.*, 2001) and Spray (Sherman *et al.*, 2001) are the simplest gliders described here. Both employ battery-powered buoyancy engines and aluminium pressure hulls that are shaped for low hydrodynamic drag. Figure 3.1 shows Slocum and the forces involved in gliding. Figure 3.6 is a photograph of Slocum. Spray is shown in Figures 3.2 and 3.3.

Slocum Battery is optimized for shallow-water coastal operation, where rapid turning and vertical-velocity changes are needed. It has a shallow pressure rating and uses a large-volume single-stroke pump to push water in and out of a port in the nose for rapid volume control. This pump is more efficient in shallow operation than are the pumps designed for deep operation. Primary pitch control is achieved by the movement of water for buoyancy control and pitch is trimmed by moving internal mass. An operable rudder controls the turning rate while maintaining a level attitude for an acoustic altimeter. Antennas are housed in a vertical stabilizer that is raised above the surface when the vehicle is pitched forward for navigation or communication. Pitch moment and surface buoyancy are augmented by inflating an airbladder at the surface. Sensors are mounted in a modular center payload bay.

Spray is optimized for long-duration, long-range, and deep-ocean use where the emphasis is on energy efficiency. The hull employs a finer entry shape than the Slocum Battery glider hull, which has about 50% higher drag (Sherman *et al.*, 2001). Spray employs a high-pressure wobble-plate reciprocating pump and external bladders in the same hydraulic configuration as ALACE floats (Davis *et al.*, 1991). GPS and satellite communication antennas are housed in a wing that is rolled vertical during navigation and communication. The vertical stabilizer houses an emergency-recovery antenna. Scientific sensors may be mounted on the hull (as is the CTD in Figure 3.2) or aft of the pressure hull in the flooded compartment that supports the vertical stabilizer. Extra room for sensors can be obtained by lengthening this compartment.

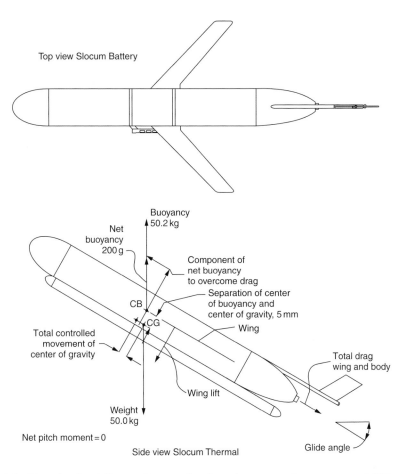

Figure 3.1 Top view is of Slocum Battery showing placement of a un-pumped CTD and the modular center payload bay. Side view is of Slocum Thermal showing the forces involved in gliding upwards, which applies to all gliders. Tubes below the Thermal model are the heat exchangers that drive the vehicle's thermodynamic propulsion cycle.

Glide control in Spray is achieved exclusively by axial translation and rotation of internal battery packs. Pitch is controlled simply by moving the center of gravity in the manner of a hang glider. Turning is initiated by rolling. This gives the lift vector a horizontal component and induces vehicle sideslip in the plane of the wing in the direction of the buoyant force. The horizontal component of lift provides the centripetal force for turning while sideslip acting on the vertical stabilizer produces the yaw moment needed to change vehicle heading. For example, to turn right during descent the right wing is dropped, like a conventional airplane, generating a lift component to the right that drives the vehicle to the right. Sideslip down and to the right acts on the vertical stabilizer causing the nose to yaw to the right. To turn right in ascent the glider is rolled oppositely by dropping the left wing.

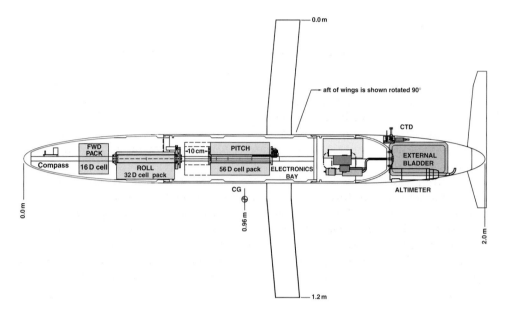

Figure 3.2 Schematic of Spray. Forward of the wings is a top view, aft is a view from the port side. The hull is three pieces. Separate battery packs are moved to control pitch and roll. Antennas are enclosed in a wing that is rolled vertical on the surface. An aft flooded section houses hydraulic bladders and some science sensors.

Figure 3.3 Spray being loaded onto a small boat in preparation for deployment. The aft flooded bay has been uncovered, exposing the hydraulic bladders. Above and forward of this a small protective covering obscures view of the Precision Measurement Engineering CTD.

3.2.2 Seaglider

Design of the battery-powered Seaglider (Eriksen *et al.*, 2001) emphasized efficient energy use to enable missions of one-year duration and ocean-basin ranges. Seaglider is enclosed in a hydrodynamic fibreglass fairing supporting wings, a vertical stabilizer and trailing antenna staff (Figures 3.4 and 3.5). The shroud is a low-drag hydrodynamic shape, with a maximum diameter at 70% of the body length from the nose, a shape that retains a laminar boundary layer forward of this maximum-diameter point. Form drag is proportional to speed to the $\frac{3}{2}$ power rather than the usual quadratic drag.

The fairing encloses a pressure hull with compressibility similar to that of seawater so that buoyant driving force is not lost as the vehicle changes depth. To achieve neutral compressibility, the hull is comprised of a series of deflecting arched panels supported by ring stiffeners. Compared with a conventional stiffer hull, a neutrally compressible hull can save pumping well over $100\,\mathrm{cm}^3$ at the bottom of a 1,000 m dive. The associated energy saving increases as dive depth squared. Seaglider efficiently maintains position in weak currents by pitching to near vertical and using minimal buoyancy forcing.

Seaglider buoyancy control is provided by a hydraulic system of the ALACE type. Movement of internal masses controls gliding and pitches the vehicle forward to raise the trailing antenna mast during communication and navigation. The wing is so far aft that the turning dynamics are opposite that of Spray. In descent, to turn right the vehicle's left wing is dropped so that lift on the wing drives the stern to left, overcoming lift off the vertical stabilizer, and initiating a turn to the right. Hydrodynamic lift on

Figure 3.4 A Seaglider recovered aboard an inflatable boat after one month in Possession Sound. An un-pumped Sea Bird Electronics conductivity cell (with plastic tubing connected) is mounted above the wing. The antenna at the end of the trailing mast is not in view.

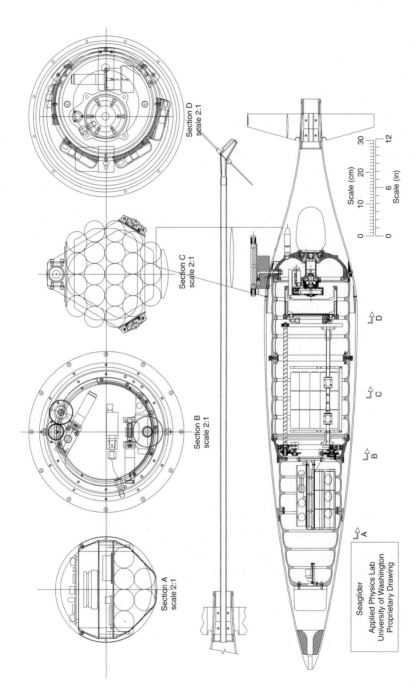

Section D
scale 2:1

Section C
scale 2:1

Section B
scale 2:1

Section A
scale 2:1

Scale (cm)
Scale (in)

Seaglider
Applied Physics Lab
University of Washington
Proprietary Drawing

Figure 3.5 Schematic of Seaglider including (from top to bottom) hull cross-sections, the antenna mast, the wing plan-form and cross-section, side view of the hull, and a scale.

the sideslipping hull produces the centripetal force to curve the course. Conversely, in ascent a roll to the left produces a left turn.

3.2.3 Slocum Thermal

Stommel's Slocum concept envisioned a glider harvesting the energy needed for its propulsion from the ocean's temperature gradient. This concept is embodied in Slocum Thermal depicted in Figures 3.1 and 3.6. In missions with electric-powered gliders, 60–85% of the energy consumed goes into propulsion, so a thermal-powered glider may have a range 3–4 times that of a similar electric-powered vehicle. Except for its thermal buoyancy system and using roll rather than a movable rudder to control turning, Slocum Thermal is nearly identical to Slocum Battery. This Slocum's wing is far enough aft that it turns, as does Seaglider, oppositely from Spray and conventional aircraft.

Slocum Thermal propulsion depends on the volume change associated with melting a material with a freezing point in the range of ocean temperatures. As Figure 3.7 describes, in warm surface waters the working fluid is heated, melts, and expands. This expansion compresses an accumulator where energy is stored. Descent is initiated by transferring fluid from an external bladder to an internal reservoir. At temperatures colder than freezing, the freezing contraction draws fluid out of the internal reservoir into the heat exchanger. For ascent, energy stored in the accumulator does the pressure–volume work and the cycle repeats. The heat exchange volume is inside tubes that run the vehicle's length (see Figures 3.1 and 3.6) and provide a large surface area for rapid heat flow.

While Slocum Thermal has yet to complete a long mission at sea, a thermally powered autonomous profiling float completed 120 profiles to over 1,250 m over 240 days (Webb, 1999) and Slocum Thermal has operated autonomously in Lake Seneca, New York.

Figure 3.6 Photographs of both Slocum Battery (above) and Slocum Thermal (below). *See* Color Plate 3.

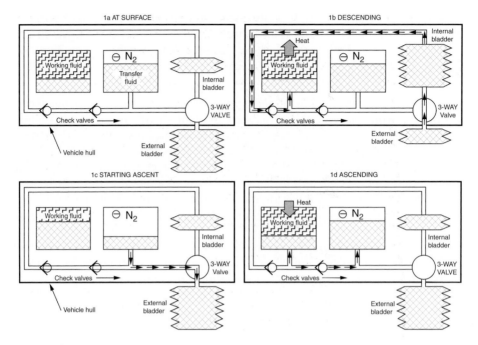

Figure 3.7 The thermodynamic cycle that powers Slocum. The left box is the heat exchange volume and the heat flow is shown with arrows. The middle box is a nitrogen-backed accumulator to store mechanical energy. The cycle is controlled by the three-way valve.

3.2.4 Operating Characteristics

Salient physical and operating characteristics of the four gliders are shown in the Tables 3.1–3.4, where U denotes horizontal velocity, W denotes vertical velocity and "Payload" indicates mass reserved for scientific instruments inside the pressure hull. Endurance figures given at a single horizontal velocity include the energy expenses of propulsion and communication of a 2 kbytes message on every cycle. Approximate costs are for a complete vehicle with conductivity, temperature and pressure sensors.

3.3 DESIGN CHALLENGES AND SOLUTIONS

The gliders described in Tables 3.1–3.4 are similar in overall characteristics because they were all designed to meet similar objectives. The ease of handling and low operating cost needed to make long time series feasible dictate small size and slow operating speed. Propulsion using buoyancy control follows Stommel's Slocum vision and wide operational experience with autonomous floats and has some engineering advantages in eliminating shaft seals and moving external parts. Ultimately, however, evaluation of buoyancy propulsion will depend on energy efficiency and the value of sawtooth trajectories for sampling the ocean. We are aware of no modern efforts to design a long-duration, efficient, slow-speed AUV using a propellor and without this,

Table 3.1 Characteristics of Spray.

Hull	Length 200 cm, Diameter 20 cm, Mass 51 kg, Payload 3.5 kg
Lift surfaces	Wing span (chord) 120(10) cm, Vertical stabiliser length (chord) 49(7) cm
Batteries	Primary lithium sulfuryl chloride, 52 DD cells in 3 packs, Energy 13 MJ at 0 °C, Mass 12 kg
Volume change	Max 0.9 l, Motor and reciprocating pump, 50(20)% efficient @ 1,000(100) dbar
Communication	Orbcomm satellite, 2-way, 0.5 byte/s net, 400 J/kbyte, GPS navigation
Operating	Max P 1,500 dbar, U_{MAX} 0.45 m s^{-1}, Control on depth + altitude + attitude + vertical W
Endurance	$U = 0.25$ m s^{-1}, 18° glide, Buoyancy 0.15 kg, Range 7,000 km, Endurance 330 days
Cost	Construction $25,000, Refuelling $2,850

Table 3.2 Characteristics of Slocum Battery.

Hull	Length 150 cm (overall 215), Diameter 21 cm, Mass 52 kg, Payload 5 kg
Lift surfaces	Wing span (chord) 120(9) cm swept 45°, Stabiliser length (chord) 15(18) cm
Batteries	Alkaline, 260 C cells, Energy 8 MJ at 21 °C, Mass 18 kg
Volume change	Max 0.52 l, 90 W motor and single-stroke pump, Efficiency 50%
Communication	RF LAN, 5700 bytes/s, 3 J/Mbyte, 30 km range, GPS navigation
Operating	Max P 200 dbar, Max U 0.40 m s^{-1}, Control on depth + altitude + attitude + vertical W
Endurance	$U = 0.25$ m s^{-1}, 20° glide, Buoyancy 0.26 kg, Range 2,300 km (estimated)
Cost	Construction $50,000, Refuelling $800

Table 3.3 Characteristics of Seaglider.

Hull and shroud	Length 180 cm (overall 330), Diameter 30 cm, Mass 52 kg, Payload 4 kg
Lift surfaces	Wing span (av chord) 100(16) cm, Vertical stabiliser span (chord) 40(7) cm
Batteries	Primary lithium thionyl chloride, 81 D cells in 2 packs, Energy 10 MJ at 0 °C, Mass 9.4 kg
Volume change	Max 0.840 l, Motor and reciprocating pump, 40% (8%) efficient at 1,000(100) dbar
Communication	Cellular 450 byte/s net, 26 J/kbyte. Iridium 180 byte/s, 30 J/kbyte, GPS navigation
Operating	Max P 1,000 dbar, Max U 0.45 m s^{-1}, Control on depth + attitude + vertical W
Endurance	$U = 0.25$ m s^{-1}, 18° glide, Buoyancy 0.22 kg, Range 4,500 km, Endurance 220 days
Cost	Construction $60,000, Refuelling $1,375

Table 3.4 Characteristics Slocum Thermal.

Hull	Length 150 cm (210 overall), Diam 21 cm, Displacement 52 kg, Payload 2 kg
Lift surfaces	Wing span (chord) 120(9) cm swept 45°, Stabiliser length (chord) 15(13) cm
Batteries	Alkaline (for instrumentation, communication), Energy 6 MJ at 21 °C, 14 kg
Volume change	Max 0.4 l, 6 kJ harvested each cycle, 10 °C minimum temperature difference
Operating	Max P 1,200 dbar, Max U 0.27 m s^{-1}, Control on depth + attitude + vertical W
Endurance	$U = 0.25$ m s^{-1}, 38° glide, Buoyancy 0.235 kg, Range 30,000 km (estimated)
Cost	Construction $70,000, Refuelling $800

one cannot evaluate the conjecture that it is easier to make wings efficient. Even if propellor vehicles can be made equally efficient, the need to span large depth ranges to adequately sample the ocean would require buoyancy control systems or significant additional drag to generate the sawtooth trajectory that gliders come by naturally.

3.3.1 Buoyancy Generation

The innovative thermally powered Slocum is capable of a remarkable 30,000 km range. While this requires temperature differences sufficient to generate enough average power to overcome the drag inevitably associated with heat transfer and to deal with ocean currents, it is remarkable that a 50 kg AUV might circumnavigate the globe. While electric power may have a broader operating region, and the reliability of all the glider propulsion systems is yet to be proven, thermal power is too powerful a concept to be ignored in either gliders or profiling floats.

Two of the present gliders that use electric-motor powered buoyancy control systems have ranges at 0.25 m s^{-1} of more than 4,000 km. A single-stroke pump with a rolling diaphragm seal is used to pump seawater into and out of Slocum Battery. This approach is simple and has superior two-way buoyancy control, but it is difficult to accomplish in high-pressure applications. Seaglider and Spray use high-pressure reciprocating hydraulic pumps to transfer hydraulic fluid between internal and external bladders. While this allows a relatively small and light hydraulic system, controlled increases of buoyancy at high pressure require special metering and cavitation at the pump inlet can induce failure (Seaglider uses a separate boost pump to overcome this and Spray has an high-compression ratio pump to handle small bubbles). While optimization will depend on gaining more field experience, electric buoyancy control appears flexible and perhaps efficient enough for most missions.

3.3.2 Hulls and Hydrodynamic Performance

Drag and compressibility are largely determined by a glider's hull. For electric gliders these characteristics are important to achieving long duration and, consequently, low cost. Seaglider has the most sophisticated hull employing a fairing that encloses a pressure hull with compressibility matched to seawater. Ports in the fairing allow any trapped air to be vented before diving. A conventional low-compressibility hull of the size of the gliders described here loses O (0.1 kg) buoyancy as it descends from the surface to 1 km depth, increasing the pressure–volume work needed to ascend. Particularly when operating with small buoyancy, the associate energy loss is significant. On the other hand, conventional (and less expensive) hulls allow a larger payload, including a larger battery pack that provides the energy for this extra work.

Discussion of hydrodynamic performance is facilitated by Figure 3.8, which describes the behavior of Seaglider (Gliding dynamics and performance are discussed at length by Eriksen *et al.*, 2001 and Sherman *et al.*, 2001). Because of drag induced by lift generation, hydrodynamic performance (e.g. efficiency, speed) depends directly on speed and angle of attack and, indirectly, on glide angle. At fixed buoyancy forcing, horizontal speed is maximized at relatively steep glide angles for which angle

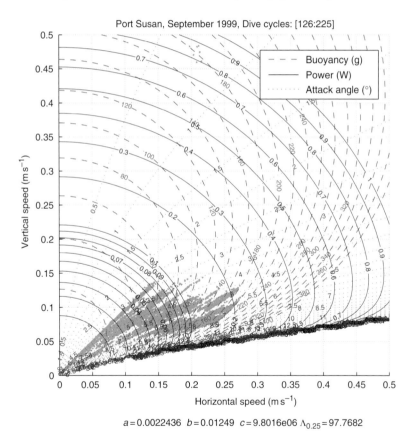

$a = 0.0022436 \quad b = 0.01249 \quad c = 9.8016e06 \quad \Lambda_{0.25} = 97.7682$

Figure 3.8 Glider performance diagram showing the buoyancy and power required to maintain a given speed and glide angle. This curve is for Seaglider but the behavior is similar for all designs. Note how at a given buoyancy, horizontal velocity U is maximized at a glide angle near 40°, whereas at fixed power U is maximized nearer a 14° glide. Green marks show observed Seaglider operating points. *See* Color Plate 4.

of attack and induced drag are minimized. At fixed power, however, speed peaks at a much shallower glide angle where induced drag is significant. All the gliders were designed with an objective of long range, which depends on the speed to power ratio, and, consequently, they use wings with high aspect ratios and high lift to induced-drag ratios.

It should be noted that the drag considerations differ somewhat between electric propulsion, where energy is limited, and thermal propulsion, where energy is unlimited but power is limited by the achievable heat transfer rate. An electric glider will typically operate near the shallow glide angle that maximizes range and at the minimum buoyant forcing to meet speed requirements. Heat flow and thermal power are, on the other hand, maximized by rapid cycling between depths with differing temperatures (overall speed affects the thermal resistance very little). For this reason, thermal gliders operate at relatively steep glide angles to increase vertical velocity. In these conditions low hull-drag is still highly desirable, but angle of attack and induced drag are low so that efficient, high-aspect-ratio wings are much less important than when energy is limiting.

Seaglider's uniquely shaped hull attempts to maximize the area over which the boundary layer in laminar. Spray has a conventional shape but efforts were made to find a low drag shape. Slocum, relying on the unlimited energy available with thermal power, uses a hull shape that simplifies construction and maximizes packing efficiency. Bio-fouling may increase vehicle drag significantly, as suggested by performance analysis on a one-month Seaglider mission in a fjord. In light of this and other uncertainties, the conditions under which each approach to hull design is to be preferred remains to be found through field experience.

3.3.3 Communication

Accurate navigation, the ability to transmit O(kilobyte) datasets quickly, and the ability to receive short messages adjusting operation are essential to autonomous operation. All the gliders described here use GPS navigation, which meets performance objectives admirably. Low-earth-orbit Orbcomm satellite (Spray), radio frequency Local Area Network communication (Slocum Battery) and Circuit Switched Cellular (Seaglider) communication have been used in the field, and System Argos is useful at least for emergency backup and locating. Low-earth-orbit systems have up to 5 orders greater speed and 3 orders better energy efficiency than Argos and additional systems (Iridium, Globalstar) are being implemented. These promise higher data rates and lower communication costs than are possible with Orbcomm. We are hopeful that at least one satellite system will survive the present economic competition.

Maintaining antennas clear of the surface in a seaway is the main technical challenge for communication and our gliders use different systems to achieve this. Seaglider and Slocums employ trailing antenna staffs that house the needed antennas. When on the surface these gliders pitch forward to raise these antenna staffs and Slocums employ an external bladder inflated by a small air pump to increase surface buoyancy and pitch moment. Spray's antennas are contained in a wing that is rolled vertical for navigation and communication. All systems are subject to loss of performance in high sea-states so adequate internal storage is necessary for several days of message buffering.

3.3.4 Gliding Control

Glider control involves monitoring performance, adjusting glide angle by controlling pitch and/or buoyancy, and adjusting heading by controlling roll or (for Slocum Battery) rudder position. All the gliders described here use Precision Navigation TCM2 attitude sensors to sense heading, pitch and roll, and they use pressure sensors to measure depth and, from pressure rate, vertical velocity. Spray and Slocum measure altitude using an acoustic altimeter while Seaglider estimates altitude from measured glider depth and a digitally stored map of water depth.

A movable rudder gives Slocum Battery the tightest turning radius (approximately 7 m) and allows turning without significant roll so that the acoustic altimeter, critical in shallow-water operations, remains accurate. The other gliders, intended for deep water, typically roll about $30°$ to achieve turning radii of 20–30 m. Because glide angle and performance are sensitively linked, gliding is generally more closely controlled than turning. In normal gliding Spray adjusts pitch around a set point using proportional control on O (60 s) intervals while infrequently adjusting buoyancy to maintain vertical velocity within an operating range. Seaglider operates similarly, controlling gliding on a longer interval of O (300 s) and uses buoyancy adjustment as a primary control. Both vehicles accelerate control at the minimum and maximum depths where buoyancy and pitch are changed significantly. Seaglider is unique in using an onboard Kalman filter to estimate currents and adjust target heading and glide angle to compensate for them.

3.3.5 Sensors

Scientific payloads for gliders are limited by size, flow disturbance, and power requirement considerations. Sensor systems must fit within the payload fraction of O (50 l) vehicle and, because gliding involves modest buoyancy forces ($\sim$0.2–4 N), ballast and trim are paramount considerations. Sensors must be hydrodynamically inobtrusive, lest they spoil gliding performance by adding drag. For example, wind tunnel tests of Seaglider demonstrated that appending a toroidal conductivity sensor with 2% of the vehicle's frontal area added more than 25% to its drag. Streamlining can be achieved by using sensors that are small or mounted flush to the vehicle hull. Outward-looking acoustic and optical sensors conveniently fit this requirement and have been used on the gliders described here.

The overall power consumption of the four gliders discussed here is O (1 W). Achieving this requires low-power electronics and sampling schemes that limit the duty cycle of sensors. Slow glider speeds allow sampling intervals of O (10 s) to achieve vertical resolution of O (1 m) but sensors with limited energy usage are still important to the overall power budget. For example, sampling temperature and salinity consumes roughly 0.1 J, dissolved oxygen about 0.4 J, and fluorescence and optical backscatter about 2 J. Glider controllers use O (0.1 W) when not in low-power sleep mode and particularly for low throughput systems, data transmission is also a significant factor in the power budget.

Like autonomous floats, gliders achieve their economy by having moderate construction costs and long operational lifetimes. Achieving this economy therefore requires scientific sensors that are stable over many months. The primary challenge to stability is bio-fouling. Compared with floats using Argos communication, gliders can reduce fouling by spending little time on the surface and in the euphotic zone. Avoidance of exposure to the sea surface itself also avoids surfactants which affect conductivity sensors, so keeping instrumentation submerged while gliders communicate is presumably helpful. Stability of temperature, conductivity (Bacon *et al.*, 2001; Riser and Swift, 2002) and optical sensors over many months has been achieved by profiling floats (Davis *et al.*, 2001), and a glider's exposure to the euphotic zone is only slightly worse.

3.3.6 Operating Costs

The principal operating costs of gliders are vehicle preparation (including energy cost), deployment and recovery, and communication. The small size and long range of the gliders described here implies low logistic overhead for operations compared to reliance on research vessels. Nearshore launch and recovery from small boats in daylight and fair weather by a crew of one or two is sufficient for glider access to most of the ice-free ocean. Communications costs depend strongly on method. Costs for global coverage range from O ($10/kbyte) for Orbcomm to O ($0.10/kbyte) for Iridium. Battery costs are of the O ($1) per deep-ocean vertical cycle. Thus even with construction costs amortized over a few deployments, the operating costs for a mission reporting hundreds of multivariable samples in each of a thousand dive cycles is about $10,000, about the same as one day of research-ship time. In perspective, gliders can collect several multivariable (e.g. temperature, salinity, velocity, oxygen, fluorescence, optical backscatter, etc.) profiles for the cost of a single expendable bathythermograph (XBT) probe.

3.4 EXAMPLES OF OBSERVATIONS

Glider technology is new and its capabilities have yet to be fully demonstrated in field experience. Nevertheless, the three battery-powered vehicles have all produced datasets that begin to sketch out how gliders can be used. They have been successfully used with, in various combinations, temperature, conductivity, dissolved oxygen, fluorescence and optical backscatter sensors. They have been used in single- and multi-vehicle arrays to collect time series of up to one-month length, time series of short sections, and a 270 km section over 13 days. This section describes some of that data.

All the gliders have also been used to measure vertically averaged currents from the difference between dead reckoning and GPS navigation. Dead reckoning is based on measured headings and speed through the water deduced from measured vertical velocity, pitch and buoyancy. A model is used to infer angle of attack from buoyancy and pitch (assuming ocean vertical velocity is negligible). With measured pitch, the angle of gliding is calculated, and from this vertical velocity determines horizontal speed through the water. Considering the main errors (vertical ocean velocities and errors in the angle of attack), depth-average current measurements are accurate to O ($1\,\mathrm{cm\,s^{-1}}$) over O ($1\,\mathrm{h}$) time intervals. When data from many depth cycles are combined, the measurement of depth-average velocity becomes quite accurate, and when coupled with sections of ocean density, allows gliders to accurately estimate absolute geostrophic currents. Thus gliders can attack the long-standing problem of hydrography: properly referencing geostrophic shear. The velocity estimates also provide accurate measurements of gliding performance. The performance figures for Spray and Seaglider given in Section 2 are based on these analyses; performance for Slocum is predicted.

3.4.1 Time Series (Virtual Moorings)

During a July 2000 field trial at the Rutgers LEO-15 research site near Tuckerton, New Jersey, a Slocum Battery completed a 10-day deployment in which it collected a 5000-dive time series of temperature, salinity and vertically averaged velocity. On average every 150 s the Slocum dived to 15 m depth, triggered on depth and altitude above the shallow bottom. Data was relayed to shore using an RF-LAN and on occasion control of the vehicle was switched between Tuckerton and Falmouth, Massachusetts using an Internet to LAN connection. The Slocum mainly maintained station but, on command, also completed one 15 km cross-shelf section.

The field trial included Acoustic Doppler Current Profilers to sample currents that provide comparisons for the depth-averaged currents measured from the Slocum by its expected and actual surfacing locations. Figure 3.9 shows a time series of the Slocum derived water velocity, and moored ADCP data any time the Slocum was within 2 km.

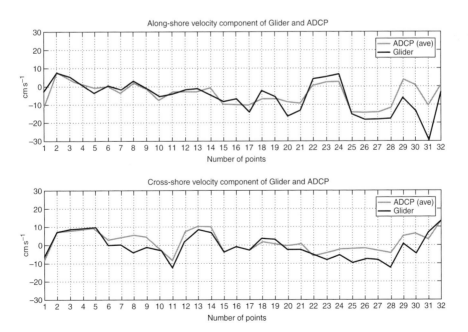

Figure 3.9   Time series of depth-average ocean velocity from Slocum (black), depth-average ADCP velocity (red) from the LEO-15 site during July 2000. *See* Color Plate 5.

The Slocum and vertically averaged ADCP compare well with an rms difference similar to that between ADCP measurements at 3 and 6 m at the same site or between sites separated across-shelf by 4 km. Some of the ADCP-Slocum difference likely also results from lateral variability. One can imagine more complex depth vs. time patterns that would allow measurement of the vertically averaged flow in vertically stacked depth ranges.

Seaglider and Spray participated in Office of Naval Research supported multi-vehicle sea trials in Monterey Bay during the summers of 1999 and 2000. In 1999 the Spray prototype was virtually moored in 450 m depth in Monterey Canyon for 10 days (see Sherman *et al.*, 2001). An acoustic altimeter established the bottom of profiles within a few meters of the bottom. The temperature and salinity time series showed internal wave motion to be bottom intensified in the canyon. In the face of internal-wave motions of O $(15\,\mathrm{cm\,s}^{-1})$ and a mean current measured to be about $3\,\mathrm{cm\,s}^{-1}$ up-canyon and to the south, surface position was maintained with standard deviation near 500 m.

Starting in June 2000, a Seaglider was virtually moored for a month in Possession Sound, a 3 km wide fjord in western Washington. In April 2001, two Seagliders were virtually moored at 1.5 km separation for a week across the Sound. Comparison of velocity computed from geostrophic shear and measured depth-averaged flow compared well with surface currents, showing that the exchange flow is largely geostrophic (Chiodi and Eriksen, 2002). These operations show how accurately positioned virtual moorings can be easily established.

3.4.2   A 270-km Section

In October 2001 a Spray was sent to sample temperature, salinity and optical backscatter along a 270 km across-shelf section southwest of San Diego. The glider was deployed not far from shore by a small boat. It dove to 500 m (or the bottom if detected by an acoustic altimeter) with buoyancy and glide angle to produce forward velocity near $0.25\,\mathrm{m\,s}^{-1}$. The course was initially west from San Diego and then south-west, crossing one shallow bank. Data and mission-control commands were relayed by Orbcomm satellite, mainly in the mode used for global communication. The section was completed in 13 days. The vertically averaged velocity (generally averaged over 500 meters) is shown in Figure 3.10 and the section of density is portrayed in Figure 3.11.

Figures 3.10 and 3.11 show how gliders could dramatically improve monitoring of the coastal environment. Figure 3.11 shows the slope of isopycnals that, through geostrophy, define vertical variation of flow through the section. The isopycnal's broad downward slope to the east demarcates the shear of the California Current – compared with deep water, surface flow is to the south toward the equator. The isopycnal upturn near the coast indicates a reverse shear that is usually indicates a shallow nearshore poleward countercurrent but might also indicate equatorward flow at depth. Cost prevents the quarterly CalCOFI shipborne survey from resolving this feature, but the glider survey's close "station" spacing makes its details quite clear. Spray's absolute transport measurements show a weak $(\sim 1\,\mathrm{cm\,s}^{-1})$ average southward flow in the upper 500 m west of 118.5°W as expected from the California Current. The structure of the

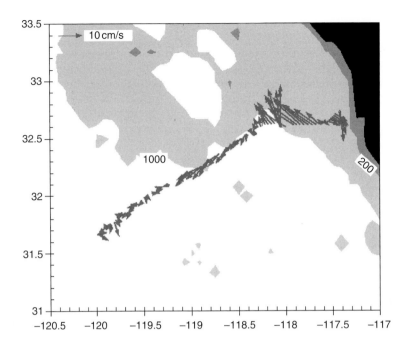

Figure 3.10 Absolute velocity averaged over the upper 500 m from a 150 nm Spray section off San Diego CA. The section was carried out to the southwest in 13 days. The 200 m and 1,000 m isobaths and the coast are shown.

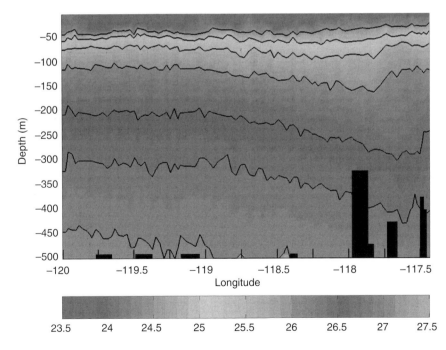

Figure 3.11 Density from the Spray section in Figure 3.10. Spacing of the temperature and conductivity profiles is about 3 km. The broad isopycnal slope downward to the east indicates the geotrophic shear of the California Current. The nearshore upward slope is associated with a near shore countercurrent. *See* Color Plate 6.

poleward flow, concentrated some 50 km off the coast, however, could not be anticipated from geostrophic shear and this emphasizes the importance of glider velocity
measurements. Highly resolved hydrographic surveys with velocity references and the
ability to identify barotropic and/or ageostrophic flows make gliders a powerful way
to observe the coastal ocean. For example, a pair of gliders could, at quite feasible
cost, produce a time series of sections like those in Figures 3.10 and 3.11 with an
average sampling interval of one week.

3.4.3 Repeated Multi-vehicle Sections

Three Seagliders were used in August 2000 Monterey Bay trials to demonstrate an
ability to gather repeated surveys using mulitple coordinated vehicles, something
that remains a rare luxury using ships (Figure 3.12). All data, as well as commands
to the vehicles, were telemetered via cellular telephone in near real time to and
from computers ashore and aboard a small vessel. The first sampling task was to
repeatedly collect sections across the continental shelf at the entrance to the bay. Two

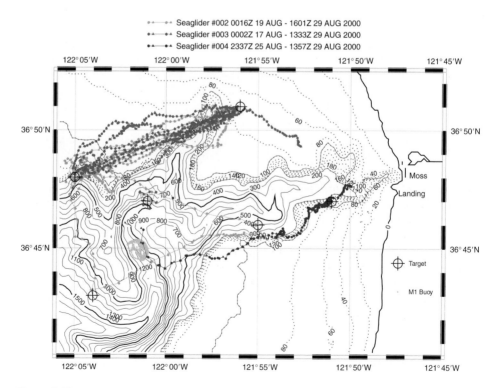

Figure 3.12 Tracks of three Seagliders in Monterey Bay (depth contours in meters). Two
gliders made a total of 13 sections along the north rim of Monterey Canyon. At the end of the
exercise, one of these (track in red) remained near a target about 2 miles north of a surface
mooring (buoy positions shown in cyan). *See* Color Plate 7.

gliders simultaneously surveyed a 15 km long transect on the northern edge of a large submarine canyon for 10 days. Temperature and salinity sections collected described the development and decay of a wind-forced surface mixed layer. The vehicles dove to within a few meters of the bottom or 250 m, whichever was shallower, demonstrating the ability to navigate over topography using a bathymetric map.

At the end of the exercise, one Seaglider was commanded to a target about 3.5 km north of the anchor position of a surface mooring maintained by the Monterey Bay Aquarium Research Institute. A tight cluster of surface positions (red symbols, Figure 3.12) demonstrated that this virtually moored glider held position at least as well as the moored surface buoy (cyan symbols, Figure 3.12). The excercise also demonstrated how a glider can, under remotely relayed commands, operate in different modes on the same mission.

3.5   THE FUTURE

Glider operations are in their infancy and the next step is clearly to use the developed technology to address scientific and environmental problems in order to develop procedures to interpret glider data, to refine and make more reliable the technology, and to assess the importance and adequacy of different technical characteristics (e.g. cost, energy efficiency, speed, endurance, reliability, and communication rate). There are many problems not yet solved. For example, how often do sharks attack these swimming aluminum fish? Can low drag be maintained for months in the presence of biofouling? Which sensors are adequately stable? These questions can only be answered from field experience. Gathering this experience can and should also advance ocean research. One would expect new technical approaches to sensors, such as those described in Chapter 15 and new combinations of existing approaches to appear in new gliders as a result of the experience gained over the next few years.

Gliders should play an important role in the emerging global ocean observing system, supplementing data from the Argo array of profiling floats particularly in regions of high interest where the Argo array has too little spatial resolution or where it is important to separate time and space variability more completely. Boundary currents, the equator, high-latitude convection regions, and continental shelves are regions where gliders in virtually moored or repeated-survey mode are likely to be valuable. Gliders will also serve as handy and efficient platforms for gathering long environmental records of variables not widely measured by operational systems. Their low cost and operational flexibility will likely also make them useful in short-term intensive campaigns. For this, new methods of communication (including underwater links) and more extensive schemes for control of networked gliders need to be developed.

The utility of all autonomous observations depends on availability of suitable sensors for a wide range of physical, optical, chemical, and biological properties. For gliders, which achieve economy through long life and low hydrodynamic drag, stability, low power and small size are key attributes. Biofouling is the primary concern for stability. It can be predicted that early successes with available sensors on gliders will provide impetus to expand the suite of variables that can be measured.

In looking back on Stommel's 1989 article anticipating autonomous gliders, we marvel at how much of what followed he had predicted. While lacking his vision, we are confident that this approach to ocean observation is just now reaching the limit of what he foresaw and that new innovation will soon carry us to areas we cannot see now.

4. SOLAR POWERED AUTONOMOUS UNDERWATER VEHICLES

D. RICHARD BLIDBERG[a] and MIKHAIL D. AGEEV[b]

[a] *Autonomous Undersea Systems Institute, Durham, New Hampshire 03824, USA and*
[b] *Institute for Marine Technology Problems (IMTP), Russian Academy of Sciences,
Far Eastern Branch, 5a Sukhanov Street, Vladivostock 690950, Russia*

4.1 INTRODUCTION

Changes to the global environment brought about by both natural forces and man's activity are a subject of concern to much of the world's population. A number of international research programs are focused on developing a better understanding of ocean processes that impact our environment. A common thread seen in all of these efforts is the need to obtain a significant increase in our ability to acquire data from the ocean. Some estimates suggest that we must increase our data gathering capability by two to three or more orders of magnitude in order to meet current needs. This limitation is reflected in the worldwide concern by research organisations over the lack of sufficient data with which to understand the dynamics of chemical, biological and physical characteristics and processes within the earth's lakes, seas and oceans.

Issues such as physical and biological coupling, biogeochemical processes and cycles both natural and human induced, fisheries and ecosystem modelling must be better understood. Spatial and temporal under-sampling in the oceans is generally recognised as one of the more important problems associated with current sampling systems.

Although more detailed monitoring of the ocean is necessary, current instrumentation does not provide sufficient capability to collect the required data from the ocean on a continuous basis. This problem of under-sampling of the ocean is a roadblock to many investigations. Few sensors exist that allow us to remotely sample large volumes of the ocean reliability. We are forced to use sampling techniques that have remained relatively the same for over a hundred years. We infer detailed processes by considering sparse data sets. If we are to meet this goal of significantly increasing our ability to acquire data and information, we must consider new technology. Low cost, unmanned, long endurance mobile sampling systems are, in many cases, a desirable alternative to the currently used oceanographic research vessels requiring large investments in manpower and equipment or by buoyed instrument strings constrained to a fixed position.

Autonomous Underwater Vehicles (AUVs) have a unique capability in that they are able to transit the ocean in three dimensions following a pre-defined path. AUV technology has evolved over many years. Many of the technological roadblocks preventing routine operational use of these systems have been overcome. Three issues

Figure 4.1   Prototype SAUV. *See* Color Plate 8.

remain as primary limitations: energy, navigation over extended times and distances, and communication of a user with the remote platform on a relatively real-time basis. Solar powered AUVs (Figure 4.1) begin to overcome all three of these limitations. They must surface to recharge the onboard energy system but the available energy is limitless. When surfacing to recharge, they are able to take advantage of Global Positioning Systems (GPS) navigation to update position (Hitchcock, 1996). They are also able to take advantage of the evolving communication infrastructure such as mature satellite based communications (Herrmann, 1997) and the newer Low Earth Orbiting Satellite (LEOS) communication systems.

When sampling areas are located near land, it is possible to take advantage of RF telemetry and communicate directly with a user. With continually easier access to the Internet, onshore receivers place data retrieval and mission control within easy reach of geographically dispersed users. If endurance is increased to a year or more, AUVs can provide the data gathering capability required to better understand global ocean processes. These autonomous sampling platforms offer the potential to acquire measurements at any point in the ocean. Current computer technology and satellite-based navigation and communications provide an opportunity to create systems that can work autonomously for long periods of time. For this reason the solar powered AUV (SAUV) has been developed.

4.2   THE SOLAR POWERED AUV

The Autonomous Undersea Systems Institute (AUSI), along with the Institute for Marine Technology Problems (IMTP) in Vladivostok, is investigating the characteristics and limitations of solar energy as an energy source for a long endurance AUV. This investigation seeks to understand the impact of the unique system components (specifically the photovoltaic array, the charging system, the energy storage system and the power management system) on the design of an AUV. It also seeks to identify constraints that an AUV system places on the solar energy system components. In parallel with these activities and experiments a small prototype SAUV testbed has been developed. This prototype vehicle is being used to evaluate the results of a number of analyses related to the use of solar energy to power long endurance, mobile data acquisition systems. The ultimate objective of the program is to develop a SAUV system for the marine community with an endurance in excess of one year.

Solar energy systems allow the endurance of AUVs to be increased dramatically thereby providing sampling systems to acquire needed scientific data over large volumes of ocean and across long time scales and to overcome the burden of recovering and recharging vehicles on a daily basis. The ability to undertake long endurance remote operations without the need for support ships and platforms and the reduced costs of acquiring that data make the development of SAUVs an important goal for today's ocean community. Inherent communication capability resulting from the need to surface on a regular basis provides the user/scientist with daily updates of data via satellite telemetry and an opportunity to adapt mission parameters based on analysis of acquired data. One can envision a scientist sitting at his/her desk studying newly acquired data and, based on the results of that analysis, modify parameters of the data acquisition task, and within minutes issuing a new command to the remote system while it recharges its energy system and updates its navigation system via GPS.

While these SAUVs have the potential to achieve these goals they, like most systems, have limitations that must be understood. These systems can store only a limited amount of energy and must be efficient in terms of energy utilisation. This limits the type of mission sensors carried onboard to relatively low power devices. While developers of scientific sensors are constantly reducing power requirements (see Chapter 15), many sensors require too much power for application on a SAUV. Most importantly the long endurance sensors must maintain calibration for extended periods of time or be capable of being remotely calibrated while at sea. Biofouling of both the scientific sensors and solar panels used to collect energy is a problem. This is particularly true when operating in the photic zone. At deeper operational depths, the problem is not as severe.

In general, a SAUV must surface on a daily basis for recharging. In some scenarios, such as shallow depth missions or those that require variable depth trajectories, this is not a problem. For missions requiring the acquisition of data from great depths the energy to ascend and descend must be part of the overall energy budget. There will also be times, depending on weather conditions, when a SAUV may not have sufficient energy to operate according to a pre-defined schedule. It may miss a day or two. If data acquisition needs to be completed during daylight hours, the system would have to operate on a 48-h cycle rather than a 24-h cycle (i.e. charge/discharge cycle period).

Certainly these limitations must be considered. SAUVs, however, offer the potential of acquiring continuous information for periods of time measured in terms of weeks to months to years. These sampling systems allow for the acquisition of data across long time scales and large ocean volumes that heretofore have proved very costly and in some cases impossible to obtain.

4.3 CAN AN AUV ACQUIRE ENOUGH SOLAR ENERGY FOR ITS NEEDS?

The amount of solar energy available on the ocean surface varies significantly with latitude, seasons and weather. The annual mean daily total solar radiation varies from less than 1 to about $12 \, \text{kWh} \, \text{m}^{-2} \, \text{day}^{-1}$ (Bahm, 1994). Conversion efficiencies for commercially available photovoltaic (PV) arrays are conservatively in the 10% range. Therefore, we can expect energy amounts in the range of 0.1–$1.2 \, \text{kWh} \, \text{m}^{-2} \, \text{day}^{-1}$. This variation in available energy will have obvious impact on possible tasks that a SAUV might perform. If we look at the latitudes roughly comprising the United States, and look at the 'worst case' numbers, which are typically in December, we see an average insolation varying from $1 \, \text{kWh} \, \text{m}^{-2} \, \text{day}^{-1}$ near the Canadian border to about $4.0 \, \text{kWh} \, \text{m}^{-2} \, \text{day}^{-1}$ in the southern US (Reineke, 1993). In order to establish some boundaries as to the range that can be expected from a SAUV, we consider two levels of solar insolation; data representing a high level of solar energy off the Hawaiian Islands in June ($6 \, \text{kWh} \, \text{m}^{-2} \, \text{day}^{-1}$) and data representing a low level of solar energy available near Vladivostok, Russia and Boston, Massachusetts in December ($1.5 \, \text{kWh} \, \text{m}^{-2} \, \text{day}^{-1}$).

It is possible to calculate the range of a small, SAUV operating in a low insolation area and in a high insolation area. With a PV array of $0.5 \, \text{m}^2$ (Figure 4.2) and a 10% conversion efficiency (PV module), in a low insolation region of $1.5 \, \text{kWh} \, \text{m}^{-2} \, \text{day}^{-1}$, this results in a PV array output of about $0.075 \, \text{kWh} \, \text{day}^{-1}$. If we consider the same vehicle, tasked to transit for 12 h and charge for 12 h in the high insolation area ($6.0 \, \text{kWh} \, \text{m}^{-2} \, \text{day}^{-1}$), we would obtain $0.3 \, \text{kWh} \, \text{day}^{-1}$ (Table 4.1) (AUSI, 1996). While the 10% efficiency for the solar arrays is an appropriate value for technology available in 2000 and has been validated by experiment using a Solarex MSX-30 Solar Module, data from a US National Renewable Energy Laboratory report (NREL, 1995) suggests that the efficiency of PV modules will increase to a level of 15–25% by the year 2010.

Table 4.1 Estimated daily range SAUV.

Location	Insolation ($\text{KWh} \, \text{m}^{-2}$)	Energy Collected (kWh)	Velocity ($\text{km} \, \text{h}^{-1}$)	Range (km)
Hawaii, year round	6.0	0.3	4.15	49.7
New Hampshire, winter	1.5	0.075	2.60	31.3

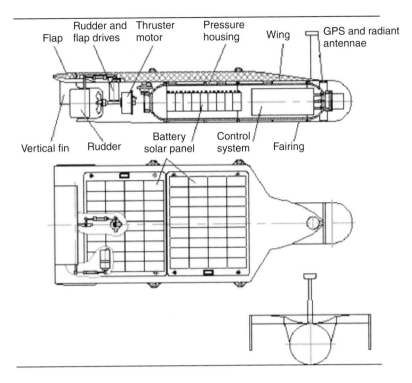

Figure 4.2 The SAUV.

4.4 SAUV SUBSYSTEMS

The current prototype SAUV, whose specifications are listed in Table 4.2, comprises a wing and a cylindrical shaped section attached to the wing. On each side of the wing are two fixed vertical stabilisers. The PV arrays are attached to the top of the wing and the electronics are inside a pressure housing within the cylindrical section of the vehicle. The pressure vessel is attached to the underside of the wing and a fibreglass fairing is attached over the pressure tube along with a hemispherical nose. The control surfaces and the thruster are integrated with the wing structure, Figure 4.2.

4.4.1 Basic Control System

Figure 4.3 describes the basic SAUV control system design. The SAUV is controlled by an autopilot mechanism. An operator is able to configure an onboard program through the external console (PC). The console can be connected to the SAUV either by the RF modem or a hard wire link. The control electronics monitor the energy in the storage batteries and ensure the energy level remains above a critical value by modifying mission parameters to match available energy. The data is also made available to the autopilot's programs. The control driver module converts system voltage to the levels necessary to drive the thruster as well as the control surface actuators.

Table 4.2   SAUV specifications.

Subsystems	
Control system and navigation	
RF modem	
Temperature and conductivity sensors	
Control surfaces and actuators	
Thruster	15 W
Electrical	
Solar boards	2 each MSX-30; 10–12% efficiency
NiCd storage battery	36 cells, 11 Wh each
Mechanical	
Working depth	1,000 m
Day time run	20–50 km
Dimensions	1.8 m by 0.73 m by 0.31 m
Mass	90 kg

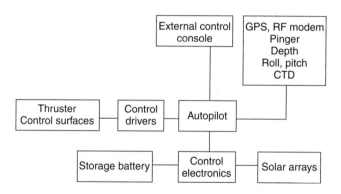

Figure 4.3   Block diagram of the SAUV control system.

4.4.2   Hydrostatics and Hydrodynamics

An AUV normally remains on the sea surface for very short periods of time. The SAUV, however, spends a great deal of time on the surface while recharging its energy system. For this reason, much thought must be put into its surface characteristics. While on the surface it should have sufficient buoyancy, when it is moving in the water column, it is desirable to be neutrally buoyant, but for safety reasons, it is desirable to be positively buoyant at all depths. One alternative is to install an active buoyancy system to add additional system buoyancy while it is on the surface. The added complexity of such a system is not attractive. An alternative that was chosen for implementation in the SAUV is to include two bladder-like tubes in each side of the upper wing. These tubes provide added buoyancy on the surface, but as the vehicle descends, the gas inside the tubes compresses and has minimal impact on the SAUV movement strategies.

Analysis was conducted to consider the effect of these buoyancy tubes both in tropical waters and the colder northern waters (Ageev, 1999).

Since the SAUV system will be at sea for extended periods of time, it must be able to withstand the surface environment. For moderate seas, the behaviour of the SAUV system is easily understood since the frequencies of the disturbances are much lower than the inherent frequency of the SAUV system. Its movements will probably coincide with the movement of the water. The most dangerous situation is falling off the crest of a wave that may cause the SAUV to overturn. It is important that the vehicle should not have some mode where it is stable in an inverted orientation. This situation does not exist in the SAUV, as it is self-righting and will return to its upright position after being overturned.

A series of tests was performed to evaluate the hydrodynamic properties of the SAUV prototype. The first test attempted to establish a drag coefficient for the overall system. At a speed of $0.4 \, \text{m s}^{-1}$ and a propeller rotation rate of 420 revolutions per minute the power required was 5.1 W. By considering the performance of the propeller and thruster along with the power, a drag coefficient of 0.16–0.17 was calculated, which was significantly higher than expected. The probable causes of this discrepancy were determined to be the lower than expected efficiency of the propeller, the voltage converter and the thruster motor. Modifications to all of these components have been made, but quantitative testing has not yet been accomplished with the new components.

Horizontal movement was first undertaken on the surface to establish how close theoretical performance matched actual data. These tests were then undertaken at depth. Figure 4.4 summarises these results. It can be seen that the actual performance was close to that expected.

Movement in the vertical plane was tested by forcing the SAUV to dive to a depth of 10 m. This test was conducted with a velocity of $0.6–0.8 \, \text{m s}^{-1}$. As seen in the Figure 4.5, the actual performance was close to the expected performance.

4.4.3 The Energy System

The SAUV relies on the energy available from the sun. A SAUV is unique in that it must include a solar array large enough to collect sufficient solar energy to accomplish

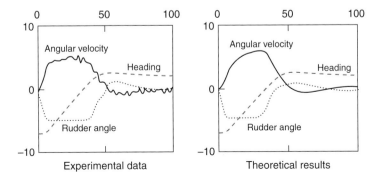

Figure 4.4   Movement in the horizontal plane.

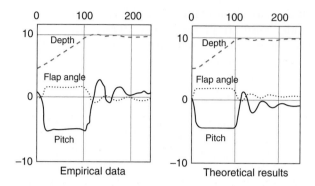

Figure 4.5   Movement in the vertical plane.

a usable sampling mission (Jalbert, 1997). An interesting analysis of the design factors associated with a SAUV can be seen in Ageev (1995). In this analysis, a conclusion is reached that the effective range of a SAUV is determined only by the efficiency of the array used to collect solar energy and not by the size of the array. Intuitively one would think that the larger the array on a SAUV, the more energy that can be acquired, hence the longer the distance that can be traversed. The problem is that as the size of the array is increased, so is the drag on the vehicle hence the larger the amount of energy necessary to push the vehicle through the water. This hypothesis, as yet unproven through in-water experimentation, suggests that the size of a SAUV can be adjusted to meet the demands of the payload (sensor size, battery size etc.). The range, however, will not be significantly affected. If, on the other hand, the efficiency of the solar array is increased, the range of the vehicle will be increased proportionately. This unique characteristic emphasises the need to clearly understand those factors that impact the range of the SAUV. If we consider the efficiency of the SAUV system, we can determine the following components of that total efficiency:

- Efficiency of the PV arrays   0.10
- Efficiency of the storage battery E_{out}/E_{in} 0.75
- Converter electronics efficiency   0.85
- Drive (thruster motor)   0.67
- Propeller   0.60
- Resulting efficiency   0.025

   This means that only 2.5% of the available solar energy is converted into thrust to move the vehicle through the water. What are the possibilities of improving this number? Currently, there are solar arrays with an efficiency of about 20%. Near term advances will probably increase converter efficiency from 0.85 to 0.90. For a larger SAUV, a drive with 0.8 efficiency is possible, and lastly the efficiency of the propeller can be increased to 0.7 (narrow, large diameter blades). This results in the following:

- Efficiency of the PV arrays   0.20
- Efficiency of the storage battery E_{out}/E_{in} 0.75

- Converter electronics efficiency   0.90
- Drive (thruster motor)   0.80
- Propeller   0.70
- Resulting efficiency   0.075

An increase of three times as much energy will result in an increase in length of transit of approximately 1.5 times for a given speed. Possibly more important is to consider that an increase in stored energy may have a greater impact on the mission of the SAUV (i.e. more energy available for mission sensors) than on its ability to transit over longer distances.

As mentioned above, increasing the battery size in a SAUV does not increase the distance it can travel. It does, however, impact the amount of energy available for sensors. For example, if a design decision is made to allow 10% of the total energy to provide for sensors, then an increase in the capacity of the energy storage system (resulting from a larger vehicle, hence large PV arrays) will result in an increase in the amount of energy available for the onboard sensor suite.

4.4.4 An Example SAUV Mission

The limiting factor when using a SAUV for any data-gathering task is the amount of available energy and the manner in which that energy is utilised. Planning must be undertaken to adjust a proposed operational scenario, and sensors to be used, to the energy storage and acquisition characteristics of the SAUV and the environment within which it is to function. If we consider the energy budget of a typical SAUV system it is possible to develop an initial estimate of subsystem energy demands as detailed in Table 4.3.

The data-gathering scenario significantly impacts the total energy requirements. It must be clearly understood that the values summarised in Table 4.4 are estimates. The SAUV velocity is assumed to be $0.5\,\mathrm{m\,s^{-1}}$ hence it would be able to travel 18 km in a 10 h period. From initial testing of the current SAUV, the 12 W of power to move the SAUV at $0.5\,\mathrm{m\,s^{-1}}$ is conservative. It is unknown how the addition of the external sensing elements, and their additional drag, will effect the power required to move the SAUV through the water. Certainly there will be an increase. The important aspect

Table 4.3 One usage profile for an SAUV gathering data over 24 h.

Each day at 8:00 am, local apparent noon and at 5:00 pm: descend to 200 m and return, three excursions 20 min each propulsion, hotel, CTD, optical sensors	1 h
At night, transit in oscillating fashion (0–200 m) out to 9 km and return: propulsion, hotel, CTD, optical sensors	10 h
Maintenance manoeuvring (to adjust position): propulsion, hotel, GPS	1 h
Data transmission, receipt of instructions, sleep, loiter	1 h
Recharging in place	11 h
Total	24 h

Table 4.4 Daily energy requirements for the SAUV.

Item	Power (W)	Duration (h)	Energy used (Wh day^{-1})
Propulsion @ 0.5 m s^{-1}	12.00	12	125.00
SeaBird SB37-SI	0.24	11	2.64
Oxygen Sensor (SeaBird)	0.10	11	1.10
Wetlabs Fluorometer	0.50	11	5.50
Wetlabs LSS light scattering sensor	0.50	11	5.50
OCI-200	0.20	2	0.40
Computers etc.	1.00	100%	24.00
Communications: Satellite	~0.1 J bit^{-1}		5.00
Navigation: GPS	2.00	5 min h^{-1}	4.00
System sensors/Actuators	1.00	12	12.00
Data logger	1.00	5	5.00
Total for basic sensors			190.14

of this summary is that the total energy usage is primarily based on the length of the nighttime transits. If the total transit is reduced, the amount of energy used will be less. If there is an excess of energy, then the velocity of the SAUV can be increased and the total transit distance increased for a specific night. The SAUV is aware of the amount of its onboard energy and can adjust its sampling strategy accordingly.

If we assume the system requires 200 Wh day^{-1}, with a solar array of 0.5 m^2 and an efficiency of 10% we must operate within an area where the average level of solar insolation is 4 000 Wh m^{-2}day^{-1}. This level of insolation exists in many parts of the world for several months during the year.

If sampling must be undertaken each day then the size of the energy storage system must be large enough to account for days when available solar energy is lower than normal. In those cases the system will use stored energy rather than acquired energy. If the sampling schedule can be modified then the system can wait until its energy system has acquired enough energy to accomplish the sampling operations (i.e. remain on the surface for an extra day). In the worst case, the system, exceeding a minimum energy reserve, can shut down most subsystems and return to the surface to await the sun and renew its energy reserves.

4.4.5 Energy Management

The SAUV can be considered as an underwater satellite whose energy comes from the sun. This is similar to space-based satellites powered by solar energy. There is, however, one difference in that, in general, space-based satellites receive solar energy constantly and SAUVs receive energy only for a period of time. The design issue with satellites then becomes to insure that the power used by all of the onboard systems does not exceed the power level available from the solar arrays. In the case of the SAUV, the concern is that the onboard systems do not use more energy than can be stored onboard between recharge periods. The design issue then becomes energy utilisation and not power limits.

The intensity of solar radiation near the sea surface experiences large seasonal and daily fluctuations. Some of these can be estimated beforehand. Other meteorological conditions result in unpredictable changes in available energy both on an hour-by-hour and day-to-day basis. To compensate for these unpredictable variations strategies must be developed that effectively utilise acquired energy as well as optimise methods by which the onboard batteries are discharged and recharged.

Seasonal and daily fluctuations of the light radiation without reference to scattering can be determined according to known astronomic formulae. Scattering and absorption of light by the atmosphere varies greatly and an assessment of those variations is best made by acquiring local data over long periods of time. To address this issue, measurements of energy generated daily were acquired by monitoring horizontally arranged solar cells at IMTP in Vladivostok, Russia for a period of one year. Figure 4.6 shows records of diurnal power in $Wh\,m^{-2}$ produced by a horizontally placed PV panel with the efficiency of 10%. Records are grouped into seasons of 100 days.

From this data it is readily observed that the solar energy system for an autonomous sampling platform must account for the variability of available solar energy. A power management strategy must effectively use onboard energy as well as adjust utilisation of that energy to account for the high variability to be expected.

The SAUV's ability to transit is limited by the amount of energy it has been able to acquire from the sun. Due to the variability of solar energy, the energy available to the SAUV each day will vary greatly. Strategies must exist to account for both the variability of solar energy and the demands of the mission the SAUV is to undertake. Three strategies were considered. The first strategy is considered the ideal case or reference strategy. In this case the SAUV travels at the same velocity during its night time transits. This implies that the average solar energy for the period of the

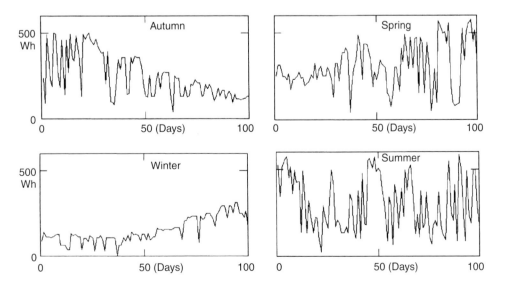

Figure 4.6   Energy acquired in Vladivostock, from July 1995 to July 1996.

Table 4.5   Effect of power management schemes on total distance travelled (Scheme 3 guarantees $20\,\mathrm{km\,day^{-1}}$).

Endurance (days) at $2\,\mathrm{km\,h^{-1}}$ and $12\,\mathrm{h\,day^{-1}}$	Scheme 1 (100%) Distance (km)	Scheme 2 (95%) Distance (km)	Scheme 3 (87%) Distance (km)
30	720	680	626
60	1,440	1,360	1,252
360	8,640	8,110	7,516

operation is known prior to the operation (this is really not possible since predicting the available energy over a given time in the future is akin to predicting the weather). Given the average amount of solar energy available each day it would be possible to determine a speed that utilised that amount of energy each night. Since the SAUV would always be moving at the same speed during the operation, its total transit would be maximised. This circumstance is defined as power management Scheme 1 in Table 4.5.

Scheme 2 is defined as a situation in which the SAUV uses all of the energy collected during the previous day while transiting during the night time hours. Its speed would be adjusted such that all of the energy acquired the previous day would be used by the time that the sun was to rise the following morning. This would mean that on some nights (when little energy was available during the previous day) the SAUV would not be able to do much more than remain on the surface waiting for the next available day when solar energy is available.

Scheme 3 is defined as a situation in which the SAUV must move the same distance each night. In this situation, the speed would be adjusted to utilise only a portion of the available energy. The remainder of the onboard energy would be kept in reserve for those days when there was little energy acquired, due to low solar insolation. The distances that the SAUV will transit are different for each of these power management schemes (see Table 4.5), however, by using such schemes it is possible to tailor vehicle performance to meet the needs of different sampling tasks. The analysis to determine the effective range presented above assumed that all of the available energy was used for propulsion. The goal of this analysis was to understand the limits associated with the effective range of a SAUV. To accurately determine this range, a number of other factors must be considered. First and foremost is to understand the amount of solar energy available at a given location and its variability during the operational period. Although, having an accurate number is like predicting the weather, it is possible to acquire from existing solar insolation databases, an average value for a region close to the region of interest. If we then assume that the variability of this energy conforms to a uniform distribution (Jalbert, 1997) we can assume that the maximum value will be twice the average value. This then allows us to determine a battery size for the SAUV (assuming that the size is such that it will fit within the size of the AUV platform). Once the battery capacity is determined, it is possible to make a number of assumptions as to the range of the SAUV. By considering different power management schemes it is possible to obtain more realistic endurance numbers for a SAUV.

For the example above, the SAUV prototype vehicle battery capacity is 420 Wh. Since the charging efficiency of the batteries (E_{out}/E_{in}) is approximately 80% then the available energy is approximately 330 Wh. If we now consider the analysis of the energy requirements of the prototype, we can determine that the system will consume approximately 200 Wh (at a velocity of 2 km h^{-1}). This suggests that the optimum range of the SAUV in a location where the solar insolation level is approximately 4,000 Wh m^{2}day^{-1} is approximately 24 km day^{-1}. Table 4.5 summarises the total endurance of the SAUV prototype for the three power management schemes discussed above.

4.4.6 Impact of the Ocean Environment

The analyses performed to date emphasise the potential of a SAUV as being an ideal sensing platform. There are, however, some questions that remain unanswered and must be resolved. AUVs have been used in the ocean for a several decades and their designs are well suited to the ocean environment. The addition of solar panels prompts a consideration of the interactions of the ocean environment and this technology. There are four potential hazards that impact the design of a SAUV: corrosion, bio-fouling of the PV arrays and mission sensors, collision with boats while recharging, and the effects of wave action on the system.

Although corrosion of system components is an important issue, since the endurance of the SAUV is so great, it is a problem that ocean engineers have been dealing with for years and is not unique to the SAUV. It is an engineering issue and best handled by appropriate system design.

Surface biofouling of PV collection surfaces by biological organisms is a serious concern for shallow-water SAUVs. There is no simple anti-fouling technique proven to be completely effective against all organisms. In order to minimise the impact of marine fouling on solar cell surfaces, a two-pronged approach is being investigated. First, any protective coating or laminate must provide 'easy-release' surface characteristics. Second, some mechanical method of 'wiping' the array during operations must be implemented.

Easy release surfaces exhibit a low surface tension to water, a quality which also significantly lowers the potential for permanent attachment by living organisms. Some commercially available solar panels, such as the Solarex MSX-Lite series, are coated with transparent ethyl vinyl acetate (EVA), a polyethylene laminate known to provide superior release characteristics to glass. The survivability and anti-fouling quality of one such panel was recently demonstrated after a one-year immersion at approximately 275 m in the Gulf of Maine.

A solar array was obtained that was part of the instrumentation buoy being used by the Ocean Processes Analysis Laboratory at the University of New Hampshire. This particular array had been on a buoy that sunk in 275 m of water in the Gulf of Maine. The Solarex MSX 10 array had been on the bottom for approximately one year. The array had not been cleaned when acquired for test purposes. Performance measurements were made prior to cleaning and immediately after it was retrieved by a fishing trawler. There was a degree of biofouling and some mechanical faults. The array was covered with barnacles, but beyond this seemed unaffected with the exception that a junction box on the back of the array had imploded due to pressure. The hard anodising had been scraped from the upper aluminium frame piece.

Measurements prior to cleaning of the array determined that, when placed in the sun, it functioned as if there had been no electrical damage whatsoever. The array was then sent to the manufacturer for a complete test and the results verified that indeed there had been little effect from being submerged. The array functioned within the original specification of 10 W ±10% and an efficiency of 11.5%. This experience provides some level of confidence as to the durability of current PV arrays in the ocean environment. Although not conclusive, it does provide a significant data point.

To further understand this issue, an in-water biofouling experiment was conducted between 23 May and 29 June 1998 at New Castle, NH to determine, in a locally worst-case scenario, how biofouling would affect the collection efficiency of solar panels as a function of time. Two commercially available solar array panels were used in these experiments. Both were identical except for the exposed surface material. Panel no. 1 – 'Lite' (Exposed material was TEDLAR, Polyvinyl Fluoride); Panel no. 2 – 'Standard' (Exposed material was tempered glass).

The solar arrays were submerged approximately 0.3 m below the sea surface at low tide such that they were exposed to the sun throughout the day. At very low tides, the panels were slightly out of the water. A report on these experiments is available (Grosholz, 1998).

Figure 4.7 depicts a plot of solar array panel area fouling and output power (as a percentage of that acquired from a reference solar panel) as a function of time. After the first 18 days, the area covered by biofouling was about 42% and the power output for both panels remained at about 80%. After 30 days, the standard panel fouled area increased to 53% and its power output dropped to 47% while the lite panel fouled area increased to 67% with its output dropping to 32%. The standard panel seems to resist fouling better than the lite panel and both systems provided 80% power output after 18 days under these conditions. A solar array panel mounted on an AUV will likely take a longer period to foul since it will be subjected to constant water washing over its surface and will be making regular excursions into deeper water.

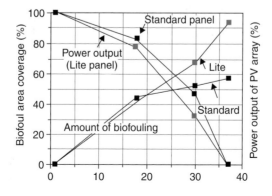

Figure 4.7   Solar panel biofouled area and power output vs. time.

SAUVs are to be at sea for extended periods of time and half of that time will most probably be spent recharging on the surface. This suggests that the possibility of collisions with surface traffic is a concern to be dealt with. There are no clear answers at this time as to the best method of mitigating this problem. It has been suggested, however, that most all of the surface hazards generate a large acoustic signature; waves splashing against a boat hull, engine sounds and other man-made noises. It may be possible to recognise incipient collisions by listening for surface traffic and, if heard, to dive beneath the surface until the danger has passed.

When the SAUV is on the surface recharging, the effects of the wave motion will impact the amount of energy that is acquired. It is important to understand what that effect will be if we are to plan effective energy utilisation strategies.

A set of experiments was designed and conducted to measure the effect on solar energy acquisition due to seawater washing over the solar panels and due to wave motion effects on the solar array in varying sea states. A SAUV body was designed and fabricated to have the same surface motion characteristics as the prototype SAUV. The system included a Solarex MSX30L solar array, a microprocessor, a battery energy gauge and charge controller (BQ2112 – Benchmarq), and a NiCd battery system. A duplicate system was housed in a land station on shore, which also monitored weather data. Data was gathered by both systems simultaneously: the SAUV body was at sea approximately 5 km off Portsmouth NH where the land station was located. Figure 4.8 depicts a plot of the results of these experiments. In general, the SAUV body floats very quietly in the wave surface and is almost constantly awash. The data in Figure 4.8 shows the percentage difference in energy acquired between the two systems, sea and land. A positive value in the Y axis indicates that the sea system acquired more energy than the land system while a negative value indicates the opposite. Two curves are shown in the plot. The top curve represents the actual raw energy acquired while the lower curve includes temperature correction. This temperature correction is used to ensure that the results measure only the effects of the wave and

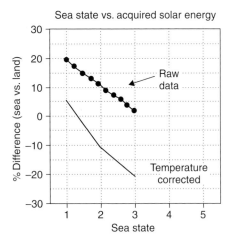

Figure 4.8   SAUV wave and wind interaction data.

water interaction with the solar collection system. The correction takes into account three different effects of temperature on charge acquisition: the battery energy gauge compensation factor, the battery self-discharge compensation factor and solar array temperature effects (PV arrays are more efficient when they are cooled by the low temperature seawater). Without temperature correction, the at-sea system collected more energy than the land system, at least up to sea state 3. The temperature corrected data indicates that the at-sea system collects slightly more energy than the land system up to about sea state 1. However, as the sea state increases, the land system begins to acquire more energy than the sea system until it is about 20% greater around sea state 3.

4.4.7 SAUV Control System

The SAUV control system is meant to be simple, as the primary emphasis is on minimising the energy usage of the computer/control system. On the other hand, the extended endurance implies that a remote user should be able to modify the onboard program to allow for insights gained from acquired data to be implemented in order to optimise the operational strategy being undertaken by the SAUV. A relatively simple set of commands has been implemented to accomplish this.

The SAUV system with its inherent low cost and long endurance lends itself to operations where multiple autonomous systems are to co-operate to accomplish a defined task. For this reason much design effort has gone in to understanding the communication issues associated with controlling multiple SAUVs. Two aspects have been addressed. First to develop a distributed simulation environment that allows multiple users to control simulated or real AUVs utilising the existing communication infrastructure. This has resulted in a Co-operative AUV Development Concept (CADCON) (Chappell, 1999). Second, to develop a communication interface to the SAUV that will allow for communication utilising various types of RF, acoustic and satellite modems.

The main idea behind CADCON is that it provides an open and flexible simulation environment for use by as many researchers as possible. Understanding that no single simulation harness could capture the full fidelity of the real open ocean environment of a deployed group of autonomous systems, nor the complexity of every sort of vehicle that might participate, AUSI has attempted to focus CADCON on one set of problems seen in multi-AUV systems: the issues associated with the interactions among multiple heterogeneous agents; be those agents simulations, real vehicles (such as the SAUV systems), or human users. To that end, the CADCON simulation environment should adhere to the following points:

- *Utilise well-known ubiquitous hardware*: All CADCON components have been developed to run on today's most available platform: the IBM-compatible computer. This allows workers to leverage well known, cheap and accessible hardware, making their participation in CADCON simulations relatively easy and flexible. Exotic hardware is not required.
- *Utilise a well-known ubiquitous communication protocol*: CADCON environment components are implemented following the client/server model and communication

between them is in the lingua franca of the Internet: TCP/IP. This non-reliance on proprietary communications protocols further leverages the system's utility and availability to other workers.

- *Be globally accessible*: Since components communicate via TCP/IP sockets, they may be distributed across intranets as well as the Internet. This opens the door to geographically distant institutions participating in joint simulation scenarios.
- *Open to any institution's AUV development style*: The client/server model fosters modular development. Coupling this with the use of the TCP/IP protocol fosters a high degree of platform/language independence for client developers. This allows great flexibility for distributed researchers to connect existing legacy simulation models or create new models on the platform and in the programming language of their choice.
- *Allow for real hardware in the loop simulation*: The modularity encouraged by clients communicating with servers via TCP/IP also provides for easy integration of real vehicles into the CADCON environment. This moves CADCON out of the strictly simulation arena. Real vehicles will react to situations presented by the CADCON simulator while safely on the bench. Given a physical communication link, this concept can be taken even further to where real vehicles situated in actual missions will begin to supply the CADCON environment with *in situ* data.
- *Access via the World Wide Web*: The current instance of the CADCON environment is now available on the Web. Users are encouraged to download example clients from the AUSI web site and try them out against a running simulator.

The current CADCON system is seen in Figure 4.9. Two of the clients can be downloaded from the web and used as tools to begin understanding how to control multiple AUVs in a semi-autonomous fashion.

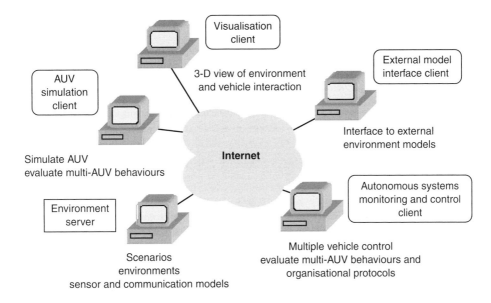

Figure 4.9 Co-operative AUV development concept (CADCON).

4.5 SUMMARY

The SAUV system has been successfully tested at sea to demonstrate many of the capabilities of the platform. This testing has identified a number of issues that have led to minor modifications of the system design. In November 1999–January 2000 a series of test evaluated the ability to communicate to remote users via the existing communication infrastructure and to make acquired data available via the Internet. This testing continues and is helping to further refine the SAUV system design and to identify the issues associated with remote users controlling and receiving data from multiple co-operating AUVs.

An important consideration is the overall system reliability. The reliability of some sensors over extended periods of time is an unanswered question especially when considering months to a year of unattended operation. Along similar lines is the question of overall platform and subsystem reliability for periods of months to years. Certainly there are examples of systems with this reliability, but few have been demonstrated in the ocean environment especially when one half of the SAUV's operational time is spent at the surface. The design of SAUV systems with extended endurance will be an evolutionary process. The potential of these systems, however, suggests that the process be endured.

5. ADVANCED MATERIALS AND THEIR INFLUENCE ON THE STRUCTURAL DESIGN OF AUVs

PETER STEVENSON[a] and DEREK GRAHAM[b]

[a] *Southampton Oceanography Centre, University of Southampton,*
Southampton SO14 3ZH, UK and
[b] *QinetiQ, South Arm, Rosyth Royal Dockyard, Dunfermline, Fife KY11 2XR, UK*

5.1 INTRODUCTION

New inventions often have to wait for appropriate materials to come about before they become a practical reality. Leonardo da Vinci's sketch showing the principles of helicopter flight remained a sketch for over 400 years until high strength steels, titanium and forged aluminium became a reality both for the airframe and the engine to drive it. Steam power did not really take a hold until iron could be reliably cast without the brittleness that limited working pressures to a few pounds per square inch and precision of manufacture improved to produce a close fit of piston to cylinder (itself partly a function of materials).

If Autonomous Underwater Vehicles (AUVs) can be classed as an invention, they too, in some measure, are dependent on material developments to make them a practical reality, Figure 5.1. Without minimising the importance of modern acoustic systems for tracking or sub-sea navigation, the advent of networked protocols easing the integration of subsystems or the godsend of the Global Positioning System (GPS); there remains the essential design compromise in the design of depth, endurance and payload.

The deeper the depth rating of the vehicle, the heavier the buoyant structure becomes to withstand the pressure. Given most AUVs are more or less neutrally buoyant, the weight of the structures, energy source and payload has to be equal to the displacement of the vehicle. Thus for a design to maximise its performance, and its usefulness, the weight of the structures have to be minimised. The key, in a word, is materials.

5.2 COMPARISON OF MATERIALS AVAILABLE TO THE AUV DESIGNER

One of the basic constraints of AUV design for oceanographic work is the corrosive nature of seawater. This chapter considers only materials that have a proven track record for being reasonably corrosion resistant. While some of these would not be considered 'advanced', for example, stainless steel, aluminium, they will nonetheless find their way into the design and being materials that engineers are accustomed to, they serve as a useful comparison against the more exotic.

Figure 5.1   The Autosub vehicle during the build phase showing a variety of materials from the ordinary to those still being developed. Polyurethane mouldings for cables, junctions, fins etc., aluminium frame work, stainless steel bulkheads, foam buoyancy in the centre section, suspended above, the main GFRP pressure vessel with titanium end domes.

5.2.1   Materials for Structural Components

Although an AUV may live an almost weightless existence beneath the waves, the dynamic loads imposed on the structure during launch and recovery are considerable when it is being lifted from the sea. The vehicle may have flooded sections, which, on recovery, may add as much as 50% to the weight of the vehicle compared to its weight in air. Engineers experienced in the design of towed bodies, landers and moorings are used to having to accommodate unexpectedly high loads but in these applications weight does not play such a large part in the performance of the design. AUVs have to be tough enough for the environment and light enough to turn in a good performance. Combine with this the necessity to accept varying payloads, access for maintenance and the design of the structure becomes a challenge.

   If the structural design is primarily driven by the need for minimum weight then specific strength and stiffness values need to be compared, where properties are normalised with respect to density. The nature of the loading is also important. If a structural member is in straightforward tension or compression, it is a simple case of comparing ratios of strength to density or Young's modulus to density. If, however, the limiting condition is that of bending stress, given that the stress is proportional to the beam section cubed and its mass proportional to section squared, it is appropriate to compare the ratio of $\sigma^{2/3}\rho^{-1}$ where σ denotes material stress, ρ the material density and E the elastic Young's modulus. Similarly, for a member constrained by a maximum deflection, we need to compare the ratio $E^{1/2}\rho^{-1}$. Table 5.1 shows a summary of properties with merit indices for materials that would normally be considered.

Table 5.1 Merit indices of materials which may be considered for structures.

Material	Yield Strength $0.2R_p$ (MPa)	Ultimate Strength R_m (MPa)	Modulus E (GPa)	Density ρ (kg m^{-3})	Specific Strength[2] $R_m \rho^{-1}$	Specific Stiffness[4] $E\rho^{-1}$	Specific Strength[5] $R_m^{2/3}\rho^{-1}$	Specific Stiffness[6] $E^{1/2}\rho^{-1}$	Specific Strength[7] $E^{1/3}\rho^{-1}$
Aluminium 6082[1]	240	280	70	2,700	1.00	1.00	1.00	1.00	1.00
Aluminium HDA 89	420	500	70	2,800	1.72	0.96	1.42	0.96	0.96
Stainless steel 316	208	540	203	7,960	0.65	0.98	0.53	0.58	0.48
Stainless steel 431	739	880	210	7,830	1.08	1.03	0.74	0.60	0.50
Titanium IMI 115	200	290	110	4,510	0.62	0.94	0.61	0.75	0.70
Titanium IMI 318	830	900	125	4,420	1.96	1.09	1.33	0.82	0.74
GFRP pultruded section (tensile)	–	290	18	1,800	1.55	See note 3	1.54	0.76	0.95
GFRP pultruded section (flexural)	–	110	14	1,800	See note 3	0.30	0.80	0.67	0.88
CFRP Unidirectional (0°)[8]	–	840	190	1,550	5.23	4.73	3.62	2.87	2.43
CFRP Unidirectional (90°)[8]	–	42	6.9	1,550	0.26	0.17	0.49	0.55	0.80

Notes
1 All specific values have been normalised with respect to aluminium 6082.
2 Applicable to members in tension or compression limited by strength, for example, struts, ties, thin wall tubes under internal pressure.
3 The flexural strength often quoted for composites is the limiting strength for a member in bending. Hence it is not appropriate to use flexural strength for members in tension/compression or tensile strength for members in bending.
4 Applicable to members in tension or compression limited by stiffness, for example, struts, ties, thin wall tubes under internal pressure (this is not a very common engineering case).
5 Applicable for members in bending, strength limited, assumes constant aspect ratio of section.
6 Applicable to members in bending, stiffness limited, assumes constant aspect ratio of section.
7 Applicable to externally pressurised tubes limited by buckling.
8 Load applied along the axis of the fibres (0°) yields the highest strength that can be achieved since the load is predominately being taken by the strong fibres. A composite is at its weakest when loads are applied across the fibres (90°) since the loads are being taken by the weaker matrix.

The merit indices have been normalised with respect to aluminium 6082 to make for an easier comparison.

This analysis throws up some interesting conclusions. The specific strength and stiffness values are dominated by the material density, hence, in some applications, titanium, normally considered to be the ultimate metal for strength and lightness can be beaten by a high-grade aluminium. Carbon fibre reinforced composite (CFRP) performs particularly well, not so much from its high strength (which in some applications can be equalled by glass fibre) but from its low density. Composites such as CFRPs deserve special mention for their anisotropic properties; their strength is highly dependent on the orientation of the fibres in relation to the direction of stress, as explained in many material text books (e.g. Hull, 1981) and further explored in Section 3.1. Strengths and moduli quoted at zero fibre orientation (as has been done in the penultimate row of Table 5.1) are unrealistic for most engineering problems where the stress condition is two dimensional if not three-dimensional. It seems more appropriate therefore to quote quasi-isotropic properties such as may be found in woven fabrics to provide a more rounded comparison.

GFRP pultruded sections have found applications for marine structural frameworks. Although lacking ductility and ease of joining compared with metals, they are tough, corrosion free and non-magnetic. The deterioration of GFRPs in the marine environment is well documented but in addition, tokens made with woven fabric subjected to a three point bend test also lost some 6–30% of strength when soaked in seawater at a combination of 70 MPa hydrostatic pressure and high applied stresses (Stevenson, 1993).

Material selection is rarely governed by weight alone. Cost, not only of the raw material but also of the manufacturing methods will clearly be a factor. The higher strength titanium alloys, for example, are difficult to form, tough to machine and if welding is required, ideally they should be electron beam welded. As a comparison, the cost of using titanium is approximately three to five times that of using aluminium or stainless steel. Corrosion due to the primary mechanism of galvanic action due to contact of dissimilar metals is another factor. A table of potential differences may be found in many sources (e.g. Myers *et al.*, 1996). Aluminium alloys, especially the higher strength grades, will corrode at the expense of stainless steel. Stainless steels can suffer from crevice corrosion after immersion times of many months, and stainless steel welds suffer due to chromium migrating away from the weld. Carbon fibre, being a very noble material, will not corrode, but will cause most neighbouring metals electrically connected to it to corrode. In this respect GFRP has the advantage of being electrically non-conducting as well being a cheaper material.

Practicality is another factor in the design process. Consideration has to be given to how a structure will be made, how subsystems will be attached and what modifications, developments or repairs at sea are likely to be needed throughout the life of the vehicle. At the time of writing, AUVs are undergoing a change from prototypes to vehicles being developed for particular roles. For example, the Autosub vehicle was designed with slotted aluminium sections allowing any number of payloads to be moved, added or removed to suit user requirements (Stevenson, 1997). This may contrast with developments in the future where vehicles could be designed for small production runs for bespoke tasks. In that case, the outer shell could be a monocoque structure designed to take fixed subsystems without any further structural members.

5.3 THE DESIGN OF PRESSURE HULLS

The design of pressure hulls for AUVs deserves special attention since it is likely that the hull will be the main source of buoyancy for the vehicle. Weight savings in this area for a given working depth have significant implications for the endurance or payload capacity of the vehicle. For every kilogram excess weight in water there will have to be some means of adding buoyancy. This would be in the form of buoyant foam of volume:

$$V_{foam} = M_{excess}/(\rho_b - \rho_{sw}),$$

where V_{foam} is the volume of foam to make up the buoyancy deficiency (l), M_{excess} the excess weight (kg), ρ_b the density buoyancy system (kg l^{-1}), and ρ_{sw} the density of sea water (kg l^{-1}).

Figure 5.2 shows the above function, illustrating how easily excess weight and dense buoyancy materials can stifle the space intended for payload. As the working depth increases, the density of suitable foam necessarily increases, requiring greater volumes of foam to keep the vehicle afloat and reducing the available space for payloads. It is better to design a lightweight hull to provide dry space and buoyancy with a minimum buoyancy deficit.

The principal failure mode for a pressure hull is either by buckling, determined by the geometry of the hull and elastic modulus of the material, or by compressive failure of the material, determined by hull geometry and compressive strength of the material. At lower working pressures, typically less than 40 MPa (~4000 m depth), failure tends to be by elastic buckling for a plain long tube, regardless of its diameter. In this case, for materials such as titanium, the inherent high strength is not being used because the tube fails elastically at stresses lower than its maximum strength, resulting in less than optimum performance. This is shown in the last column of Table 5.1, where it is seen that aluminium will out-perform titanium when stiffness

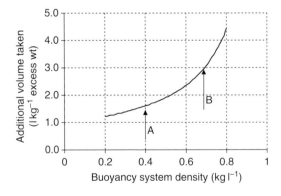

Figure 5.2 The additional volume of buoyancy required for each kg of excess weight. For example, foam rated for 500 m (point A) will take up 1.5 l for every 1 kg of buoyancy required, whereas foam rated for 6,000 m (point B) will take up 3 (see also Section 4).

dominates the mode of failure. This situation can be improved by adding circular stiffening frames along the tube's length, in effect making it a series of short tubes where
the skin is more highly stressed and the frames resist buckling. In this instance, the
mass of the hull is dominated by the thickness of the shell and high strength materials
can be used to good effect. The design and analysis of hulls for isotropic materials
has been extensively covered in other works (e.g. Bickell and Ruiz, 1967; Myers *et al.*,
1969; Ross, 1990). Bickell and Ruiz (1967) has a useful chapter on the design guidelines on roundness, factors of safety and rule of thumb methods of determining the
section of stiffening frames. Myers *et al.* (1969) helpfully shows in graphical form the
functions predicting buckling in terms of material modulus, thickness to radius and
length to diameter ratios. Since the calculations to determine the number of lobes
formed when buckling are iterative, these graphs give a particularly quick solution.
Using this data, an approximate guide of mass to displacement ratios for hulls in
different materials is shown in Figure 5.3. The graphs need to be used with some cau-
tion, especially for anisotropic materials since some sweeping assumptions have been
made as to the mode of failure. These assumptions are described in the following

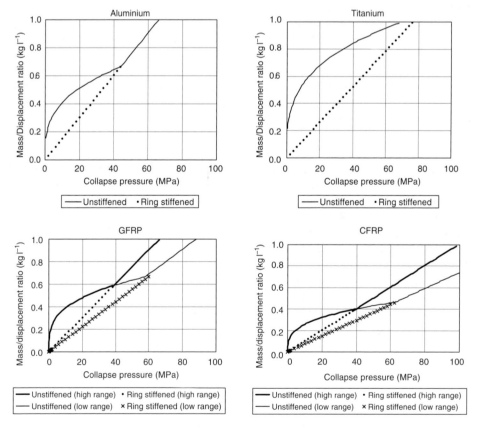

Figure 5.3 Summary of mass to displacement ratios of long tubes (length/diameter >10) for
different materials. These exclude the mass and displacement of end domes and fixings.

sections. For this reason, the graphs show a high and low range to give an indication of performance. Finite element (FE) stress analysis is increasingly used to analyse the failure of structures but such methods still require an initial design as a starting point. The older, classical methods still have their use in generating designs likely to be close to the desired performance.

5.3.1 The Use of Composite Materials

The previous section showed the potential of composite materials such as glass fibre and carbon fibre in the use of lightweight hulls. This section considers in more detail their nature and how they may be applied as pressure hulls.

5.3.1.1 Material anisotropy

Composite materials offer great scope to the structural designer, giving the freedom to tailor materials for specific applications. Choice of constituent fibres and resin, fabrication route, format of individual plies and lamination or winding patterns can all affect the final properties of the material. However, this flexibility has a cost. The anisotropic properties described in Section 2 significantly complicate the design of composite structures, as traditional solutions based on isotropic elasticity, such as Roark's Formulas (Young, 1989), cannot be relied upon.

The case of a pressure hull is a good example where this tailoring of properties can be used to good effect. For a thin cylinder, the axial stress due to pressure acting on the ends is half the circumferential stress due to pressure acting on its diameter. Thus, the fibres may be orientated to share the circumferential to axial stresses in a $2:1$ ratio. For a filament wound tube where continuous fibres are helically wound over a mandrel, simple netting analysis, where all the stress is assumed to be taken by the fibres, shows the optimum winding angle to be $55°$ with respect to the polar axis (Hull, 1981, pp. 182–191). The prediction of composite strengths, methods of analysis, their uses and modes of failure are described in a variety of materials textbooks (e.g. Hull, 1981; Agarwal et al., 1980).

The elastic properties of an isotropic material are completely defined by two constants, Young's modulus and Poisson's ratio, but a fully anisotropic material requires 21 independent elastic constants. Most practical composites are not anisotropic but are orthotropic, having three orthogonal planes of material symmetry. This reduces the number of elastic constants to nine: three Young's moduli, three shear moduli and three independent Poisson's ratios. The interested reader can find more information on the equations of anisotropic elasticity in textbooks such as Reddy (1997) and Smith (1990).

The effects of material anisotropy on stress analysis can be illustrated in the simple, but quite relevant, case of a thick cylinder under external pressure. Figure 5.4 shows the circumferential and longitudinal stresses through the thickness of three different cylinders under a hydrostatic load: steel, hellically wound ($\pm55°$) GFRP and polar wound ($90°$) CFRP. The stresses in the steel cylinder obey the familiar Lamé theory given in any strength of materials textbook (e.g. Hearn, 1977). The GFRP cylinder is not highly orthotropic, the circumferential modulus is approximately 50% greater

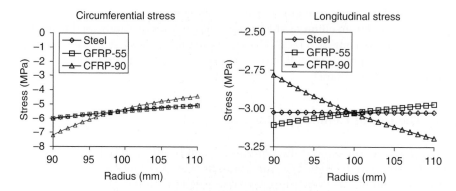

Figure 5.4   Stresses in a thick cylinder illustrating how the anisotropic properties affect the stress distribution through the thickness of a thick wall cylinder.

than the axial modulus, and the distribution of circumferential stress is virtually identical to the steel cylinder. However, although the axial stress is no longer constant through the thickness, the discrepancy is no greater than about 3%. The CFRP cylinder is an extreme example; the peak circumferential stress is almost 20% greater than in the equivalent steel cylinder.

Although unidirectional carbon fibre composites can be as stiff as steel in one direction, the nature of most practical composites, which have fibres oriented at various angles, means that they are normally significantly less stiff than steel. The in-plane modulus of a quasi-isotropic GFRP composite can be an order of magnitude lower than that of steel. Because of this, design against excessive deformation and elastic buckling is essential.

Once the displacements, buckling response and stress distributions have been assessed the designer will be faced with the problem of determining the strength of the composite structure. Unlike metals, which require only one parameter, the yield stress, to be specified, composites require a number of parameters to account for the variety of modes in which failure can occur. Composite materials are strongest in the direction parallel to the fibres but there can be significant differences between tensile and compressive strengths. The former is governed by the fibre strength but the latter depends on the stability of the fibres in the resin with compressive failure resulting from fibre micro-buckling. The failure behaviour transverse to the fibres is governed by the matrix and can vary widely between tension and compression and intra- and inter-laminar modes. The photograph of a collapsed tube in Figure 5.5 shows the complex pattern of fibre and graphically illustrates the complex state fibre and matrix interaction to maintain equilibrium.

A variety of strength criteria have been developed for composite materials that account for the anisotropy of the material strengths (e.g. Thom, 1998). Perhaps the most widely used are the quadratic criteria such as the Tsai–Wu (1971) criterion. These criteria, basically modifications of the familiar von-Mises criterion for isotropic materials, have been used successfully in many applications and are available in many commercial FE analysis packages. However these methods fail to differentiate between failure modes and have been less successful in predicting failure in the thick section

Figure 5.5 Collapsed carbon fibre pressure hull, the cross over of the fibre tows is seen in the lower section, the fracture follows the ±55° wind angle.

composites that are required for deep diving pressure hulls (e.g. Graham, 1995). Criteria that predict the failure mode, as well as load (other than the simple maximum stress and maximum strain criteria) have been developed, for example, Hashin (1980) and Hart-Smith (1992, 1996) but these do not appear to be so readily available in commercial software.

5.3.1.2 Thick shell behaviour

The through-thickness shear moduli, G_{13} and G_{23}[1] of a general composite laminate are normally substantially lower than the in-plane modulus, E_{11}, sometimes by as much as an order of magnitude. Because of this, thick shell behaviour can occur in structures that would be described as 'thin' using normal criteria for isotropic materials, for example, length to thickness >10 for a plate. The consequences of ignoring these effects include under-predicting displacements and over-predicting elastic buckling pressures.

Most commercial FE packages account for these effects by means of specially formulated thick shell elements that include through-thickness shear deformation. These elements normally use the first-order shear deformation theory (FSDT) which assumes a constant transverse shear strain through the thickness. This theory necessarily introduces shear correction factors, which although readily shown to be 5/6 for an isotropic plate, depend on lamina properties and lamination scheme for a general

---

1 The double suffix notation defines the direction of the normal to the plane on which the stress acts (first suffix) and the direction of the property in question (second suffix).

laminate. The FE code will evaluate these shear correction factors internally and, by returning to the 3-D equilibrium equations, calculate an accurate distribution of interlaminar shear stresses through the thickness, which meets the requirements of being continuous at layer interfaces and zero at the surfaces.

Higher order shell theories have been developed (Reddy, 1997), but currently these are more of academic interest as the improvements in accuracy are modest in most practical situations while the increase in computational effort is considerable. The theory of shells and shell finite elements is a specialised area with a vast literature but, at a minimum, the analyst should understand the formulation of the elements being used and the consequent capabilities and limitations.

5.3.2   Unstiffened Monocoque Construction

The simplest pressure hull configuration consists of an unstiffened monocoque cylinder with flat or domed end closures. A monocoque cylinder is simple to manufacture, for example, by filament winding or forming a prepreg (matrix impregnated woven fibre) on a cylindrical mould and similar methods can be used to produce hemispherical or torispherical end domes. The resulting structure will be free from interlaminar shear stresses, with the possible exception of a region near the ends if there is a stiffness mismatch between the cylinder and the end closures. If there is a mismatch, it is possible to eliminate it by suitable design (Graham, 1995).

However, as has already been mentioned, the monocoque cylinder is not a particularly efficient design for most practical depths because of the large thickness required to prevent buckling. For example, Graham (1996) describes a monocoque CFRP thick walled pressure vessel, closed with hemispherical titanium domes, that was designed for a pressure of 60 MPa and which failed at 61 MPa by overall buckling. The cylinder was wound at ±55°, which is easy to manufacture, strong in the 2 : 1 stress field which exists (Kaddour *et al.*, 1998), but results in low circumferential and axial Young's moduli, and hence poor resistance to buckling. The stiffness can be significantly improved by using a $[0°/90°_2]^2$ configuration but winding fibres close to the axial direction is difficult and can lead to loss of strength. Optimisation of winding patterns offers considerable potential for improving designs and is currently an active area of research (e.g. Keron *et al.*, 2000).

5.3.3   Ring-stiffened Construction

Efficient design of cylindrical pressure hulls, particularly for shallow water applications, requires the use of circumferential stiffening to provide sufficient resistance against buckling. Through experience with military submarines the behaviour of ring stiffened, isotropic thin-shell pressure hulls has come to be fairly well understood and various attempts have been made to extend this to include material anisotropy and thick shell effects (McVee, 1994).

---

2 The suffix '2' denotes the number of layers and their fibre directions are symmetrical about the mid section.

Ring stiffened composite cylinder design can be compromised by the low inter-laminar shear strength of the material. However, there is potentially more scope for development than with traditional isotropic materials. A traditional pressure hull consists of a cylinder with 'T-shaped' frames and the principal design parameters are frame size and frame spacing. Additional variables offered by composites include vary-ing fibre angles to modify properties; for example, the shear stiffness of webs can be increased by adding fibres at $\pm45°$. Smith (1991) provides a useful introduction to the design of monocoque and ring stiffened composite pressure hulls.

5.3.4 Ceramic Pressure Hulls

Modern ceramic materials with high toughness properties have been considered as a potential material for pressure hulls since the mid 1960s. This work has been pursued principally by the US Navy and in 1993 a paper described the manufacture and test of a 635 mm diameter, 2.29 m long alumina–ceramic hull with a working depth of 6,100 m and a mass to displacement ratio of $0.6 \, kg \, l^{-1}$ (Stachiw et al., 1993).

The manufacturing process involved a number of stages:

(1) The ceramic grains were mixed with water and binding additives to form a slurry, spray dried to form agglomerates and screened to the desired size.
(2) The hull was cast into shape and isostatically pressed at 60 MPa. (Pressure vessels to carry out this operation, especially on the size quoted above are scarce.) At this stage, the component has some handling strength but has the consistency of chalk.
(3) The component is machined, and sintered in a refractory kiln to form the mole-cular bonds, consolidate the component and remove the binder material. This process results in approximately 15% shrinkage.
(4) The component is final machined using diamond grinding wheels. Assuming the isostatic press and refractory kiln to be available at a sensible cost, it is this process where most of the costs lie. The benefit is a precision component with negligible imperfections that could initiate the onset of buckling.

As well as the high manufacturing costs, the main disadvantage of the material is its relatively low fracture toughness, or brittleness. There is no plastic deforma-tion, which, in metal designs, accommodates highly localised stress concentrations prior to ultimate failure. In a brittle material such features initiate cracks which will propagate as a function of the fracture toughness and the geometry of the crack. Titanium, for example, has a fracture toughness of $50–60 \, MPa \, m^{-1/2}$, for ceramics it is $3–10 \, MPa \, m^{-1/2}$. In practical terms, even the corners of O ring grooves on sealing faces could give unacceptably high stress concentrations and in the work quoted, the design used a gasket for sealing in order to avoid such features. For those converted to the use of ceramics this is an inconvenience overcome by careful design, impact protection coatings and careful handling. Other engineers might consider it a serious problem.

Composites can be subject to similar problems. Impacts can cause hidden delami-nation damage mid thickness, seriously weakening the structure. On this point glass

Figure 5.6   Pressure hulls from the Autosub vehicle about to be proof tested. The carbon fibre hulls have been overcast with foam buoyancy to provide additional buoyancy and mechanical protection to the carbon.

fibre possesses another advantage: being translucent it is possible to see the damage. One solution adopted by the Autosub project is to cast the CFRP pressure cases in syntactic foam buoyancy to form the outer diameter of the vehicle, the knocks are then absorbed by the relatively tough foam shown in Figure 5.6.

A more well established ceramic is the Borosilicate P40 glass in the form of 43 cm diameter spheres, used by the MIT Odyssey range of AUVs for buoyancy and payload space at depths to 6,000 m (Bellingham, 1993). These spheres have the advantage of being affordable and readily available but provide limited and inconvenient payload space and can prematurely implode if their brittle nature is not taken into account when designing the support structures.

5.4   FOAM BUOYANCY SYSTEMS

Most AUVs do not achieve all the necessary buoyancy from the pressure vessels alone. Also, there tend to be spaces that are difficult to fill with payload or subsystems but that could be put to good use with additional foam buoyancy. Table 5.2 shows the typical performance of foams with different depth ratings. These materials are cast as sheets and can be laminated and shaped or they can be cast as a solid with macro spheres embedded to reduce the mass to displacement ratio (moulded plastic spheres typically 65 mm diameter). Developments on these materials include the casting of smaller carbon fibre spheres (typically 10 mm) that are lighter, and by virtue of their small size, produce a better packing factor in awkwardly shaped pieces. Buoyancy systems for submerged pipelines are being developed with CFRP tubes buried in the foam to further reduce the mass to displacement ratio. Since AUVs tend to be long and cylindrical, this new technology may prove useful in the future.

Table 5.2 Densities of foam buoyancy systems for a range of operating depths.

Description	Maximum Operating Depth[1] (m)	Density (kg l^{-1})
CRP Marine polyurethane foam (intermittent immersion[2])	200	0.20
CRP Marine polyurethane foam (long term immersion[3])	200	0.33
Balmoral BF/SE/60 Syntactic composite foam[4]	300–600	0.32–0.38
CRP Marine co-polymer foam (long term immersion[3])	500	0.38
CRP Marine co-polymer foam (intermittent immersion[2])	800	0.38
Balmoral BF/SE/62 Syntactic composite foam[4]	1,200–1,800	0.45–0.51
Balmoral BF/SE/72 pure syntactic foam	2,100–3,000	0.51–0.54
Balmoral BF/SE/75 pure syntactic foam	5,400–6,700	0.60–0.64

Notes
1 Operating depth is given as an indication only. Different applications demand varying factors of safety, stability of density and creep, etc.
2 Intermittent immersion might be for ROVs or AUVs but in the case of an AUV, the material might be de-rated in order to achieve a more stable density value with changing depth.
3 Long term immersion is for applications such as moorings deployed over a number of years.
4 Composite construction comprises of pure syntactic foam with larger spheres (e.g. moulded plastic or carbon) cast in the body of the foam.

5.4.1 Effects of Compressibility and Contraction on AUV Performance

AUVs are generally ballasted and trimmed to have a small positive buoyancy to provide a failsafe method of bringing the vehicle to the surface should power systems fail. In the interests of simplicity, AUVs generally dive and control depth by means the sternplanes, this places limits on the permitted buoyancy: too little buoyancy and the AUV is in danger of not floating due to changes in water density in different geographical locations, too much buoyancy and the vehicle may struggle to dive beneath the wave zone and will need excessive pitch and sternplane angles to overcome the buoyancy forces, so creating additional drag.

The limited range of acceptable buoyant force raises a number of different scenarios as far as buoyancy is concerned, illustrated in Figure 5.7. When a free-flooding vehicle first dives, it is taking with it surface water trapped inside. This trapped water is less dense than water at depth. Thus, for the early part of the dive, the vehicle gains buoyancy. As the water conditions inside and outside the vehicle equalise, the vehicle is in more dense water but the hull and buoyancy will have compressed due to pressure and may have contracted due to a lower temperature. In the case of the pressure hulls,

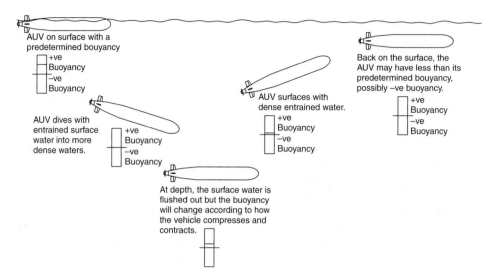

Figure 5.7   Illustration of the changes in buoyancy an AUV may undergo on a mission.

if the strain vs. pressure is known, a reasonable estimate of the buoyancy change can be found. The compression of the foam, however, is more difficult to estimate. Some manufacturers quote a bulk modulus for their materials but since they have some inherent compressive strength such a value as would apply to a fluid is questionable and dependant on the shape of the form. One of us (Stevenson) has made attempts at measuring the compressibility by hydraulically testing the total volume of foam in a test vessel and measuring how much water it takes to be bled from the vessel to reduce the pressure to zero. However, the expansion of the test vessel and the compressibility of the water needs to be taken into account, two values which, depending on the relative size of vessel being used, tend to swamp the foam compressibility measurement.

The extent of a vehicle's buoyancy change as a result of the above factors can be estimated from vehicle pitch and dive plane data whilst the vehicle is in steady state, constant depth flight. Changes in steady state pitch angle with depth indicate that the downward force generated by the hydrodynamic lift of the body has to change with depth to compensate for changes in buoyancy. An example of this characteristic is described in the account of Autosub-1 diving to its maximum depth of 500 m (Griffiths et al., 1999).

5.5   THE FUTURE?

Without a doubt, metals will continue to play a role in the construction of hulls particularly small subsystem housings; the design and failure mode analyses are more straight forward, the performance more predictable and they lend themselves more readily to penetrations for connectors and lead throughs. For the main source of buoyancy new material developments have a big part to play. Composite materials offer many advantages for AUV pressure hulls and can provide considerable design flexibility. However, specialist knowledge is required to take full advantage of their properties and to exploit their potential fully. Work on thick section composites is progressing apace with improvements being made in both performance and consistency. Ceramics have proved to be effective and may provide the ultimate mass to displacement performance where budgets are more generous.

At the time of writing AUVs are creating serious interest in the fields of offshore energy, naval data gathering and ocean science. The AUV equivalent of the Ford Model T has yet to come (it may seem strange to equate new technology with such an ancient machine but in 10–20 years time AUVs in existence today will possibly seem quite crude). To continue the analogy of the motor car, successive designers adopted the internal combustion engine for power, four wheels for stability, steel construction for the bodywork etc. These decisions were by no means obvious at the outset. It may be that AUVs will develop to take on a *de facto* style of design; particular materials and designs being adopted as an industry practice, perhaps composites for pressure vessels, rechargeable lithium ion batteries for energy, foam buoyancy for the form and so on. This is a new technology, different avenues are being explored and designs are evolving in order to optimise the basic compromise of depth, endurance and payload.

6. DOCKING SYSTEMS

ROBIN GALLETTI DI CADILHAC and ATTILIO BRIGHENTI

SATE s.r.l, Santa Croce 664/a, 30135 Venice, Italy

6.1 INTRODUCTION

This chapter describes a general approach to designing docking systems for Autonomous Underwater Vehicles (AUVs), with emphasis on a systems design approach, aimed at identifying the needs of the AUV to be serviced and the main aspects to be addressed in designing the system.

Underwater docking systems have the aim of permitting AUVs to perform repeat missions without being retrieved to the surface, and in particular to be able to perform data downloading and battery recharging activities underwater. Docking systems have the potential to overcome two of the principal current restrictions on the use of AUVs; their limited independence from expensive surface support, and the risk of loss or damage inherent in repetitive surface launch and especially recovery. A small AUV with a docking station can operate in an area over an extended duration and provide an extensive search or survey of that area cost effectively.

Docking systems may therefore be defined as physical interfaces located underwater that permit AUVs to replenish their power supply and exchange data in a secure fashion. The technology and philosophy of docking stations are applicable to small (<50 kg), medium and large (>1,000 kg) AUVs and to vehicle applications in support of science research, commercial surveys and defence.

6.2 MISSION REQUIREMENTS FOR DOCKING SYSTEMS

Prior to analysing the actual design of docking systems, it is instructive to give a brief description of the types of missions where these systems can be applied.

6.2.1 Ship Supported at Any Water Depth

This scenario foresees the use of the dock as a support to AUVs that are deployed and recovered from dedicated or chartered ships. It is foreseen that the dock, with its integrated AUV, would be launched from the ship. The AUV would commence its mission once the dock has reached a suitable depth. Within this scenario the AUV may perform repeat missions without the requirement to be recovered on board the ship, as once docked all the activities of recharging and data downloading may be performed at the deployment and docking depth.

6.2.2   Open Water Missions from Coastal Waters

This scenario foresees the use of the docking system for missions in open waters as in the case above, however in this case the dock will be placed on the sea floor and not be deployed from a ship. The AUV will treat the dock as a subsea home base from whence it will emerge to perform its mission, and to which it will return at the end of the mission for data downloading and recharging. The initial hypothesis is that the location on the seabed be in relatively shallow waters and therefore this scenario has been defined as from 'coastal or continental shelf waters'. The main challenge for this solution will be reliability, while the main goal is for significant operational cost savings, as a consequence of the need for minimal use of a surface ship.

6.2.3   Ice-infested Waters

One of the promising applications for AUVs is considered to be their use in ice-infested waters. In particular, the development of AUVs is considered one of the essential 'technological packages' for the exploration and exploitation of Arctic and Antarctic waters. A number of AUVs have already been developed and deployed to this end, however, currently, they are restricted to the use of a surface support vessel. This scenario will have obvious cost and availability advantages.

6.2.4   Deep Waters

Until recently this was seen as a distant application, however increasing interest in AUV technology from the offshore oil and gas industry, in which operational water depths have passed 2,000 m, has increased the attention given to this application (see Chapter 8). Apart from the industrial applications, scientific use in support of the installation and maintenance of large sea floor observatories is also being considered. AUVs are also seen as central to the data gathering from autonomous ocean sampling networks, including event-triggered investigation and survey from long-term docking stations (Singh *et al.*, 2001).

6.3   DESIGN APPROACH

The suggested design approach for the identification of the most suitable docking system for a specific AUV is to adopt a systems design methodology. This entails performing a design analysis divided into a number of discrete steps as described below.

The first step is to define the basic premise for the overall project, based on the general specifications of the AUV(s) to be catered for, the general environmental conditions foreseen and the global mission parameters for the AUV. The basic premise data forms an environment within which the conceptual design may be carried out.

6.3.1 Definition of the Mission Requirements and Operational Phases

The mission requirements for the docking system should be defined based on an analysis of the mission parameters of the AUV's foreseen activities. These should be developed taking into consideration:

- the missions for which current AUVs developed or operated by the designer are utilised;
- the objectives for AUVs under development, to identify the near term goals for future missions of AUVs;
- analysis of the requirements of clients and industry to define possible mission requirements as a longer term goal;
- consensus discussions with all parties involved, in order to define the mission requirements for the docking system, bearing the above in mind.

Once the mission requirements have been established, it is possible to identify the various operational phases that will be required, which will lead to the selection of a suitable design solution. In the case of underwater docking systems identifying the operational phases will entail some or all of the following:

- homing, including relative position identification, navigation control, and target detection;
- docking, including mechanical interfacing, dissipation of the vehicle's momentum, fine alignment, and electrical or electromagnetic and perhaps optical connection;
- garaging, including mechanical blocking, protection, and release;
- the reverse functions of the above for when the AUV leaves its dock;
- launch and recovery of the docking system, with or without a docked vehicle.

6.3.2 Concept Generation

The generation of the concepts for the principal operational phases defined above may then be carried out based on the following:

- literature analysis of the methods currently in use in the industry (state-of-the-art analysis);
- discussions with operators as to how the main operative phases are currently carried out, and which developments they foresee;
- technology transfer from other industries (e.g. aerospace, nuclear, defence, etc.);
- brainstorming activities with other members of the dock design team and the AUV designers.

Once the concepts have been developed, a screening activity may be performed based on the selected scenarios and the basic premise data. This screening activity will assist in the selection of the most promising concepts.

The final step of the conceptual design is the development of the most promising solution. This will explore the feasibility of the most promising concept identified by analysing the principal critical issues, and will give clear indications on how the further design activities should proceed. The following sections describe technologies and practical solutions for each of the principal operational phases necessary for successful docking.

6.4   DOCKING TECHNOLOGIES AND SOLUTIONS

6.4.1   Homing

As outlined above, the homing operational phase includes a number of aspects, the principal ones being relative position identification, navigation control, and target detection. In some cases the solutions will be predicated by the capability of the AUV(s), while in others specific solutions will have to be identified.

Recent developments in the inertial navigation systems aboard AUVs have permitted high reliability and accuracy in position identification, and therefore the approach from a long distance of an AUV to a docking station is generally no longer considered a critical aspect – see for example, the results with the Maridan AUV using the MarPos inertial navigation system described in Chapter 12. In the event that the AUV positioning system is considered insufficient for long distance tracking, a Long Base Line (LBL) acoustic navigation system with a set of repeater stations may be placed on the sea floor in the operational area of the docking system. Combined Global Positioning System – LBL buoys have been developed by one manufacturer, suitable for first circle re-entry (Coudeville and Thomas, 1998).

Once the AUV is within close range to the docking station a tracking system will be utilised to assist in guiding the AUV into the dock. The tracking system will require components within the AUV and within the docking station. Three main alternatives have been proposed for short distance tracking:

Acoustics: The acoustic solution consists of an Ultra Short Base Line (USBL) system, which is suitable for close range approach. This will improve the relative position accuracy between the vehicle and the dock to typically ±0.2 m, at a rate of two to four fixes per second, with improving accuracy as the range reduces. The USBL receiver could be on the AUV if already part of its equipment or could be on board the docking system itself. In this latter case the AUV would only carry the transponder part of the system. Alternatively a low cost ROV type transponder grid may be used for sea bottom based scenarios.

Electromagnetic: An electromagnetic (EM) homing system is an alternative that can provide accurate measurement of the AUV position and orientation relative to the dock during homing (Feezor *et al.*, 2001). The EM system consists of a dock with EM coils that provide the homing signal, and a set of sensing coils that are mounted in the AUV. Such an EM system operates with a dipole magnetic field generated at the dock and sense coils in the AUV to measure the bearing and orientation relative to the dock entrance. In operation, magnetic field sensors in the AUV identify a magnetic field line emanating from the dipole on the dock, and the AUV simply follows the field lines into the dock entrance. The range of the system currently on the market is limited to 25–30 m.

Optics: This approach for terminal guidance of an unmanned underwater vehicle is roughly analogous to that which is employed by a heat-seeking air-to-air missile when locked onto a target. In this case the target is a light emitter which is located at the underwater dock. By using a signal-processing algorithm that calculates the difference between the light intensities that are received by adjacent quadrants a great deal of the forward scattered light can be rejected when the signals from the four quadrants are subsequently processed. The distinct spot of light on the detector face formed by the un-scattered light generates a high-quality control signal that can be employed to aim the nose of the vehicle at the light source, as the illumination is equal in all four quadrants only when the tracker's optical axis is exactly aligned with the source.

Two solutions are possible for an optical system: either the AUV is fitted with a complete recognition system (including an image processor), that searches for the pattern of the recovery device, or the AUV steers toward the bearing of highest intensity. Although an image-recognition system can be of great use on board an AUV, it is expensive, difficult to operate and requires significant calculation time, which makes it incompatible with the need for real time navigation. In addition, the effective range of an image-recognition system is considered limited.

A light source tracking system has a better range and is easier to use. It should be noted that all optic systems have the limitation of only being suitable for clear waters.

As a consequence of the limitations of EM and optical systems, acoustic communication is generally selected as the technology used for homing, and is featured by most AUVs. A typical implementation is illustrated in Figure 6.1 based on the scheme

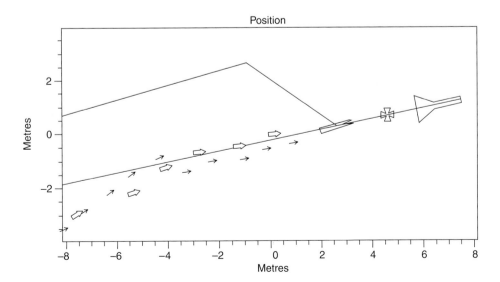

Figure 6.1 REMUS makes its final approach to the dock. Open arrows show the actual position of the vehicle, thin arrows show where the navigation system 'thinks' the vehicle is.

used in the REMUS project (Stokey *et al.*, 2001). The first step is for the AUV to get into a position where it can interrogate the transponder located on the dock. Once the AUV receives a response it obtains an updated position fix, and thus the AUV position jumps, correcting the accumulated dead reckoning errors, for example due to a substantial current flow. The AUV now turns toward the track line. Once it nears the track line, it turns towards the dock, which it is now able to interrogate regularly, and hence receive position updates as it closes on the dock.

6.4.2 Docking

A number of systems for AUVs to mate to the docking component have been considered by researchers and industry. Among these are:

- *Articulated arm*: This solution foresees placing the AUV interface, which in this case will be a cone for interfacing with the nose of the AUV, on an articulated arm similar to the manipulators currently used by ROVs and submersibles for deepwater work. The arm may either be attached to a ship deployed tow fish for near-surface use or attached to a bottom based structure (Figure 6.2).
- *Guide wire cone*: This solution foresees the location of an interface nose cone on a guide wire that is deployed utilising buoyancy elements from a bottom based structure.
- *Garage*: In this solution the AUV if foreseen to swim into a garage, rather than a nose cone. The garage may either be positively buoyant if deployed from a bottom-based station, or negatively buoyant if launched from a vessel.
- *Stinger*: This system is based on the concept used for aeroplanes landing on an aircraft carrier. The AUV is supplied with a hook that latches onto a suitable slot or

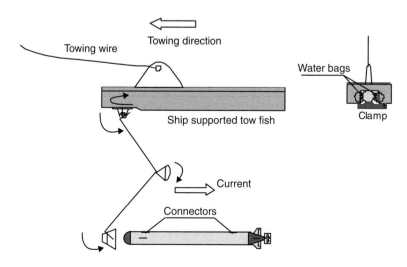

Figure 6.2   Typical articulated arm concept.

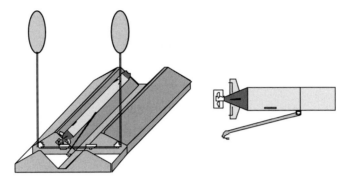

Figure 6.3   Typical stinger concept.

catch wire during a close fly-by, following which the vehicle may be manoeuvred
for final docking (Figure 6.3).
- *Free swimming*: In this solution an ROV is deployed from the vessel or the bottom based structure and collaborates with the AUV to interface the vehicle to
  the dock.

Generally, in order to select the most suitable mating system it is necessary to identify the other operative functions that must be catered for by the mating solution,
principally:

- fine guidance during docking, with particular reference to attitude in a strong
  current;
- the system for manoeuvring the AUV, especially at slow speed, when a nose cone
  is used;
- protection for the AUV during docking and launch and recovery;
- the system for connecting the electrical connector to the AUV;
- and the manner in which the AUV is released for its next mission.

6.4.3 Garaging

Garaging is the function of protecting the AUV after docking, during recharging and
during launch and recovery. The garaging function can be independent or combined
with this latter operation. It is important to bear in mind that the garage concept
selected should minimise the required supplementary vehicle hardware, which might
affect adversely the vehicle hydrodynamics.

A disadvantage of having a garage is that it may result in a more difficult navigation
control problem, since the vehicle cannot merely head towards the docking point.
Rather, it must orient itself on a proper glide path, much as an aircraft must align
itself over a runway. This disadvantage can be overcome by maintaining a forward
motion to the ship in the surface support case, and the possibility for the garage to

Figure 6.4   Example of a framed modular garage.

orient itself with the current, with the garage entrance downstream, when bottom supported. In this way the vehicle can more easily align itself towards the garage entrance by maximising the flow over the vehicle control surfaces.

Of the many concepts for AUV garages, two variants are considered below:

- *Framed modular garage*: This concept consists of a prismatic frame constructed from tubular beams forming the corners, connected as an open structure (Figure 6.4). The advantage of this solution is a low added mass relative to the internal storage volume enveloped. However, in the simplest implementation this solution would not allow a smooth AUV re-entry and would need to be designed specifically for each vehicle cross-section.
- *Tubular garage*: This concept consists of an open tube of a diameter larger than that of the AUV. It is convenient for cylindrical or torpedo shaped AUVs, for which the lost volume is minimum and the added inertia associated with the trapped water and the hydrodynamic added mass is limited. However, it is not suitable in this regard with flat horizontal shapes, such as the Maridan vehicles (Chapter 12), or with streamlined, low drag AUVs such as Hugin (Chapter 11), as the added inertia would become unacceptable. This concept could have two add-on variants to make docking easier and to allow adaptation to a variety of AUV diameters. First, a propeller could be added to help the deceleration of non-hovering AUVs just prior to and during mating with the garage in the case of bottom support, generating an artificial current against the approaching vehicle. Second, by a pumping-out effect, a guiding bottle would help stop non-hovering vehicles after they had entered the garage.

In both basic garage architectures identified above a number of supplementary elements can be considered to assist in the mating and securing of the AUV:

- *Front bumper*: In order to dissipate the residual kinetic energy a shock absorber may be placed at the rear end of the garage. A spring loaded mechanical system or an inflatable water bag are possible solutions.
- *Locking mechanisms*: In order to ensure that the AUV does not move once it is in the garage, securing elements can be included. Suggested solutions are piston depressors or water bags.
- *Guidance bumpers*: To assist the AUV in reaching a well-defined position in the garage to ensure the possibility of performing the electrical connection. These are typically low friction material rails shaped to cope with the AUV cross-section. Connecting the bumpers to the structure with resilient elements will provide shock absorption, while giving the bumpers a low angle conical configuration will perform braking of the AUV within the garage. The use of special alignment and support elements can be considered as an alternative.

As illustrated above the supplementary elements may be active or passive. It is clear that the choice to use active components will provide higher reliability for the actions they are aimed at but will also necessitate using hydraulic or electric systems that may reduce the robustness of the overall system.

Whichever solution is adopted for the garaging of the vehicle, an important aspect to consider when the garage is used for underwater docking is the release and expulsion of the vehicle from the garage for redeployment.

A number of solutions are possible, including:

- using a piston to force out the AUV via a nose cone interface;
- using pulleys with elastics in the place of the piston;
- permitting the garage structure to open up or down (when sea bottom or ship supported respectively) so that the AUV can swim away under its own power.

In selecting the preferred solution, simplicity and operational reliability must be born in mind.

6.4.4 Launch and Recovery

Although the launch and recovery system is not strictly part of the docking system, it is considered to be a critical system to the design and development of the docking solution, in particular if ship deployed operations are envisaged. Several different concepts are commonly used for launch and recovery, they include:

- *Crane deployment*: Although this is the most common form of launch and recovery system, in which the vessel crane is utilised to lower the AUV into the water and, following the mission, to lift it back on board, consideration should be given to the sea state in which this operation can be performed and the amount of human intervention that is necessary. Recent systems have automatic mechanical

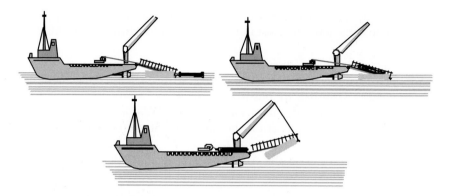

Figure 6.5   Operational sequence for a stinger system.

connection systems that reduce the need to deploy a tender to assist in recovery operations, although the reliability of these systems should be carefully analysed. Motion-compensated cranes have also been used for AUV launch and recovery. Dedicated crane-like launch and recovery systems designed specifically to handle AUVs have proven successful, for example the telescopic cradle fitted with winches used with Autosub (Stevenson, 2001).

• *Stinger or slipway deployment*: In order to reduce the dependence on human assistance, autonomous systems such as that depicted in Figure 6.5 have been suggested. The system used with the Hugin vehicles (see Chapter 11) is a variant of the slipway method, where the slipway is an integral part of the vehicle support container (Vestgård, 2001).

• A-frame crane solutions should also be considered, in particular for operators that have a dedicated vessel. Operation in up to sea state 5 can be achieved, for example by using a latching head mount derived from an ROV tether management system as used for the Maridan-600 vehicle (Baunsgaard, 2001).

In order to overcome the limitation imposed on the launch and recovery system due to the sea state, consideration should be given to recovering the AUV from a distance below the sea surface. Typically this would be performed by the AUV swimming into a garage type structure, which would be deployed from a crane and subsequently retrieved to the surface containing the AUV. This solution is clearly applicable to many of the docking solutions discussed previously, thus illustrating the synergy that can be had from a surface deployed docking system and a launch and recovery solution.

The main aspects that must be born in mind when developing such a solution are the effect of the current on the orientation of the garage, the heave motions induced on the garage by the vessel, and the horizontal motions of the garage while passing through the splash zone. Use of a dead weight situated under the garage and attached to guide wires has been suggested as a suitable solution, as illustrated in Figure 6.6.

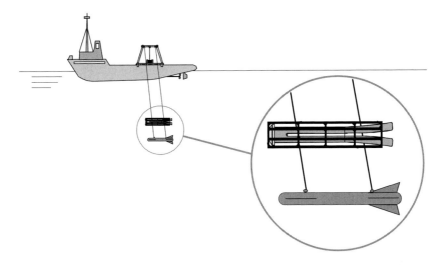

Figure 6.6   Underwater launch and recovery with dead weight.

6.4.5   Special Requirements for Connectors and Power Supplies

Wherever possible the development of a docking system for an existing AUV will be relatively non-intrusive to the actual AUV design. However, the addition of the electrical connector, and possible modification to the battery charging system may be necessary and these are described below.

In order to perform data downloading and power recharging underwater, it will clearly be necessary to perform an electrical or electromagnetic connection between the docking system and the AUV. For an electrical connection this will entail including either the female or male part of an underwater mateable connecter to the AUV itself. This component has already been developed within the offshore industry for the connection of umbilicals to the control systems of underwater production wells, but will require careful selection in order to guarantee reliability in an autonomous system. The precision mating required for direct electrical connection might be relaxed somewhat if an electromagnetic coupling is used. Such a system has been demonstrated on the Odyssey AUV (Singh *et al.*, 1997).

Recharging AUV battery packs is generally performed in a controlled environment on the surface. In some instances, the vehicle's battery packs are physically exchanged for charged units to enable the discharged packs to be handled under controlled conditions. These conditions will be more difficult to provide when recharging batteries in an autonomous underwater docking system. Modifications will be dependent on the type of batteries on the AUV and on the sophistication of the battery recharging management system in use, however some important considerations to bear in mind are:

- Charging cells in a closed environment will require particular consideration to hydrogen gas formation as the by-product of the recharging sequence of lead acid batteries in particular. Other battery chemistries are not immune from this problem. Vincent and Scrosati (1997) give an overview of charging requirements for many different types of secondary cells. Particular care needs to be given to balance the charge among cells connected in series-parallel arrangements.

- Recharging duration should be kept to a minimum, in particular in the case of ship deployed docking systems, otherwise it may turn out to be more convenient to retrieve the AUV to the surface and physically change out the battery pack. Charging at high rates may require careful thermal management, again dependent on the electrochemistry of the chosen cells.
- Energy on board the AUV should be more than adequate for the mission tasks, as prudence requires that the docking be performed with sufficient energy on board to repeat the homing sequence in the event that docking is unsuccessful.

During recent years the batteries used for AUVs have undergone a significant evolution, taking advantage of the advances inspired by the consumer electronics markets (see Chapter 2 for a detailed review of AUV power sources). Fortunately for underwater docking technology, current attention is being placed on the use of secondary lithium ion and lithium polymer batteries, which are more suitable to the requirements of underwater recharging than lead–acid or nickel–cadmium batteries. Secondary lithium batteries have a higher specific energy and energy density than silver–zinc batteries. An advantage of lithium–polymer batteries is that they are solid state and can operate at ambient pressure, therefore they do not have to be contained inside a pressure hull. Other advantages include a high cell voltage (3.8 V), long endurance with over 600 cycles, and a battery life of over 5 years. The *in situ* charging of a lithium pack on an AUV has been demonstrated (Bradley *et al.*, 2000). Cost currently remains high and the cells must be charged and discharged between a precise range of voltage in order to prevent damage from excess charge or discharge, however it may be assumed that in the future this type of battery will increasingly enter the AUV market.

6.5   STATE OF THE ART IN DOCKING SYSTEMS

A number of underwater docking systems have already been developed by the research community and industry, although they are at present at the prototype stage of development. The most significant systems, which utilise a number of different architectures from among those described in previous sections, are discussed here.

6.5.1   Nose Cone Docking Systems

With accurate homing and positioning being one of the principal concerns for underwater docking, many of the early studies looked at the use of nose cone docking systems. This type of docking system has been used successfully with Woods Hole Oceanographic Institution's REMUS vehicles (Stokey *et al.*, 1997). REMUS is only 1.35 m long with a body diameter of 0.19 m, and is therefore suitable for use with the relatively large cone size that can be seen in Figure 6.7(a). Homing is performed with

(a) 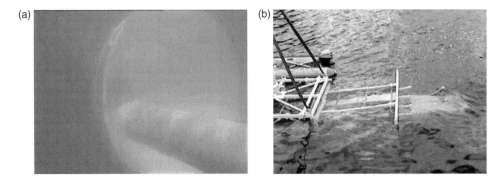 (b)

Figure 6.7   (a) REMUS entering in the docking cone and (b) the MARTIN 600 AUV entering the EURODOCKER garage.

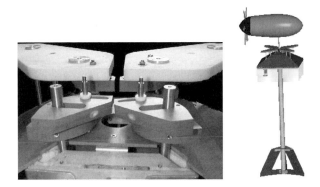

Figure 6.8   Florida Atlantic University's Ocean Explorer stinger docking system.

an acoustic system that has a range of 50 m, which is considered necessary in order to ensure that the AUV flies directly down the centre of the cone to achieve docking.

6.5.2   Stinger Docking Systems

Florida Atlantic University developed an alternative docking system for its Ocean Explorer series of AUVs, based on the stinger concept (Figure 6.8). Current Ocean Explorer AUVs are designed to operate with 8–12 h mission duration at $1.5 \, \text{m s}^{-1}$. Through multiple missions over the period of deployment, the AUV can provide coverage of over $1,000 \, \text{km}^2$ surrounding the docking facility. After each mission, the AUV returns to the dock to 'refuel', download the mission data and upload any mission changes.

The purpose of the docking system is to move the vehicle towards and eventually into the dock by providing successive fuzzy goals for the AUV. As the AUV approaches the dock it continually slows down and aligns itself with a docking axis (Rae and Smith, 1992).

Most docking methods for AUVs use a nose-first approach, however due to the modularity of the Ocean Explorer this is highly undesirable, as nose-cones are mission-specific while the rear section is standard, containing the propulsion and navigation systems. Instead, a docking method using a belly-mount stinger has been developed. The stinger interfaces with the dock itself through a four-petal configuration that allows the AUV to approach from any direction. Once the stinger is locked into place at the centre of the dock, the electrical connection can be performed.

Odyssey AUVs have been adapted successfully to dock with moored stations by using a scissors-like stinger on the nose (Bowen, 1998; Bowen and Peters, 1998).

6.5.3   Garage Docking Systems

Another prototype docking system that has been constructed is based on the garage concept and is called the EURODOCKER (Brighenti *et al.*, 1998; Reinhardt *et al.*, 1998; Galletti *et al.*, 1998, 2000). The EURODOCKER project has selected a system that both guides the AUV in the final approach and provides a suitable location during the period that the AUV is docked. The garage also protects the AUV after docking, during recharging and/or launch and recovery.

The concept consists of a frame made of tubular beams (Figure 6.7(b)), segmented transversely and longitudinally, hosting four curved guidance bumpers, shaped to cope with the AUV cross section. Connecting the bumpers to the structure with resilient elements provides shock absorption, while giving the bumpers a low angle conical configuration assists with braking the AUV during entry into the garage.

The release and expulsion of the vehicle from the garage is performed by permitting the garage structure to open up or down (when sea bottom or ship supported respectively) so that the AUV can swim away under its own power.

Homing is performed with an USBL system that guarantees the AUV to position itself within the 0.4 m misalignment that is permitted by the funnel of the garage. The velocity of the vehicle prior to docking is designed to be between 1 and $2.5\,\mathrm{m\,s^{-1}}$, although lower velocities can be handled, depending on the thruster power of the AUV. The active components on the garage are driven by a 2.2 kW seawater pump that drives the door actuation cylinders and water bags for shock absorption and positioning. The positioning system lodges the AUV within the garage to with 0.01 m accuracy, although the connector is mounted on a floating seat that permits misalignments of 0.025 m. The connector permits both battery charging via a dedicated battery charging management system and data downloading.

The fully functional prototype was constructed and successful sea testing was carried out in May 2001 utilising Maridan's Martin AUV. Tests were performed both on the surface and in a fully submerged configuration. Following docking into the garage the electrical connector was mated to the AUV and data was exchanged, prior to the vehicle being redeployed via the opening of the garage doors. Other facilities that compose the garage are permanent and variable buoyancy systems, instrumentation, blocking devices and the electrical connector.

6.5.4 Submarine Docking Systems

Within naval military research and development, one aim over the last few years has been to develop the capabilities to both deploy and subsequently recapture AUVs from submarines, in particular from torpedo tubes (see Chapter 10). The docking and retrieval concept, as used in the Long-Term Mine Reconnaissance System (LMRS) autonomous vehicle follows the following basic sequence:

- following its mission the AUV swims alongside the submarine;
- an articulated retrieval arm is deployed out of one torpedo tube;
- the AUV swims into a nose cone positioned on the retrieval arm;
- handling mechanisms are deployed to pull the AUV into an adjacent torpedo tube;
- the AUV is retrieved into the submarine.

Even the first stage of this procedure, positioning the AUV alongside the submarine, involves complex hydrodynamics as the (relatively) small AUV comes alongside the (relatively) large submarine and enters its boundary layer (Huyer and Grant, 2000; Ramamurti and Sandberg, 2000). The retrieval mechanisms contain complex electromechanical and hydraulic components, yet are constrained to fit within the 0.53 m diameter of a torpedo tube. There is also the operational issue of committing two torpedo tubes to the AUV and its docking system. For these and other reasons more innovative solutions have been proposed that do not involve docking into existing torpedo tubes. One example is the US Naval Underwater Warfare Centre's MANTA concept, where the AUVs hull shape would allow conformal docking with the hull of a submarine.

6.6 THE FUTURE

The various underwater docking systems developed to date have illustrated the technical feasibility of the solutions discussed in this chapter and recent developments in on-board power systems and inertial positioning systems have overcome remaining doubts as to the operability of these systems. It is therefore likely that an increasing number of AUVs will operate with underwater docking in the future, both in order to perform repeat missions and in order to launch and recover the vehicles in more severe meteo-oceanographic conditions.

However, the most promising future for underwater docking is envisaged to be in the area of subsea intervention (Taylor and McDermott, 2001). A number of research projects are underway, or have been proposed, that aim to use AUVs in place of tethered remotely operated vehicles for inspection, maintenance and repair operations subsea, particularly within the oil and gas industry. In most cases these projects foresee docking the AUV to a sea bottom anchored connection point in order to provide the energy and signal bandwidth required to perform such operations. This technology will not only vastly increase the capabilities of the AUVs, in particular for manipulation, but will also open up other tasks in the offshore industry. A project of note in this sector is represented by the Subsea Operations with Autonomous Vehicles (SOAVE) project commenced at the start of 2002 by the partnership formed by Maridan, SATE, Shell International, Oceaneering, Fugro UDI and the University of Hannover.

7. PROPULSION SYSTEMS FOR AUVs

SULEIMAN M. ABU SHARKH

School of Engineering Sciences, University of Southampton,
Southampton SO17 1BJ, UK

7.1 INTRODUCTION

Almost all AUVs use screw-type propellers because they are in general the most efficient propulsion devices at the operating speed of AUVs, although some vehicles, for specific purposes, may use other propulsors, for example the pump-jet as in the Marlin submarine-launched AUV (Tonge, 2000) and the gliders described in Chapter 3. As these are very much the exception currently, this chapter is concerned primarily with screw propellers driven by electric motors. It begins by defining some of the terms used in propeller design. A simple theory of the propeller action is used to develop a set of equations relating its driving motor torque and power requirements to its basic dimensions, output thrust, and thrust efficiency. Methodical test series data is used to compare the mathematical model to existing propeller designs and to illustrate the critical design parameters.

Two critical design parameters are identified: the propeller diameter and its pitch, which determine its torque and power requirements and hence the size and the cost of the AUV. Other parameters, like the number of blades, blade thickness, hub size, rake etc., have little effect on the power and torque requirements of carefully designed propellers. It is also shown that for AUVs with low advance speeds (below $2\,\mathrm{m\,s^{-1}}$) it pays, in terms of propeller thrust efficiency, to have a duct or Kort nozzle, named after Ludwig Kort who did some pioneering work on ducted propeller for use in ships (Kort, 1934). He showed that the profile of the cross-section of the duct and its thickness were critical to the performance of the ducted propeller.

The torque and power requirements of the propeller, which need to be minimised to minimise the size and the cost of the AUV, are shown to be in conflict with each other; minimising one results in maximising the other. This suggests that there should be an optimum combination of propeller basic dimensions, that is, diameter and pitch, which results in minimising the size and cost of the AUV, for certain performance requirements. Optimisation strategies for propeller drive systems are discussed; emphasising that the optimisation criteria will have to take into account the size, cost and performance of all the drive system sub-components, from the supply, through the inverter to the propeller, in addition to other operational and physical constraints.

The chapter concludes with a short review of submersible electric motor technology, concentrating mainly on the available engineering solutions for housing and protecting motors operating in the seawater environment.

7.2 PROPELLER NOMENCLATURE

A marine propeller is a device that converts the greater part of the driving power into a thrust force to propel a vessel. The screw propeller is the most common form of marine propulsion device; in general it is also the most efficient. Most common types of propeller used in AUVs are hub driven, as illustrated in Figure 7.1. The propeller hub is fitted onto the shaft of an electric motor, or alternatively, the motor might have an outer rotor on which the blades can be fitted, and thus the motor forms the hub of the propeller. Sometimes, hydraulic motors (a less efficient system) may be used, but usually only where electric power is supplied through a tether. Hydraulic motors are much smaller that conventional electric motors, thus resulting in a considerably more compact unit, at the expense of overall poorer system efficiency. A detailed comparison between hydraulic and electric thruster systems was presented in Abu Sharkh *et al.* (1994).

The alternative to driving the propeller through its hub is to use a tip drive. Tip driven propellers using a gear drive were proposed by Sounders (1957) and Radojcic (1997), but these are not common due to gear system complexity and inefficiency. However, recent advances in electric motor technology and manufacturing techniques have made the tip drive concept that is illustrated in Figure 7.2, an attractive option for ducted propellers that can offer a compact and more efficient alternative to hub driven propellers (Holt and White, 1994; Abu-Sharkh *et al.*, 2001). The integrated electric thrusters are now available commercially with a wide range of sizes and power ratings from 50 W to 11 kW. Figure 7.3 is a photograph of prototype integrated thrusters that were developed at the University of Southampton.

The motion of a marine screw (Figure 7.4), as suggested by the name, combines a rotation with a translation along the axis of rotation. It consists of a number of identical twisted blades (usually from two to five) equally spaced around a hub or boss, or inside a ring in the case of a tip driven thruster. The inclination and the twist of a blade are defined by the angles between a datum plane normal to the axis of rotation and a number of datum lines fixed relative to the blade. A set of pitch datum lines is taken at a series of constant radii from the axis of rotation, and the angle

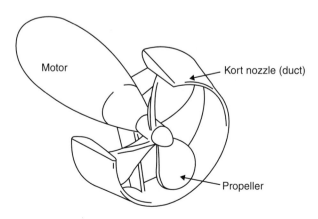

Figure 7.1   A hub driven ducted propeller.

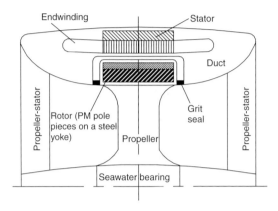

Figure 7.2   The integrated electric thruster concept.

Figure 7.3 Prototype electric integrated thrusters developed by the University of Southampton.

between the datum plane and the pitch datum line is termed the pitch angle ϕ. The pitch angle varies radially: typical values are $\phi = 50°$ near the boss to $\phi = 15°$ near the blade tip. The surface of the blade facing aft is called the face of the blade, the forward side being the back.

The face of each blade is part of a helicoidal-like surface, which can be generated in its simplest form by a radial line which rotates about its axis and also moves along this axis a distance directly proportional to the angular movement. The axial distance corresponding to one complete revolution of angular movement is termed the pitch P, which is related to the pitch angle ϕ_r at a distance r from the propeller centre by the following equation:

$$\tan \phi_r = \frac{P}{2\pi r}. \tag{7.1}$$

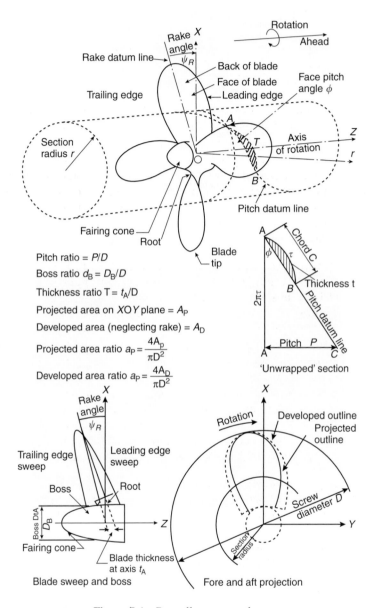

Figure 7.4 Propeller nomenclature.

The pitch for a true helicoidal surface is the same at all radii, however, in practice the pitch is not always constant, and it is fairly common to have a reduced pitch towards the hub to reduce vibration (Lewis, 1988). The ratio of the pitch/propeller diameter, P/D is called the pitch ratio, and usually P is taken to be the pitch at $0.7\,R$, where R is the radius of the propeller. The pitch ratios for marine propellers range from 0.6 for highly loaded propellers, such as those on tugboats, up to two or more on high-speed motor boats.

Other terms commonly used in propeller design are defined in Figure 7.4.

7.3   THEORY OF PROPELLER ACTION

Theories of propeller action are reviewed by Lewis (1988) while a more comprehensive review of the theoretical studies of ducted propellers has been given by Wessinger and Maass (1968) and Küchmann and Weber (1953). There are three basic theories of the propeller action, the momentum theory, the blade element theory and the circulation theory. The treatment presented in this section is based on the momentum and a simplified version of the blade element theories, since, in general, these two theories are simpler than the circulation theory and are more amenable to analytical treatment. In addition, the circulation theory is usually only needed in the final design stages of a propeller where more detailed refinements to the blade geometry are required.

In the simple momentum theory the propeller is regarded as a disk mechanism capable of imparting a sudden increase of pressure to the fluid passing through it, the method by which it does so being ignored.

It is assumed that:

(a)   The propeller imparts a uniform acceleration to all the fluid passing through it, so that the thrust thereby generated is uniformly distributed over the disk.
(b)   The flow is frictionless, thus thermal energy will be neglected.
(c)   There is an unlimited inflow of water to the propeller.
(d)   The region of fluid that is acted upon by the propeller forms a circular column. The centreline of this column is assumed to be horizontal. The propeller race (also sometimes termed the slipstream) is accelerated, to have an axial motion only, thus ignoring the angular motion and turbulence effects.
(e)   Placing a duct around the propeller has the effect of modifying the streamlines (like an airfoil) of the flow in order to reduce or increase the contraction of the flow thus changing its velocity profile. Drag forces and turbulence generated by the presence of the duct, as well as forces produced due to the differential pressure between the outside and the inside of the duct are neglected.

The first assumption involves a contraction of the race column passing through the disk, and since the contraction cannot take place suddenly at the disk, the actual acceleration must occur outside the disk and be spread over a finite distance fore and aft.

Consider a propeller disk of an area A_o and a diameter D advancing with a uniform velocity V_A into undisturbed fluid. The hydrodynamic forces will be unchanged if the system was replaced by a stationary disk in a uniform flow of the same velocity V_A, as shown in Figure 7.5.

In accordance with Newton's second law, the thrust force will be given by:

$$S = U_2 \dot{m}, \tag{7.2}$$

where $U_2 + V_A$ is the velocity of the water in the slipstream, and $\dot{m}$ is the mass of water flowing through the propeller in one second which can be shown is given by:

$$\dot{m} = \rho A_o (U_1 + V_A), \tag{7.3}$$

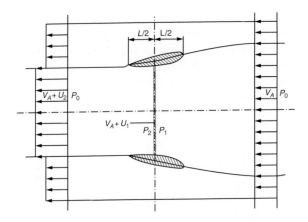

Figure 7.5   Changes in pressure and velocity at propeller disk, momentum theory.

where ρ is the density of water, and $U_1 + V_A$ is the velocity of water at the propeller disk. From Equations (7.2) and (7.3) the thrust produced by the propeller is given by:

$$S = \rho A_o U_2 (U_1 + V_A).$$
(7.4)

Similarly, the power expended, which is equal to the increase of the kinetic energy of the flow, can be shown to be given by:

$$P = \tfrac{1}{2}\rho A_o U_2 (U_1 + V_A)(U_2 + 2V_A).$$
(7.5)

The thrust efficiency is defined as the ratio of the output thrust to the driving power;

$$\eta_t = \frac{S}{P} = \frac{2}{U_2 + 2V_A}.$$
(7.6)

Clearly, to maximise η_t for a given output thrust and advance speed, the relative slipstream velocity U_2 must be minimised while increasing U_1 in equation (7.4), that is, by minimising the ratio of U_2/U_1. This can be achieved by placing a duct around the propeller which modifies the water flow stream lines (Figure 7.6), accelerating the flow around the propeller or decelerating it depending on the shape of the duct. In the first case the propeller efficiency is increased (by minimising U_2/U_1) and in the second case the cavitation properties of the propeller are generally improved.

For an open water propeller (equivalent to a neutral duct) it can be shown (Harvald, 1983) using Bernoulli's equation that U_1 is related U_2 by the following equation:

$$\frac{U_2}{U_1} = a,$$
(7.7)

where $a = 2$.

Assuming no flow separation at the duct, in addition to the earlier assumptions, it can be shown that a in Equation (7.7) is a constant for a propeller with a duct.

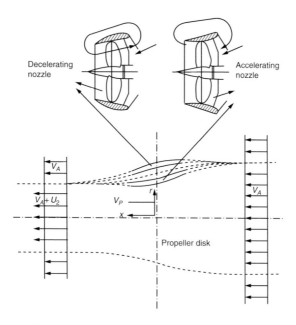

Figure 7.6   Streamline forms produced by different ducts.

The value of a is 1 for an ideal accelerating duct. For actual accelerating ducts a lies between 1 and 2, while for decelerating ducts a is greater than 2. Cavitation is not a problem at the usual operating depths of an AUV propeller, and the main interest is in accelerating ducts, to increase propeller efficiency.

The static thrust efficiency η_{ts} at the bollard pull condition can be found from Equation (7.6), by substituting $V_A = 0$, as

$$\eta_{ts} = \frac{2}{U_2}. \tag{7.8}$$

Substituting Equation (7.8) in Equation (7.6) produces

$$\eta_t = \frac{\eta_{ts}}{1 + \eta_{ts} V_A}. \tag{7.9}$$

The thrust efficiency is therefore directly related to its value at the static bollard condition and it is sufficient to consider the static thrust efficiency to compare the performance of different propellers, which is the case for highly loaded propellers with small advance speeds and large thrust. However, for lightly loaded propellers with low output thrust and large advance speeds this may no longer be the case especially with ducted propellers where the drag losses on the duct result in a worse performance than open propellers at these conditions (Küchmann and Weber, 1953).

By arranging the above equations, it is possible to derive an expression for the static thrust efficiency in terms of the propeller diameter and the output static thrust as follows:

$$\eta_{ts} = \sqrt{\frac{\pi\rho}{a}}\frac{D}{\sqrt{S}}. \tag{7.10}$$

Another useful quantity which is independent of the propeller dimension and the output thrust and may be used as a figure of merit to compare different types of propellers, usually at bollard pull conditions, is the efficiency coefficient η_d (Gongwer, 1984; Lewis, 1988) which is defined as

$$\eta_d = \frac{S}{P}\left(\frac{S}{\rho A_o}\right)^{1/2}. \tag{7.11}$$

The efficiency coefficient can be shown to be proportional to the ratio of the axial kinetic energy imparted to the water to the power driving the propeller shaft, that is, it is a measure of the efficiency of the propeller in converting the input power into useful thrust.

When the thrust and power are those at the bollard pull conditions, the static efficiency coefficient η_{ds} can be found by re-arranging Equation (7.10) as follows

$$\eta_{ds} = \sqrt{\frac{4}{a}}. \tag{7.12}$$

So far the propeller was considered as a mechanism for increasing the momentum of the race, but no attempt was made to explain how this is done, and drag forces were neglected. It is possible to account for the effect of the actual geometry of the propeller and the drag forces using the blade element theory in which the propeller is considered to be made up of a number of separate blades which in turn can be divided into successive strips of elements as shown in Figure 7.7.

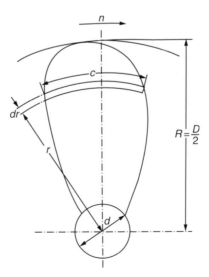

Figure 7.7   Blade element theory.

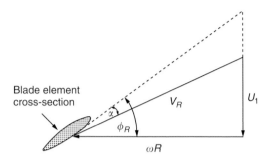

Figure 7.8 Blade element velocity diagram at bollard pull.

The forces acting on each element are evaluated from the knowledge of the relative velocity of the element to the water and the characteristics of the element section shape.

The velocity components on a blade element at the tip of the propeller, a distance R from its centre, at the bollard condition are shown in Figure 7.8 (neglecting the rotational induced velocity of the water). In this figure, ω is the angular velocity of the propeller and ϕ_R is the pitch angle at the tip of the propeller. The angle α between the resultant water velocity V_R relative to the blade element direction and the face pitch line, is known as the angle of incidence (or attack), and it is always small in an efficient propeller, usually in the region of 4° to 6° to maximise the lift to drag ratio (Lewis, 1988). From the velocity diagram in Figure 7.8 it is possible to find an expression for the propeller angular speed ω as follows:

$$\omega = \frac{2U_1}{D \tan(\phi_R - \alpha)}. \tag{7.13}$$

For given advance speed, propeller diameter, and output thrust, ω is maximised by reducing the pitch angle ϕ_R in Equation (7.13). However, there is a limit on the minimum propeller pitch angle (about 10° in practice), below which the propeller efficiency is impaired due to interference between the blades and restriction of water flow.

From Equations (7.7), (7.8) and (7.13) another useful expression for the static thrust efficiency can be derived as follows:

$$\eta_{ts} = \left(\frac{4}{a \tan(\phi_R - \alpha)}\right)\frac{1}{\omega D}. \tag{7.14}$$

The size of the motor required to drive the propeller is directly proportional to the driving torque Q (Say, 1983) that can be shown to be related to the propeller geometry and the output thrust by the following equation which is obtained by re-arranging 7.14):

$$Q = \frac{P}{\omega} = \frac{a \tan(\phi_R - \alpha)}{4}SD. \tag{7.15}$$

The driving power can be also found in terms of the output thrust and the diameter of the propeller by re-arranging Equation (7.10):

$$P = \sqrt{\frac{a}{\rho\pi}}\,\frac{S^{3/2}}{D}. \tag{7.16}$$

7.4 EXPERIMENTAL INVESTIGATIONS OF PROPELLER ACTION

Experimental data on existing propeller designs is available in the form of design charts, which give the results of open-water tests on a series of model propellers (e.g. Lewis, 1988). These cover variations in a number of design parameters such as pitch ratios, blade area and section shapes. A propeller that conforms to the operating characteristics of any particular series can be rapidly designed to suite the operating conditions.

The open-water test conditions should be set such that advance speed of the model is scaled to satisfy the Froude number. However, in this case the Reynolds number will not be the same for the model and the actual designed propeller. In practice, tests are conducted such that the value of the Reynolds number is as high as possible, the requirements to run at the correct Froude number being ignored. However, provided that the propeller is run with adequate immersion, so that there is no wave making on the surface, the lack of Froude-number identity will not have any important effect.

The pressure should also be scaled in these tests. In practice, this is not usually the case. However, since the forces on the propeller blades are caused by differences in pressure, they will not be affected by this fact unless cavitation occurs, in which case other types of tests must be made to correct the open-water test results.

Typical curves produced by these open-water tests are shown in Figure 7.9 in which the parameters K_T, K_Q, η_0 and J are defined as follows:

$$\text{Advance ratio} \quad J = \frac{2\pi\,V_A}{\omega D}, \tag{7.17}$$

$$\text{Thrust coefficient} \quad K_T = \frac{4\pi^2 S}{\rho\omega^2 D^4}, \tag{7.18}$$

$$\text{Torque coefficient} \quad K_Q = \frac{4\pi^2 Q}{\rho\omega^2 D^5}, \tag{7.19}$$

$$\text{Propeller efficiency} \quad \eta_0 = \frac{J}{2\pi}\frac{K_T}{K_Q}. \tag{7.20}$$

From these equations it is possible to derive expressions for the thrust efficiency as follows:

$$\eta_t = \eta_d\sqrt{\frac{\rho\pi}{4}}\,\frac{D}{\sqrt{S}} \tag{7.21}$$

and

$$\eta_t = \frac{K_T}{K_Q}\frac{1}{\omega D}, \tag{7.22}$$

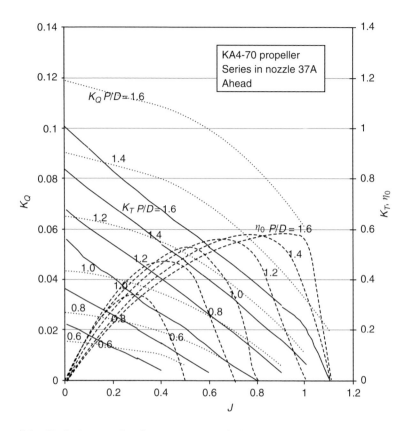

Figure 7.9 Typical curves for thrust, torque and efficiency obtained by open-water tests.

where the efficiency coefficient is given by

$$\eta_d = \frac{(K_T/\pi)^{2/3}}{K_Q}. \tag{7.23}$$

The torque, speed and power required to drive the propeller can also be given by the following equations:

$$\omega = \frac{2}{\sqrt{\rho}}\sqrt{\frac{S}{K_T}\frac{1}{D^2}}, \tag{7.24}$$

$$Q = \frac{P}{\omega} = \frac{K_Q}{K_T}SD, \tag{7.25}$$

$$P = \frac{1}{\eta_d}\sqrt{\frac{4}{\rho\pi}\frac{S^{3/2}}{D}} = \frac{2\pi}{\sqrt{\rho}}\frac{K_Q}{K_T^{3/2}}\frac{S^{3/2}}{D}. \tag{7.26}$$

The theoretical estimates of performance based on momentum theory in equations (7.10)–(7.16) can be compared to the actual performance of a given propeller with known K_T, K_Q versus J characteristics:

$$K_T = \frac{\pi^3 a}{4} \tan^2 (\phi_R - \alpha), \tag{7.27}$$

$$K_Q = \frac{\pi^3 a^2}{16} \tan^3 (\phi_R - \alpha). \tag{7.28}$$

The theoretical expression provides an upper limit on the achievable performance as it neglects the losses occurring in practical propellers. Other useful information on propellers including those with large hubs and unconventional duct shapes can be found in Andrews and Cummings (1972), Harvald (1983), Gongwer (1984, 1989), and Sparenberg (1984).

7.5   MATCHING THE MOTOR TO THE PROPELLER

Figure 7.10 shows the variation of torque and power requirements (based on Equations (7.25) and (7.26)) for the combination of a propeller style Ka4-70 in a 37-A duct, whose particulars and characteristics are published in Oosterveld (1972, 1973) and shown in Figure 7.9. The curves are for a propeller with a pitch ratio of 1.0. Figure 7.11 shows the variation of thrust, power and speed as a function of pitch ratio for the same propeller with a diameter of 300 mm and producing 1800 N of thrust (using Equations (7.24)–(7.26)). The curves in Figures 7.10 and 7.11 have the same general shape for other values of pitch ratio, diameter and thrust.

For other types of propeller, these curves also have the same general shape. In general, for a particular pitch, diameter and thrust output, the power and torque requirements are not significantly affected by other parameters such as number of blades, blade thickness, hub size, rake etc., for well designed propellers. However, the shape of the duct wall cross-section can have a large effect on power and torque, especially in the astern direction.

The size of an electric or a hydraulic motor can be shown (Say, 1983) to be proportional to propeller torque, while the size of the power supply power (battery/cable) is proportional to propeller power. The fluid friction loss in a motor with a radial gap is proportional to the cube of the speed and the fourth power of the diameter (Belgin and Boulos, 1973). Ohmic loss in the motor increases roughly as the square of motor torque for fixed motor dimensions.

As shown in Figure 7.10 (and Equations (7.25) and (7.26)), for constant thrust and pitch, as the propeller diameter increases, torque and motor size increase while power and power supply size reduce. Hence an optimum diameter can be selected to minimise the overall size or cost of the thruster system. In the case of a hub driven propeller the optimum choice of diameter may be biased towards reducing torque and hence motor diameter. For hydraulic systems, the optimisation may be biased towards reducing the size or cost of the large power pack.

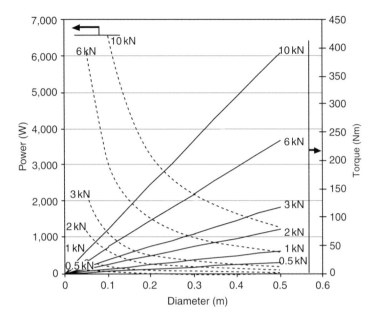

Figure 7.10   Effect of diameter of power and torque requirements of a Ka4-70 propeller with a pitch ratio of 1.0 in 37-A duct.

The required speed of the thruster is inversely proportional to the square of its diameter, and hence by increasing diameter the fluid friction loss decreases. But this increases torque, and so for fixed motor dimensions ohmic loss increases. Again an optimum propeller diameter can be selected to maximise overall efficiency.

On the other hand, power and hence power supply size is not significantly affected by pitch ratio (for the practical range of pitch ratios of 0.6–1.4), as illustrated in Figure 7.11. However, torque and hence motor size increase with increasing pitch ratio. In general, a small pitch ratio would therefore be preferred to reduce the size of the overall system, and reduce motor ohmic loss. But speed increases by increasing pitch, hence increasing fluid friction loss in the motor which can be large for motors of large diameter.

To summarise, the size and efficiency of the drive are determined by the design of the propeller. In general, attempting to improve overall efficiency results in increasing the size of the motor and vice versa.

Depending on the type of drive, the optimisation objective may be either to minimise the size or cost of the system (or one or more of its sub-components), or to maximise its efficiency. This of course should be subject to constraints (or boundaries) on acceptable size of motor, power supply and propeller, and acceptable efficiency of the system. The maximum diameter of a hub motor is typically restricted to 50% of the diameter of the propeller. The diameter of propellers used in AUVs is generally less than 500 mm, restricted by available space on the vehicle. Possible optimisation strategies of motor/propeller/supply configurations are discussed in more detail in Abu Sharkh *et al.* (1995b).

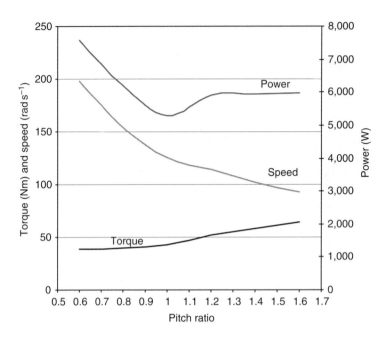

Figure 7.11   Effect of pitch ratio on power, torque and speed of a 300 mm diameter Ka4-70 propeller in 19-A duct producing 1,800 N of thrust.

7.6   SUBMERSIBLE ELECTRIC MOTOR TECHNOLOGY

The main engineering problem that needs to be solved when using an electric motor to drive a propeller is protection against water ingress which can weaken the dielectric strength of the winding insulation and can cause corrosion of laminations, magnets and other motor and drive electronic components. One solution is to house the motor in a pressure vessel, and to use rotating shaft seals to prevent water ingress. However, at the operating depths of AUVs, this solution can be very expensive. Rotating shaft seals have large friction losses thus reducing the efficiency of the motor.

Another solution is to use a magnetic coupling to transmit mechanical power from the motor to the propeller as illustrated in Figure 7.12 for commercial thrusters produced by Tecnadyne. This eliminates the need for seals. But the magnetic coupling adds to the cost and size of the thrusters. Synchronous coupling (e.g. using permanent magnets) also suffers from the problem of losing synchronism when subject to shock loads or sudden acceleration, which can require the motor to be restarted. However, reliable operation of this type of motor at pressures of up to 600 bar has been reported (Stevenson *et al.*, 1997).

The most common solution, however, is to house the motor (and sometimes the electronics) is an oil filled and pressure compensated housing as shown in Figure 7.13. The oil inside the housing is kept at a slightly higher pressure than that of the seawater by a spring-loaded diaphragm. This insures that oil leaks out of rather than seawater leaking into the motor housing. The oil is selected to be compatible with the winding

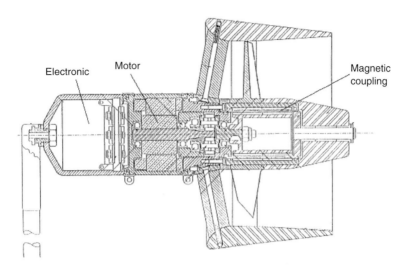

Figure 7.12 Magnetic coupling used to transmit power from motor in a pressure vessel to propeller in water, without the need for a shaft penetrating the housing (courtesy of Tecnadyne).

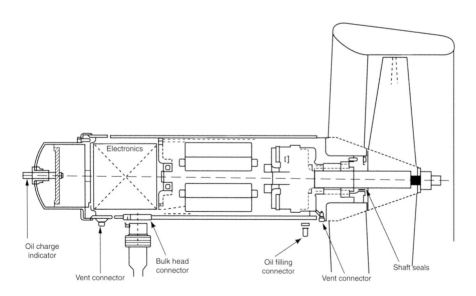

Figure 7.13 Oil filled and pressure compensated electric motor.

insulation and other parts of the motor. One disadvantage of this arrangement is the higher friction (stirring) loss in the motor.

Water filling is also used in some motors. Large induction motors produced by Hayward Tyler in the United Kingdom are examples of this. Water is less viscous than oil (less friction loss) and is a better coolant, but it causes corrosion. A special polymer

Figure 7.14   Polymer windings of water filled motor (courtesy of Hayward Tyler).

insulated winding is used (Figure 7.14), which reduces the slots copper fill factor and hence increases the size of the motor and or reduces its efficiency. The integrated thrusters developed at the University of Southampton encapsulate the motor and components in epoxy as protection from water ingress.

7.7   CONCLUSIONS

Using a simple theory of the propeller action, in addition to methodical test data, it is possible to predict the hydrodynamic performance of a propeller. Two critical propeller design parameters are identified, the propeller diameter and pitch, which determine its torque and power requirements and hence the size of the AUV. Other parameters, such as the number of blades, blade thickness, hub size, etc. have little effect on these requirements for carefully designed propellers. For AUVs with low to moderate advance speeds, it may pay in terms of thrust efficiency to have a duct or nozzle shrouding the propeller. The profile of the cross-section of the nozzle is critical, especially to the astern performance of the propeller. The optimum length/diameter ratio of the nozzle has to be selected to maximise the effect of the nozzle on improving the thrust efficiency of the propeller while minimising the effect of drag forces on its walls. For the illustrated propeller designs, the optimum length/diameter ratio of the nozzle lies between 0.5 and 1.0.

The torque and power requirements of the propeller, which are related to the size and cost of the AUV are shown to be in conflict with each other, minimising one maximises the other. This suggests that there should be an optimum combination of propeller diameter and pitch, which results in minimising the cost and the size of the AUV, for certain performance requirements. By careful matching of the propeller to the motor, it is possible to maximise the efficiency of the thrusters/motor system, while keeping the size of the motor to acceptable level.

Symbols

ϕ	pitch angle
ϕ_r	pitch angle of blade element at radius r
ϕ_R	pitch angle of blade element at radius R
A_o	propeller disk area, m^2
D	diameter of the propeller, m
m	mass flowing through the propeller disk in 1 s, $kg\,s^{-1}$
N	angular speed of the propeller, $rev\,min^{-1}$
P	power, W
ρ	density of water, $kg\,m^{-3}$
r	radius of a blade section, m
R	radius of the propeller, m
S	thrust, N
U_1	axial velocity of water at the propeller disk at static bollard condition, $m\,s^{-1}$
U_2	velocity of water in the slipstream at static bollard condition, $m\,s^{-1}$
V_A	propeller advance speed, $m\,s^{-1}$
P	pitch of the propeller, m
η_t	thrust efficiency, $N\,W^{-1}$
η_{ts}	static thrust efficiency, $N\,W^{-1}$
η_d	efficiency coefficient
η_{ds}	static efficiency coefficient
a	ratio of U_2/U_1
K_T	thrust coefficient
K_Q	torque coefficient
J	advance ratio
η_0	propeller efficiency in open water
ω	propeller angular speed, $rad\,s^{-1}$

8. AUV TASKS IN THE OFFSHORE INDUSTRY

EDWIN F.S. DANSON

Edwin Danson Associates, 14 Sword Gardens, Swindon SN5 92E, UK

8.1 INTRODUCTION

In the offshore industry, where survey work is primarily of a commercial nature, the Autonomous Underwater Vehicle (AUV) is just beginning to make an impact. Technical dependability is a high priority in this sector, even more so are cost-efficient solutions, especially in those regions where conventional acquisition methods have reached maturity or are unable to deliver the crucial information upon which engineers depend.

The commercial importance of the offshore industry to the interests of many nation states, both developed and developing, is without doubt considerable. Key to these activities is surveying and offshore engineering, both of which can benefit from the capabilities of autonomous underwater robotics. Even so, the offshore industry is not the largest sector that could, in the future, take advantage of the new technology.

Of the 1,400+ survey and research vessels in national government service world-wide, some 25% are engaged in hydrographic[1] surveying of one sort or another. Their work is principally focused towards charting for navigation, for military and defences applications, for environmental assessments and monitoring and for national economic exploitation. Some 52% of the planet's entire marine survey effort is concentrated within this range of activities and it is not surprising that much of the early development of AUVs was funded by this sector. Of the remaining 48% of the world's hydrographic activities, 28% are for, or within, the ports, harbours and coastal engineering sector. Traditionally, this is a fragmented and conservative industry and is unlikely, at least for some time, to embrace advanced technology. The last 20% represent the surveying activities that support the offshore industry. These divisions are shown in Figure 8.1.

The activities comprising the hydrographic tasks that support the offshore industry are multitudinous. This sphere is overwhelmingly of a commercial nature and primarily performed by organisations within the private sector. At this time (2001) the broad divisions of the offshore industry are shown in Figure 8.2.

---

1 Hydrographic surveying – in the context of this chapter, means the activities associated with the physical depiction of the sea and ocean floors, the charting of natural and man-made artefacts and structures, and the geophysical examination of the first 10 m or so of the sub-seabed. Deeper penetrating geophysics and seismic exploration is excluded.

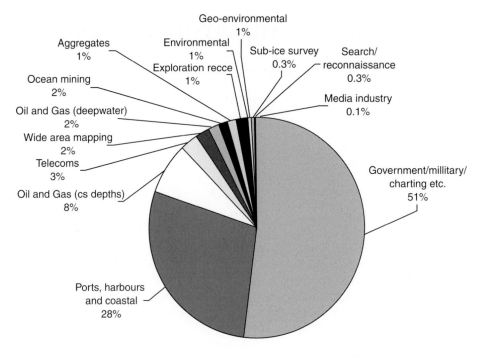

Figure 8.1 Hydrographic surveying by sector (2001).

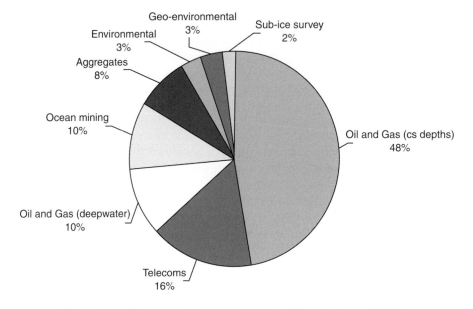

Figure 8.2 The sectors that comprise the offshore industry (2001).

The oil and gas sector accounts for the greatest proportion of the offshore industry with about 83% of its activities taking place in water depths less than 300 m while the remainder is focused in the so-called deep-water areas such as off West Africa, South America and the Gulf of Mexico. In the future, this ratio will change as the exploitation of hydrocarbons moves into ever-deeper water.

Telecommunications is another sector of the offshore industry that is evolving rapidly. Its current 16% share of activities is on the increase as the demand grows for secure, robust and high-capacity telecommunication networks to link the nations of the world. Ocean mining, for precious minerals and metals, is also expected to be a growth area in the future as technologies are developed to harvest these products from the sea.

The oceans are still, for the most part, a mystery. While scientists for decades have known that the ocean drives our climate, it is only recently that politicians and the general public have come to accept the importance. Climatic change and the need to understand and monitor the ocean, as well as to assess the effects of our polluting industries and population, has already led to an increase in environmental monitoring as a market for the private sector. This is expected to increase from the current levels of about 6%. The technologies and tools developed and used within the private sector are well suited to many of these vital tasks and the commercial efficiency inured in this sector will provide cost-efficient solutions in the future. Without doubt, the AUV will be one of those tools.

8.2   THE ECONOMICS

A paper by Chance *et al.* (2000) has been quoted on a number of occasions as demonstrating and illustrating the cost-effectiveness of AUV operations. In essence, the technical (and commercial) arguments for replacing 'towed system' technology with AUVs are:

- *Line running direction*: Towed systems cannot easily accommodate sharp changes in direction whereas an AUV can accommodate sharp changes.
- *Line running position*: Keeping a deeply towed system on line is difficult – AUVs run on line.
- *Line running altitude/aspect ratio*: Keeping a towed system at correct depth is not easy; an AUV can fly at precisely the correct depth or height above the seabed.
- *End of line turns*: Towed systems, especially in deep water, require a long line turn (anything up to 2 h) – AUVs turn on a dime.
- *Run-ins/run-outs*: Towed systems require a run-in to a line and a run-out (in deep water the run out can be six or more kilometres) – AUVs need neither a run-in or run-out.
- *Positioning*: In deep water, positioning a towed system is problematic and, if any degree of accuracy is needed, calls for a second USBL vessel – AUVs can run (a) autonomously; and (b) (if a UUV) be positioned by USBL from a single vessel.

- *Survey speed*: Survey velocity of a towed system is a function of water depth and length of cable; the deeper the water, the slower the speed. The speed of a deep-tow system can drop to a half a metre a second whereas AUVs can travel at speeds of between one and a half and two and a half metres per second, irrespective of water depth.

Undoubtedly, AUVs can offer timesavings over conventional survey techniques. To what extent these savings translate into economics is a more complex issue than is often supposed. Even compared with some of the more exotic survey instrumentation, an AUV can appear to be quite expensive. To the company amortising assets with which an AUV could, or might, compete, or even supersede, the wisdom of an investment in the technology can appear questionable. For a new-start, or a company considering expanding capability, the AUV offers an interesting and commercially attractive alternative to traditional technology.

C&C Technologies of Lafayette, La., quote freely $55,000 (2000) as the day rate for a survey spread equipped with its Hugin vehicle (Northcutt *et al.*, 2000). From extensive comparative studies, this rate appears to be in line with other vendors, and reflects the investment, the associated development and integration costs and, of course, the rapidity of data acquisition and reduction in exposure risk.

The commercial economics for AUV operations have also been evaluated independently for a 'typical' vehicle and its systems. Although the 'savings' are generalised, they nevertheless demonstrate that the efficiencies of AUVs cannot be dismissed. In the examples it is assumed that:

- transit time to site and return is 2 days;
- the 'conventional survey' consists of multibeam and a towed sidescan and sub-bottom profiler;
- line run-out and line turn times are included for the 'conventional' survey's towed systems (layback is 3× water depth);
- weather downtime is 20% of the time running survey lines.

The AUV operation includes:

- simultaneous multibeam, side-scan and sub-bottom profiling;
- time for diving to depth and resurfacing;
- battery change out and data download on deck.

The first example, Table 8.1, is the type of survey that might be undertaken for oil and gas activities in a deep-water (2,000 m) 10 km by 10 km block; a total of 520 survey line km. By conventional methods, the total time for the project is 21 days. In all cases of AUV speed and endurance the cost savings are significant.

The next example, Table 8.2, is of a simple 200 km by 800 m wide cable or pipeline route corridor in an average water depth of 300 m and having one change of direction; a total of 665 survey km. In this instance, the conventional survey ship is operating at 2.0 m s^{-1}. In addition to the geophysical survey, an allowance of 150 h has been made for geotechnical sampling, a very typical offshore survey activity. This example demonstrates that the choice between an AUV and a conventional vessel towed system

Table 8.1   Comparison of survey time for a 10 km by 10 km block.

AUV Speed	AUV Mission Duration		
	12 h	24 h	48 h
$2.5\,\mathrm{m\,s}^{-1}$			
No. of missions	6	3	2
Project duration	6.8 days	6 days	5.8 days
$2.0\,\mathrm{m\,s}^{-1}$			
No. of missions	7	4	2
Project duration	7.7 days	7 days	6.5 days
$1.5\,\mathrm{m\,s}^{-1}$			
No. of missions	9	5	3
Project duration	9.4 days	8.4 days	7.9 days

Table 8.2   Comparison of survey time for a 200 km route survey in 300 m of water.

AUV Speed	AUV Mission Duration		
	12 h	24 h	48 h
$2.5\,\mathrm{m\,s}^{-1}$			
No. of missions	7	4	2
Project duration	8 days	7.2 days	6.7 days
Cost savings are significant			
$2.0\,\mathrm{m\,s}^{-1}$			
No. of missions	9	5	3
Project duration	9.4 days	8.4 days	7.9 days
Cost savings are marginal			
$1.5\,\mathrm{m\,s}^{-1}$			
No. of missions	12	6	3
Project duration	11.7 days	10.2 days	9.5 days
Cost savings are negative			

requires care. Only in the case of the AUV travelling at $2.5\,\mathrm{m\,s}^{-1}$ are the cost savings significant. It should also be noted that the choice is the operator's and not necessarily the customer's.

8.3   THE VEHICLES

It was during the latter years of the 1990s that the survey community first began to investigate the possibility that AUVs could be employed to solve some of the problems then facing the oil and gas industry. The impetus came from the upstream oil and gas operators where a number of enlightened individuals were considering how best to approach surveying tasks in the deep-water blocks then being explored in the Gulf of Mexico, West Africa and offshore South America. Their contention was that, in these difficult regions, traditional surveying methods and existing technology were approaching their practical limits and that the AUV appeared to solve many of the difficulties.

It soon became apparent that the world was not short of excellent AUV development centres such as MIT's Sea Grant Laboratory, the Woods Hole Oceanographic Institution and Florida Atlantic University in the United States. In the United Kingdom, the University of Southampton had developed and built its first vehicle, the Autosub. In all, there were nearly a hundred vehicles in existence or under development. There was also the surprise that AUV technology was quite mature and had a track record dating back to the 1950s. However, as surveying platforms, there was much work to be done. Nevertheless, these and many other vehicles around the world were repeatedly demonstrating their extraordinary capabilities. MIT's Odyssey had shown that a small vehicle could perform complex scientific programs in the hostile environment of the deep ocean (e.g. Bellingham, 1993) and Southampton's Autosub had demonstrated its prowess by performing complex scientific work programmes (e.g. Millard et al., 1998). Of the commercial AUV manufacturers, International Submarine Engineering's ARCS vehicle had over 900 successful dives to its credit, Thomas (1986), while its mighty Theseus had shown that an autonomous vehicle could lay 175 km of fibre optic cable beneath the polar ice (see Chapter 13). In Norway, Kongsberg Simrad AS and their military partners had developed the first generation of their successful Hugin[2] vehicles and were holding regular customer demonstrations (see Chapter 11).

At about the same time the oil and gas industry began to investigate AUVs. The De Beers mining group in South Africa was also studying AUVs as a possible improvement to their survey methods for offshore diamond prospects. They turned to Maridan AS, a small entrepreneurial Danish company who was then developing a new model from an earlier experimental vehicle (see Chapter 12). Elsewhere in Europe, Japan and around the Asia-Pacific, private enterprises collaborating with university projects were also developing autonomous vehicles for use within the private sector. AUV technology continues to attract great interest in academic circles, and will continue to do so as military and commercial applications advance; more importantly for industry, the vehicles have unquestionably reached maturity for commercial exploitation (Wernli, 2000).

From the start, the major oil and gas operators were reluctant to fund directly the development of autonomous technology, arguing that AUVs were surveying tools and therefore the surveying companies should acquire and develop the technology themselves. There was, at first, considerable reluctance to pursue the technology. One of the main, and reasonable, arguments against an early adoption of AUVs was a financial accounting one. It was argued that investing in a new, untried and very costly technology would place an even greater financial burden on a small, low-margin, survey industry. Additionally, such an investment would limit the scope to recoup the investment the sector had made in deep-water acquisition systems, the so-called deep-tow systems. These deep-tow systems were, and remain still, costly items and can compare in capital cost with many AUV models. However, while many argued

---

2 In the context of this essay, the term AUV incorporates vehicles that are both fully autonomous and those that operate under permanent, or semi-permanent acoustic command from the surface, the so-called Unthethered Underwater Vehicle or UUV.

against the vehicles on the grounds of cost, none could disagree with their potential to change the way seabed data would be collected in the future.

AUVs are costly items and probably will always be; offshore surveying systems have always been expensive. Despite predictions that equipment costs should reduce over time, history does not concur. While models of non-evolved systems have come down relatively in price, such as the older sort of side-scan sonar, such are the demands of the offshore industry and its customers that new technology is constantly required. Apart from the enterprising De Beers mining group, three of the larger private survey companies embarked on the road to AUV acquisition. The US company C&C Technologies was the first in the offshore sector to acquire and field a vehicle. Its choice was Kongsberg Simrad's Hugin 3000, a deep-water vehicle targeted initially at the Gulf of Mexico market. About the same time, the UK based Thales GeoSolutions announced its intention to acquire a small fleet of vehicles from the MIT spin-out company Bluefin Robotics of Cambridge, USA (Willcox, 1999). Its vehicle, Sea Oracle, was developed in a partnership program combining Thales' expertise in sensor technology with Bluefin's autonomous vehicle skills and know-how. The Netherlands-based Fugro group was also very active and, through an open competition, selected to partner with the Canadian manufacturer, International Submarine Engineering (ISE) of Vancouver. However, after a shift in focus, Fugro's development program moved towards where they saw their main market, the Gulf of Mexico. Its US company, Fugro Geoservices, and its partner, Oceaneering, took advantage of the Boeing corporation's inter-company innovation scheme to develop a large survey AUV.

Alone of the first four advocates, Thales chose that its vehicle should be small, arguing that it would be easier to handle and that, by using multiple vehicles, it could meet its deep-water acquisition needs. The other companies chose much larger configurations to cater for extended operational duration, or mission time, while carrying a full complement of survey sensors.

In 2001, Halliburton's UK subsidiary, SubSea, a major offshore construction contractor, joined the race by announcing its intention to build survey class AUVs based on the University of Southampton's Autosub design. A net purchaser of survey services, SubSea recognised that in the Autosub it could perform its own seabed surveys independently, reduce its costs and enhance its product.

Time and again it has been shown that a good survey, to a sensible specification, saves project time and money. Even civil engineers are beginning to understand the economics. Optimal structural design versus over-design saves the oil and gas industry millions every year; however, this is only workable when the engineers have the essential data they need. The AUV appeared to solve both problems – excellent data acquisition capabilities at an affordable price.

There was considerable debate within the survey community on what tasks AUVs were capable of performing. Most agreed that multibeam echosounders, side-scan sonars and sub-bottom profilers should be carried; some added magnetometers to the inventory while others included some of the more exotic systems. There was the issue that centred on how long should an AUV mission last. Twelve hours? Twenty four? Forty eight or even longer? When Fugro set out its requirements in 1998 through an invitation to tender, speed and endurance were key commercial issues. While no one envisioned running a vehicle for more than a few hours in the first instance, it was recognised that a 12-h mission would not be that long in coming. However, to be

Table 8.3 Comparison of Fugro's 1998 desired AUV specification with that of Hugin 3000 (Hugin data from the C&C Technologies Inc. web site).

Item	Fugro 1998	C&C's Hugin 3000 Vehicle
Length	<6 m	5.2 m
Working speed	2–2.5 m s^{-1}	2 m s^{-1}
Depth capability	3000 m	3000 m
Endurance (range)	24–48 h (170 km min to 430 km max)	40 h (288 km)
Sub-sea navigation	USBL/LBL, DVL plus INS to achieve <0.05% of distance travelled between updates	USBL/LBL, DVL plus inertial guidance (fibre optic gyro)
Collision avoidance sonar	Yes	Yes
Bi-directional acoustic transmission	Yes	Command and control/low bit rate data
Mission planning software	Yes	Yes
Payloads survey suite	Multibeam echo sounder	Simrad EM2000 multibeam echo sounder
	Sub-bottom profiler	Chirp (2–16 kHz) sub-bottom profiler
	Side-scan sonar	Chirp (120 kHz or 410 kHz) side-scan sonar
	Option of a magnetometer	Option of a magnetometer

commercially viable, 48 h appeared to be the optimum and the vehicle would have to be able to perform, if not immediately, then within a short time.

The other questions centred on equipment and sensor packages and operating configurations. In this respect, it is interesting to compare the similarities in the following table of Fugro's 1998 specification with the Hugin 3000 delivered to C&C Technologies in 2000 (Table 8.3).

Running such a spread of survey sensors, the command and control systems as well as the vehicle propulsion, consumes significant power. Hugin 3000's power system is an aluminium–hydrogen peroxide semi-fuel cell; other power sources are chiefly batteries such as primary manganese alkaline or secondary cells such as silver–zinc or, increasingly, lithium–ion.

Another question on AUV economics centred on the efficiency of vehicle operations, especially whether it was best to have one multi-role or several single-role vehicles. Several economic evaluations, based on manufacturers 1998 prices, seemed to point towards 'bigger is beautiful'; however, there is more to the answer than simple economics. Smaller vehicles are easier to handle and are more portable, can be operated from vessels of opportunity and offer some flexibility for sensor packages.

Another vehicle class, akin to the AUV, is the so-called Hybrid vehicle. This centaur-like vehicle is half AUV and half ROV. As a class, the concept was stimulated by the work, among others, of the French company, Cybernetix and its SWIMMER vehicle (Cybernetix, 2001), and the AutoROV designs of the tripartite of ISE, Fugro-UDI and the Stolt Comex group (Gallett, 1999). The industry's need for such a hybrid vehicle was, at first, in some dispute. While oil and gas companies saw benefit and calculated huge savings, ROV operators remained sceptical, and with good reason. However, the idea of the hybrid vehicle stimulated more thought, and design, and its applications became clearer. The best, and most economically resilient use, was seen to be where ROVs were used as observation vehicles and where some 'light weight' intervention was required. The market best defining these characteristics is offshore construction support where ROVs are routinely used to monitor pipeline installation (touchdown monitoring) and the placement of sub-sea structures such as template and manifolds.

Very significant cost-savings are available to any customer who employs vehicles that do not have to rely upon specialist ROV support vessels, or whose missions are limited by weather conditions. Indeed, certain hybrid vehicle designs point towards these vehicles remaining on station, or in an oil field complex, rather than being tied to the capabilities and availability of specialist vessels. In late 2000, J. Ray McDermott, one of the major offshore construction companies, announced a collaboration with ISE for such a vehicle, SAILARS, aimed specifically to meet its construction support needs (McDermott, 2001). SAILARS employs a variant of the Dolphin semi-submersible AUV (see Chapter 14) as a stand-off vehicle from which is deployed a deep-water ROV. Other designs, such as SWIMMER and AutoROV, use an AUV-type shuttle to take an ROV to the seafloor construction site where the vehicle is mated to a power/optical fibre relay point.

8.4   THE TASKS

AUVs are, or soon will be, capable of performing many of the tasks traditionally associated with ship work. Whether they will or not depends upon the sense and sensibility of the industry and the operators. Assigning to AUVs tasks that could be performed more efficiently using traditional tools could cause more harm than good. Nevertheless, the most obvious application for AUV operations is that most traditional of ship-based science – surveying.

Equipped with a multibeam echo sounder, sub-bottom profiler and side-scan sonar, the AUV is the ultimate surveying tool. Detractors of this survey data acquisition method point out that the real-time quality control available from the vehicle while underway, from limited acoustic bandwidth, is not sufficient to ensure the data collected is adequate. However, this should not be regarded as an impediment to progress and, already, artificial intelligence is being developed that promises results better than any unaided human could maintain. It should of course be noted that UUVs (vehicles in direct acoustic link with the mother vessel) provide a substantial amount of real-time data.

8.4.1   Oil and Gas

The main driver for introducing the survey AUV has been the oil and gas industry's deep-water blocks off the Americas and Africa, where the costs associated with surveying, using traditional technology, appeared untenable. The alternative method for imaging the seafloor, and advocated by many in the industry, was to use reprocessed 3D seismic exploration data (Rutledge and Leonard, 2001). It has been argued that

the results compare more than favourably with multibeam echo sounder data; fortunately, for design engineers and the environment, most professionals disagree and
uphold the value of surveys conducted using appropriate tools and quality-assured
procedures. The AUV bridged the gap and offered the prospect of proper surveys
performed at costs comparable to surveys conducted in more shallow water. AUV
economics, like AUV technology, required a paradigm shift[3] in thought by contract
tenderers and evaluators alike.

AUVs are not just limited to deep-water and Maridan, for example, produce a version of their vehicles specifically for shallow water. It is anticipated that AUVs will one
day be a viable alternative for inshore survey and have the capability of going places
that are less accessible to, or could endanger, a traditional survey launch operation.

For the oil and gas industry, survey AUVs are suitable alternatives for:

- geohazard/clearance surveys;
- rig site surveys;
- acoustic inspection of pipelines and sub-sea installations;
- pipeline route surveys;
- construction site surveys.

It has been noted above that most of the vehicles entering the market are quite large,
five or six metres overall. Handling such large vehicles is not without its difficulties
and, as a result, specialist launch and recovery systems are needed. This tends to
delimit the vessels capable of operating an AUV to the larger type of vessel or survey
ship. The techniques for AUV operations are only just developing; for site surveys,
it has been suggested that the mother ship will continue to perform the digital data
acquisition leaving the so-called analogue acquisition to the AUV.

The so-called hybrid vehicle, described above, is an evolutionary line of AUVs considered by some in the oil and gas industry as one of its most exciting developments.
At this time, the greatest limitations for autonomous underwater robotics (as work-
class vehicles) are power and artificial intelligence. The tasks performed by ROVs
are complex indeed and, as yet, no one has been successful in replacing a human
operator. It is of course only a matter of time and money and if the commercial case
for autonomous robotics is demonstrated it is very probable that hybrid vehicle developments will produce a viable alternative to the ROV. In the meantime, these sorts of
vehicles, if they enter the industry, have an interesting future. In oil and gas pipeline
construction, the traditional method of monitoring pipe lay is to use a specialist ROV
vessel. The ROV monitors the pipe's touchdown point and can also be used for light
intervention tasks such as wire cutting etc. The hybrid vehicle can replace the ROV for
much of the critical touchdown monitoring. In deep water, this aspect of surveillance
takes on even more importance, as there is the added difficulty that pipes tend to
buckle under the extreme loads and then collapse with the higher water pressure.
With a hybrid vehicle monitoring the operation, its support ship can be employed

3 'Paradigm shift', much used to describe the thought and attitude changes necessary to maximise the
benefits of AUVs, in this context has been attributed to Jens Pind of Maridan AS.

in a much more efficient way providing construction support to the lay barge and checking the route ahead.

Unless the autonomous vehicle has a primary power source (such as SAILARS), the finite electrical power available has meant that the use of power-hungry video and camera systems is, at this time, limited. Similarly, heavy consumers like laser-scanners cannot be readily incorporated into today's AUV. The oil and gas industry has come to depend on visual observation for much of its tasks and will not readily, or speedily, change its ways. Unless AUVs can be fabricated that cater for this need, then it is unlikely that they will ever replace ROVs for observational tasks or as work-class vehicles.

Where AUVs do offer very tangible benefits is in pipeline route surveys, where the primary sensors are multibeam echo sounders, side-scan sonars and sub-bottom profilers. These surveys range in length from a few score metres to many hundreds of kilometres. Here the AUV has a distinct advantage because, while it collects the acoustic imagery, the mother vessel is free to perform the time-consuming and heavy geotechnical investigation. Similarly, the vehicles can gather the shallow geophysical data for construction site surveys while the mother ship gathers the more difficult deeper digital geophysics and *in situ* testing and site investigations for the all-important foundation studies.

The AUV has been studied by the seismic exploration industry as an alternative seismic streamer tow vehicle, as a seismic source and as a 'rear guard' data collector for far-field seismic signatures. Where it is expected the AUV has a new role is as an environmental data acquisition tool. In the deep-water regions, where much is still unknown, the vehicles can study and monitor environmental conditions, observe shallow water flows and measure the strength of currents throughout the water column. It can also be used as for environmental protection detecting and observing protected benthic populations and other ocean floor phenomena.

8.4.2 Telecommunications

In common with the oil and gas industry, route surveys are also the principle interest of the submarine telecommunications sector. Here the AUV is particularly well suited to continental shelf operations where fielding a specialist survey vessel to a remote location can be expensive. The cable industry, attracted by the prospects of rapid deployment and efficient data collection, has showed considerable interest in AUVs. It has even been suggested that an AUV could perform short cable crossings and inter-connector route surveys entirely autonomously, launched from a shore facility.

While it is less likely that AUVs will find a role off the continental shelf where, traditionally, surveys are conducted by multibeam echo sounder from surface vessels at moderately high speed, there is an opening for the more difficult shore approaches. In this zone, typically water depths less than 30 m and where the ocean going vessels cannot manoeuvre, surveys are generally conducted from a third party in-shore vessel. AUV performance competitions in the US have shown that some vehicles are particularly good operating in these difficult, close to shore, conditions.

The other obvious use of the AUV is as a cable layer, which has already been shown to be feasible by ISE, whose Theseus vehicle laid some 175 km of fibre optic cable beneath the polar ice (see Chapter 13). A number of organisations are looking at this option for small diameter communication cables.

8.4.3   Minerals and Mining

In the search for seafloor minerals, the AUV has already made its appearance. The DeBeers company's acquisition of a Maridan vehicle was for just this purpose and their arguments for going the AUV route were purely commercial. Augmenting their surface fleet with AUV capability has significantly increased the amount of ground they can cover each day in their search for diamonds.

The sub-sea mining industry will undoubtedly be a growth area for the future. The AUV is a logical tool for mapping, for example, manganese nodule fields and, perhaps, versions of the hybrid class of vehicles will be employed for the mining and monitoring.

8.5   CONCLUSIONS

In conclusion, autonomous underwater and semi-submersible vehicles, be they AUVs, UUV or hybrids, have important roles to play in the offshore environment. They may not be a substitute for many traditional ship-based methods of data acquiring, or replace the ROV as a construction support and intervention vehicle, but they do hold the prospect of becoming an indispensable tool for the engineer and surveyor.

9. MULTIDISCIPLINARY OCEAN SCIENCE APPLICATIONS OF AN AUV: THE AUTOSUB SCIENCE MISSIONS PROGRAMME

AUTOSUB SCIENCE AND ENGINEERING TEAMS*

*c/o Nicholas W. Millard, Southampton Oceanography Centre,
University of Southampton, Southampton SO14 3ZH, UK*

9.1 INTRODUCTION

In December 1995 a group of marine scientists and technologists met to define the scope of a thematic programme proposal that would use Autosub to demonstrate the utility of an Autonomous Underwater Vehicle (AUV) for ocean science. Further, the programme would aim to tackle questions that could only be answered using the unique features of such a vehicle. The ideas were presented at an open forum of the UK marine science community in May 1996. After incorporating the views raised at the forum, a proposal was submitted to the Natural Environment Research Council's (NERC) Marine Science and Technology Board in June 1996. The £5 m proposal was ambitious in its aims. It foresaw applications covering hundreds of kilometers in the open ocean, work in ice-covered areas, terrain-following missions with a host of sensors for physical, geophysical, biological and chemical oceanography. At the time the proposal for the Autosub Science Missions programme was submitted the longest mission achieved by the vehicle was some 300 m; the deepest depth achieved was some 9 m. Autosub had only been tested in Empress Dock, Southampton.

* **Science Teams**
Steven A. Thorpe[1], Thomas Osborn[2] and David M. Farmer[3]
George Voulgaris[4], Michael B. Collins[1], Eugene Terray and John H. Trowbridge[11]
Andrew S. Brierley[5], Paul G. Fernandes[6] and Mark A. Brandon[7]
Peter Statham[1], Julian Overnell[8] and Christopher German[1]
Alex Cunningham[9], Peter H. Burkill[1] and Glen Tarran[10]
Kate Stansfield[1] and David Smeed[1]

Engineering Teams
Nicholas W. Millard[1]
Steven D. McPhail[1] and Peter Stevenson[1]
Miles Pebody[1] and James R. Perrett[1]
Andrew T. Webb[1] and Mark Squires[1]
David Meldrum[8] and Gwyn Griffiths[1]

[1]Southampton Oceanography Centre, Southampton; [2]Johns Hopkins University, Baltimore, USA; [3]Univerity of Rhode Island, Narragansett, USA; [4]University of South Carolina, Columbia, USA; [5]University of St Andrews, St Andrews; [6]FRS Marine Laboratory, Aberdeen; [7]Open University, Milton Keynes; [8]Scottish Association for Marine Science, Oban; [9]University of Strathclyde, Glasgow; [10]Plymouth Marine Laboratory, Plymouth; [11]Woods Hole Oceanographic Institution, Woods Hole, USA.

At its February 1997 meeting the Council of the NERC took a bold step by rec-
ommending funding for the programme to begin in April 1998. However, Council
decided that, because of its ambitious targets, the programme should first demonstrate the utility of Autosub in shelf seas and the open ocean. Council suggested that
a programme proposal for beneath ice-shelf work with Autosub should be submitted
later, if the vehicle proved itself. The Autosub Science Missions Scientific Steering
Committee met for the first time on 8 December 1997. An Announcement of Opportunity for the "Autosub Science Missions" programme was issued on 8 January 1998
with a closing date of 14 February 1998 for outline proposals. An "Open Day" was
held at the Southampton Oceanography Centre on 27 January 1998 so that the scientists who were potentially interested could meet up with the technical team to discuss
their ideas. In all 24 outline applications were submitted. The Scientific Steering
Committee approved 19 of the outlines to proceed to full proposals with a submission deadline of 27 March 1998. Following international peer review, the Committee
ranked the proposals at their second meeting on 29 June 1998. Sufficient funds were
available to award six research grants and a small grant for data quality assessment.

The six projects supported were:

- Dissolved and particulate Mn and O_2 in the benthic boundary of hypoxic sea lochs:
  Overnell of the Dunstaffnage Marine Laboratory and Statham and German of the
  Southampton Oceanography Centre.
- Flow through a strait and over a sill in the Strait of Sicily: Smeed, Southampton
  Oceanography Centre.
- Sonar and turbulence studies in the upper ocean: Thorpe, Southampton
  Oceanography Centre.
- Spatial variability of bottom turbulence over sandbanks: Collins of the
  Southampton Oceanography Centre and Voulgaris, University of South Carolina.
- Subsurface single cell and particle analysis using flow cytometry: Burkill of the
  Plymouth Marine Laboratory and Cunningham, University of Strathclyde.
- Under sea ice and pelagic surveys: Fisheries and plankton acoustics in the North
  Sea and Antarctica: Fernandes, Fisheries Marine Laboratory, Aberdeen, Brierley,
  British Antarctic Survey and Brandon, Open University.

In addition, a small grant was made to Brian King of the Southampton Oceanography Centre to assess the data quality of the CTD and ADCP instruments on the vehicle
in support of the other projects.

Preparing and installing the sensors and using and interpreting the data from these
sensors, for several disciplines, has required close collaboration between the engineers
supporting the AUV and the scientists relying upon their measurements. The AUV is
still a young sensor platform. While it has excellent characteristics, such as low self-
noise and vibration coupled with high stability – all of which are desirable for many
measurements – speed, payload and power constraints impose restrictions on the
measurements that can be made. In this chapter we focus on three issues concerning
the use of sensors on an AUV undertaking multidisciplinary ocean science, drawing
upon our results from over 100 Autosub missions. First, how can data from several
sensors on an AUV be integrated to improve our understanding of the process under
investigation? Second, how can the combination of sensor and AUV characteristics

provide novel information that would be near impossible to obtain using conventional methods? Third, data quality: what are (some of) the advantages and disadvantages of employing sensors on an AUV?

9.2 STANDARD SENSORS USED ON AUTOSUB

The standard sensor suite on Autosub provides for the basic measurement needs of the physical oceanographer. These sensors also provide measurements of parameters of interest to biological oceanographers and biogeochemists. Over the 100+ missions of the Autosub Science Missions programme, many of the nuances of operating these sensors within an AUV have been discovered. Changes in placement or sampling position have been made in the light of this experience.

In the past, the vehicle has carried a Chelsea Instruments Aquapack CTD and fluorometer and an FSI CTD (Griffiths *et al.*, 1999). More recently, the Seabird SBE-9+ has become the instrument of choice. On the port side of the vehicle a water-inlet protrudes to allow the temperature and conductivity sensors to be as close to the external environment as possible, Figure 9.1. The sensors are pumped. The pump is the last module in a chain that starts with the inlet, followed by the platinum resistance thermometer; conductivity cell; electrode for dissolved oxygen concentration; beam transmissometer (that also provides beam attenuation); and a fluorometer. On the starboard side of the vehicle an identical inlet leads to a second set of temperature and conductivity sensors. Replicating these essential sensors is especially important as the vehicle has a limited ability to collect and store water samples for calibration.

This series arrangement of sensors, working from a pumped supply, provides a convenient and compact arrangement that minimises temporal and spatial anomalies in the measured quantities. There is a need to minimise the pressure difference between the inlet and outlet tubes to ensure constant flow rate past the sensors. Hence the inlet and outlet tubes are placed near each other. This is quite different to the early practice with the Aquapack CTD in Autosub. Then, the inlet and outlet locations were chosen to provide a pressure differential that forced water to flow over the sensors without the need for a pump. One weakness of that arrangement was the flow rate depended on the vehicle forward speed.

The RDI 300-kHz Workhorse ADCP makes its current profile measurements remote from the vehicle and so might be thought to be less susceptible to the details of its mounting within the AUV. However, for best performance, the transducers do need mounting slightly proud of the vehicle's hull. As well as providing current profiles, the downward-looking ADCP provides the vehicle with bottom-track navigation when at altitudes of less than 200 m. The minimum value of slant range to the bottom from the ADCP is used by the navigation system as an additional collision avoidance sensor, augmenting the forward-looking sonar.

Since June 2000 the vehicle has carried a second, upward-looking, ADCP (Figure 9.1). Its primary function is to extend the range of current profiles when operating in deep water, but it has also been used to measure the draft of icebergs. In addition, the instrument has shown that "bottom track" velocity can be obtained from backscatter from the undersides of icebergs. This is an important factor in improving

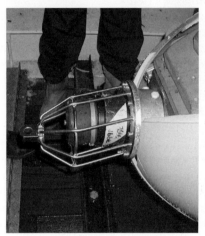

Figure 9.1   Standard sensors on the Autosub AUV. Top left: inside the nose of the vehicle showing the PAR sensor, CTD and associated sensors; top right: the upward-looking 300-kHz ADCP; bottom left: a bent acoustic transponder protective cage after collision with an iceberg with bottom right: the CTD sensors shield remaining in place after the collision, even though the gel coat on the side panel had been shattered.

navigation accuracy for future AUV missions under ice shelves, from which icebergs were formed.

9.3   COMBINING SENSORS TO STUDY PROCESSES

It is often the case that scientists need to observe and measure a number of parameters when studying ocean processes. Constraints on the use of suitable combinations of sensors on conventional platforms can limit the effectiveness of these measurements, for example, due to surface waves or ship wakes, due to spreading of acoustic beams from surface instruments and interference from sidelobes. The Autosub Science Missions programme encouraged scientists to design experiments that used multi-sensor measurements, taking particular advantage of the platform characteristics of the AUV. In the two examples that follow the AUV characteristics that were most important were first, operation near the sea surface with low self-noise or self-motion and second, following varying terrain at a constant altitude within 5 m of the seabed.

9.3.1 Upper Ocean Physics

There are few published observations of turbulence just beneath the water surface, and none that relate simultaneous measurements of dissipation, breaking waves, bubble clouds and Langmuir circulation, measurements that can be used to investigate the processes of momentum and gas transfer in the upper ocean and their fluxes between air and the water.

In January 1988 efforts to measure turbulence and acoustic scattering from the US Navy research submarine *Dolphin* were frustrated by a major failure in the submarine's generators at an early stage of the experimental period (Osborn *et al.*, 1992). Plans were therefore made to repeat some aspects of the *Dolphin* study using the Autosub vehicle as a platform for sensors. During missions in March–April 2001, Autosub carried a two-beam sidescan sonar package, AIRES II (Thorpe *et al.*, 1998), in addition to turbulence probes (see below) and a bubble resonator (Farmer *et al.*, 1998), as shown in Figure 9.2. Upward-pointing 250-kHz transducers were mounted pointing ahead and to the side of Autosub. The former detected waves breaking ahead of the vehicle, allowing the age of the consequent turbulence to be estimated when the Autosub subsequently reached, and sampled the turbulence, at the breaker location. The second sonar was used to detect bands of turbulence formed in Langmuir circulation, and hence provided information about the location and scale of the phenomenon to which the other measurements could be related (Thorpe *et al.*, 2002).

Although greatly modified (e.g. by greater storage and size reduction of some components) the sensor suite carried by Autosub was very similar to that carried by *Dolphin* 12 years earlier. Autosub provided a stable and steady platform to support the sensors in operations at 2–10 m depth in winds from calm to $14\,\mathrm{m\,s^{-1}}$ and obtained a set of unique data at relatively low cost.

The turbulence package has nine channels of data: two shear probes (one measuring the cross-stream horizontal velocity fluctuations and the other the vertical velocity component), temperature, temperature derivative, three accelerometers (for pitch, roll and heave), pressure and pressure derivative. The shear data are used to calculate the rate of dissipation of kinetic energy in 1.3 m horizontal bins. The temperature and temperature derivative data are combined and digitally filtered (Osborn *et al.*, 1992, using the method of Mudge and Lueck, 1994) to produce high-resolution temperature traces with resolution of millidegrees.

The pressure data from the turbulence package have been processed to produce a high-resolution time series to estimate wave spectra (Osborn *et al.*, 1992). Since the body is almost neutrally buoyant, the wave signal also appears in the accelerometer data and these can also be used to estimate the period of the wave field.

Figure 9.2   The three primary sensors used to study Langmuir circulation, turbulence and gas transfer in the upper ocean. Upper panel: forward- and starboard-looking sidescan sonars; lower left panel: velocity shear and temperature microstructure probes on a spar fitted to the nose of the vehicle; lower right panel: a free-flooding resonator for estimating bubble size and gas fraction.

Bubble size distributions provide useful clues as to the redistribution processes at work near the ocean surface. Measurements appear to indicate that immediately following a wave-breaking event, the size distribution approximates a power law, with slope between -2 and -3. Thereafter the bubble spectrum is eroded by buoyancy sorting at the larger radii and dissolution at smaller radii. These competing factors lead to a transition in the slope of the bubble spectrum at an intermediate radius of $\sim 100\,\mu$m. Since the bubble size distribution is an evolving property, its measurement leads to insight on turbulence and other near surface processes. A particularly important variable in this connection is the age of the bubble cloud, which might be deduced from simultaneous sonar measurements ahead of the moving platform.

Bubble population densities were determined as a function of bubble radius with a free-flooding resonator. The operational principle (Farmer *et al.*, 1998) involves the excitation and detection of broad band noise between two circular steel plates. Thirty or more resonant modes are excited and detected. The presence of bubbles within the resonant cavity alters the acoustic properties of the medium, modifying both the amplitudes and frequencies of the resonant modes. Resonant peak amplitudes and frequencies are found by fitting the data with an appropriate Lorentzian, derived from a theory for the resonator. This information is used in an inversion to yield the bubble population as a function of radius. The frequency range is 15–200 kHz, encompassing a radius distribution range of 20–220 µm. In principle, the information derived from the peak frequencies provides a separate means for determining the size distribution; in fact, if linear acoustical response is justified, the real and imaginary components of acoustic dispersion are conveniently linked through the Kramer–Kronig relationship. The optimal inversion procedure exploits the resonant peak amplitudes, which are subsequently compared with the measured peak frequencies, so as to verify internal consistency of the data.

We encountered some practical problems in deploying the resonator from Autosub. The resonator needs to be mounted on the upper surface of the vehicle so as to be better positioned for near surface bubble observation. The added weight of the steel resonator plates provided ballasting problems. Moreover, the resonator had to be placed near the aft end of the vehicle, thus separating it from the turbulence sensors. Some flow distortion must also be expected from flow along Autosub's hull ahead of the sensor. Bubble populations decay almost exponentially with depth, with an e-folding depth of 1–2 m. Thus the depth of measurement is crucial to data interpretation. Maintaining a steady vehicle depth this close to the surface was a challenge, but the evidence from the deployments indicates the operation was successful. There is some suggestion that the lowest acoustic frequencies on the resonator were contaminated by transmissions from an acoustic beacon mounted on Autosub and not all of the electrical noise sources were completely eliminated. Preliminary analysis of the bubble population shows that generally clean and high quality data were acquired, with the integrated air-fraction varying in concert with simultaneous acoustic measurements of bubble clouds (Figure 9.3).

9.3.2 Physical Processes in the Bottom Boundary Layer

The morphological evolution of the coastal ocean is the result of gradients in sediment transport, which in turn depend on the gradients of the hydrodynamic forcing. For example, theoretical work on the development and maintenance of linear ridges and data on peak tidal flows in the sandbanks of the southern North Sea, suggest that the bottom stress expected in the swales between the banks is of the order of $1 \, \mathrm{N \, m}^{-2}$ and increases by a factor of order two over the crest of the bank. Despite this large spatial variation of stress, previous measurements were either limited to single points or averaged over the entire wavelength of the shoals. Thus, observations sufficient to constrain the spatial variability of flow in general and bottom turbulence in particular produced by models are virtually non-existent and our understanding of the development and maintenance of such linear features is limited. Further, the validation and/or calibration of numerical models is based on comparisons with stationary time-series. However, the accuracy of these models is controlled on their ability to predict the gradients that are actually present in the marine environment. Measurements of flow and turbulence from moving platforms such as the Autosub provide a unique opportunity for attempting the spatial mapping of flow and turbulence over uneven topography, as is the case over ridges. These types of data can be used both for process studies as well as for data assimilation into numerical models.

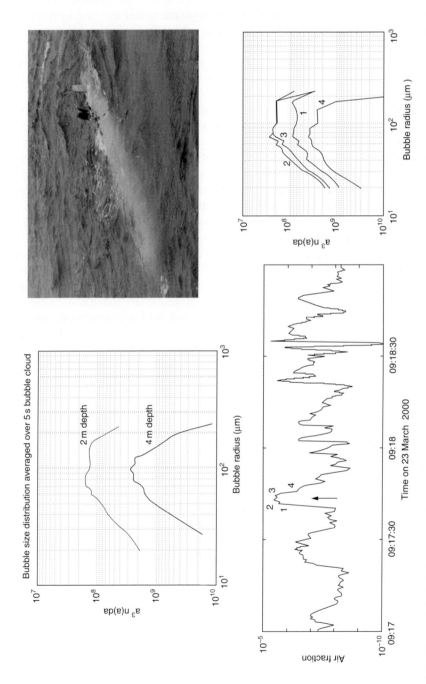

Figure 9.3 Upper left: bubble size spectra at 2 and 4 m depth from the acoustic resonator; upper right: resonator mounted on the upper rear surface of the Autosub; lower left: time series over 2 min of the air fraction at 2 m depth with; lower right: bubble size spectra at the four instants marked on the time series.

The AUV was used during August 1999 to measure the spatial variation of turbulence over a sandbank in the southern North Sea. The scientific program consisted of collection of stationary, ship- and AUV-borne observations during a 10-day cruise over the Broken Bank; a tidally dominated sand ridge situated 100 km off the East Anglian coast of the United Kingdom. The bank is typical of those found in the region; it is 41 km long and 5 km wide and has an asymmetric cross-section with steep northeastern slopes and more gently sloping stoss sides. The operations took place onboard the RRS *Challenger* during the period 17–27 August 1999 (Voulgaris *et al.*, 2001). During this survey Autosub was deployed on three missions of which only two were successful. The vehicle was programmed to collect data along a 5 km long transect running perpendicular to the main axis of the bank and at a height 5 and 4 m (for the second and third missions,respectively) above the seabed in terrain following mode (see Figure 9.4). The ambient tidal flow speed varied between 0 and 60 cm s^{-1}. On average the AUV covered 11 transects per tidal cycle for a total period of over 70 h with one change of batteries between data collection periods.

The repeatability of the transects under the strong flows encountered was within 700 m in the across-transect direction. For AUV applications in very dynamic environments and in process studies that require high accuracy in positioning this might be a limitation. Shorter missions contribute to better accuracy in positioning. Altitude was maintained within 0.75 m, attributable largely to the large bedforms that were found in the area with wavelengths of 10–15 m and wave heights of the order of 0.2–0.5 m (Velegrakis *et al.*, in preparation). Combination of pressure data from the vehicle and bottom range revealed that the vehicle was following the large bedforms.

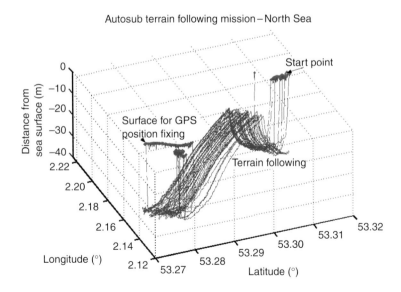

Figure 9.4 Three-dimensional representation of the AUV transect during one of the missions in the North Sea under terrain-following mode. Note the scatter in AUV position across the main transect direction (from Griffiths *et al.*, 2000).

However, the existence of the surface waves on the pressure signal does not allow the use of these data to calculate the bottom elevation in relation to the sea-surface.

The platform was equipped with sensors to estimate the main source terms (production and dissipation) in the simplified Turbulent Kinetic Energy (TKE) equation. Shear near the bed was measured using a downward-looking 1,200 kHz RDI broadband ADCP that provided current measurements with a vertical resolution of 0.25 m. Dissipation estimates were obtained by applying the "inertial dissipation" technique to spatial velocity spectra using a pulse-to-pulse coherent, 1.75 MHz Doppler sonar (Veron and Melville, 1996; Zedel et al., 1997) and to frequency velocity spectra obtained using a standard 10 MHz Acoustic Doppler Velocimeter (Voulgaris and Trowbridge, 1998) that provides point measurements of the 3-D current structure. The 1.75 MHz Doppler sonar was installed on the nose of the vehicle so as to provide turbulence measurements without any distortion from the vehicle, while the 10 MHz ADV was fitted near the stern of the vehicle looking downward.

The data reduction was based on spatial averaging defined by the variation in bottom elevation. Segments along the survey transect were identified where the mean water depth was within 1.5 m around a mean value. Mean water depths varied from 37 to 15 m and the horizontal distances corresponding to each (3 m) vertical elevation bin varied from 250 m (in the swales of the bank) to only 40 m near the crest, where the bottom slope is greatest (see Figure 9.5(c)). Flow data from the ADCP and the ADV were averaged for the time period the AUV was over the particular segment. In total 15 segments were defined. A time-series of flow variables was constructed for each segment, consisting of approximately 70 unevenly spaced points spanning a period of four days.

Harmonic least-square analysis was used to examine the spatial variability of the tidal and sub-tidal component of the flow and the results are shown in Figures 9.5(b) and (c), respectively. The interaction of the flow with the topography and in particular the veering of the tidal M_2 ellipse clockwise to the main axis of the ridge as predicted by the theory is clearly shown in Figure 9.5(b) at a high spatial resolution compatible to those of numerical models. Also, the typical clockwise circulation is shown where the along-ridge subtidal flow (u) changes sign on either side of the ridge. The across-ridge component shows that the transverse flow is away from the crest of the ridge, a result that was predicted by theory.

Also, for the first time we were able to accurately map the spatial variability of the subtidal flows (see figure 9.5(b) for along the bank (u) and across the bank (v) flows). The clockwise subtidal flow around the bank was confirmed and the location of flow reversal was found to be located on the stoss side of the bank some 200 m before the bank crest. The subtidal flow across the bank (v) was found to be away from the crest of the bank in accordance to the theory of Johns and Dyke (1972). This is the first data showing this agreement between theory and measurement in the across ridge direction. The same data from the ADCP and the log-profile theory were used to estimate bottom shear stress and bottom drag coefficients across the bank at all 15 positions.

Experience showed that although the installation on instruments on the nose of the vehicle provided distortion-free flow measurements, the sensors were vulnerable to damage during unexpected collisions at deployment and/or recovery during high seas. During this expedition, the Doppler sensor was slightly damaged during a test

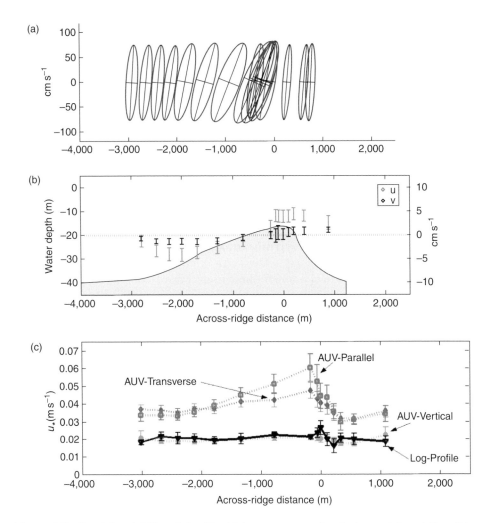

Figure 9.5 Spatial variability of the M_2 tidal component of turbulence (a) and flow (b), as well as the subtidal flow along (u) and across (v) the main axis of the ridge; (a) Comparison of the semi-diurnal amplitude of bottom shear velocities estimated from the 3-components of flow measured with the ADV using the inertial dissipation technique and from the downward looking ADCP using the log of the profile. Notice the agreement between the ADCP-derived values and those derived from the vertical component of the ADV; (b) Spatial variability of the near-bed, semi-diurnal tidal ellipse across the bank as revealed by the analysis of the ADCP data; (c) Spatial Variability of the nearbed subtidal velocities along (u) and across the ridge (v), (from Voulgaris et al., in preparation). The bottom topography along the AUV transect is shown on (c).

recovery, which resulted in poor performance for the remainder of the cruise. On the other hand the protected ADV was functional and operational during the whole period. This reinforces the intuitive notion that the optimum location for instruments that are sensitive to the details of the flow is forward of the nose of the vehicle. However, sensors at this location are exposed and are often damaged. Modelling the flow on the boundary layer of the vehicle can be used to guide the choice of more protected locations for the installation of off-the-self sensors like the ADV.

9.4   EXPLOITING SYNERGY BETWEEN SENSOR AND AUV

The AUV should not be thought of merely as a substitute for existing data-gathering methods such as research ships, drifting buoys or moorings. With suitable sensors, the AUV can make measurements that would be either very difficult, inordinately time consuming or next to impossible to make with traditional methods. This section describes three different applications, where the combination of AUV and special sensors can provide new insight.

9.4.1   Scientific Echo Sounding from Autosub

Acoustic techniques provide a powerful and non-invasive means for quantitatively sampling large bodies of water in short periods of time. Scientific echosounders are therefore used routinely to conduct surveys of commercially and ecologically important species (e.g. North Sea herring, Simmonds *et al.*, 1997; Antarctic krill, Brierley *et al.*, 1999). Despite the major benefits that acoustic techniques provide, however, they are restricted in scope by limitations in conventional sampling platforms. Research vessels, for example, typically have echosounder transducers mounted several metres below the sea surface (on the hull, in a drop keel or in a submerged towed body) and, as a consequence, leave the upper water column unsurveyed. This area may be of biological importance. Furthermore, there is a common perception that fish exhibit avoidance behaviour to the noise from research vessels, thus biasing abundance estimates (Magurran, 1999). Physical constraints also restrict the regions in which acoustic surveys can be conducted. Sea ice, for example, plays a major role in global climate and circulation processes (Johannessen *et al.*, 1994) and is believed to be a habitat of major ecological importance (Loeb *et al.*, 1998).

The under ice realm remains virtually unknown (particularly in the Southern Ocean) because of sampling difficulties: whilst ice strengthened research vessels can penetrate ice, in so doing they disrupt the very environment of interest. Deep-sea regions are also problematic for acoustic surveys from surface vessels because of the physical attenuation of sound, particularly at higher frequencies, by seawater (cf. Kloser, 1995).

The Under Sea Ice and Pelagic Surveys (USIPS) project was conceived to address some of these issues. A Simrad EK500 scientific echosounder (Bodholt *et al.*, 1989) was modified for autonomous data collection and reassembled to fit inside a pressure vessel that could be accommodated in Autosub (Figure 9.6). This was connected to two transducers (38 and 120 kHz) that had the capability of being orientated either upwards or downwards (or one on each side) for specific mission types.

In July 1999 a total of 13 scientific missions were successfully completed by Autosub during the course of a North Sea herring acoustic survey carried out by the RV *Scotia* (Fernandes and Brierley, 1999). Of these, eight were totally autonomous, where, for the first time, the AUV was sent beyond the communication range of its support

Figure 9.6   Autosub fitted with an EK500 system for fisheries and under sea-ice studies. Left: The 38-kHz transducer nearest the nose with the 120-kHz transducer immediately behind. Right: The EK500 electronics pressure vessel being lowered into Autosub.

services (Fernandes *et al.*, 2000a). Using combinations of transducer orientation (one up, one down) acoustic data were gathered for the first time on the distribution and abundance of herring throughout the entire water column (Figure 9.7, upper panel). Analysis of these data revealed that surface schools were not common in the western North Sea, and less than 1% of the total number of herring in the area were in the upper 20 m and hence remained un-sampled by the research vessel. Thus, research vessels are able to detect the vast majority of fish in this area. On the subject of avoidance, the research vessel and Autosub did not detect significantly different quantities of fish, suggesting that avoidance was not a source of bias (Fernandes *et al.*, 2000a). Our findings, which are valid only for noise-reduced vessels such as RV *Scotia* (Fernandes *et al.*, 2000b), justify the additional expense that noise reduction measures incur in vessel construction.

In January and February 2001 Autosub was deployed under sea ice in the northern Weddell Sea, Antarctica, equipped with the same scientific echosounder used in the North Sea. With the transducers mounted on the AUV's dorsal surface, we were able to collect data on the distribution and abundance of Antarctic krill beneath sea ice. Autosub ran for a total of over 275 km under ice-covered seas collecting the most extensive data set yet available on both krill distribution and abundance, and on Antarctic sea ice thickness (Figure 9.7, lower panel). Comparison of krill acoustic data collected by Autosub with data collected using the same type of acoustic system by RRS *James Clark Ross* along reciprocal transects in adjacent open water has enabled

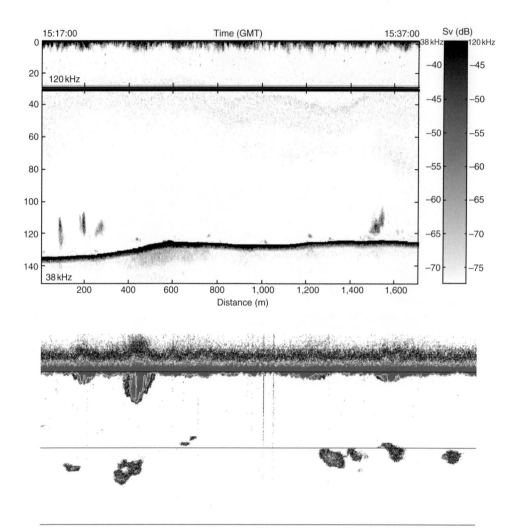

Figure 9.7 Upper panel: A composite echogram section from Autosub: 120-kHz transducer orientated towards the surface (upper image); 38-kHz towards the bottom (lower image). The horizontal bands at 30 m depth correspond to the transmit pulses of each of the transducers and indicate the position of the vehicle. The intensity scale is in decibel (dB) units of the volume backscatter strength (S_v). Four large herring schools are evident in the lower image, close to the seabed. Internal wave activity is evident from the scattering of plankton (diffuse layer) at 40 m. Bottom panel: A 38-kHz echogram from Autosub looking upwards under ice showing swarms of krill and downward-projecting sea-ice keels. The horizontal lines are 50 m apart and the total horizontal distance is 1,400 m.

hypotheses on the relative importance of ice covered and open waters to krill to be addressed, Brierley *et al.* (2002). Krill is the keystone species in the Southern Ocean and, with concerns that sea ice extent is changing because of climatic change, it is vital to understand the importance of sea ice to krill (Brierley and Watkins, 2000). In the Southern Ocean, data on the thickness of sea ice has previously been restricted to point measurements from holes drilled into the ice (Wadhams, 2000). The echosounder system on Autosub was able clearly to detect the sea ice/water interface (Figure 9.7, lower panel) and will enable sea ice thickness along the entirety of the 275 km of transect to be determined. A PAR sensor was also mounted on Autosub for its under ice forays and will enable us to evaluate the impact of ice cover on light attenuation, a measure vital for estimation of primary production under sea ice.

There is no doubt that Autosub has proved to be an effective platform for gathering fisheries and environmental acoustic data. Data quality is very high because of Autosub's very low noise characteristics (Griffiths *et al.*, 2001). The flexibility of depth operation that Autosub provides enables the echosounder to be targeted to depths more appropriate to the biota under observation, overcoming problems of sound attenuation, and the ability of the vehicle to dive beneath an obscured surface enables it to collect data from environments that would otherwise be impenetrable. In future, when identification of species becomes possible on the basis of acoustic data alone (cf. Brierley *et al.*, 1998), and the requirement to catch samples of the species under investigation in nets is alleviated, we can envisage acoustic surveys being conducted in their entirety by AUVs.

9.4.2 *In Situ* Chemical Measurements: Manganese and Methane

Conventional approaches for obtaining data on dissolved trace metals and gasses rely on collection of samples with water sampling bottles, and subsequent laboratory analyses. Whilst our understanding of key biogeochemical processes involving trace metals has developed using these cumbersome and time consuming approaches, we are now limited by their poor spatial and temporal resolution. This is particularly a problem in many near-shore and upper ocean systems where time and space scales are short. AUVs provide an excellent platform for continuous methods of metal analysis, and thus providing quasi-synoptic distribution data that can be related to processes. An *in situ* dissolved manganese analyser system was deployed on Autosub in the fjord-like system of Loch Etive in Scotland, where development of sub-oxic conditions in deeper isolated basins between water replenishment events are concurrent with increases in concentration of dissolved manganese. This is, to the best of our knowledge, the first application of an *in situ* metal determination system in combination with an AUV.

The *in situ* analyser was based on the spectrophotometric detection of the Mn-1-(2-pyridylazo)-2-napthol (PAN) complex at 560 nm, as used by Chin *et al.* (1992). The coloured complex was formed after mixing of the seawater sample and buffered reagent in a continuous flow manifold, in which the solutions were propelled through the tubes by a peristaltic pump (Figure 9.8(a)). The coloured complex was detected by a solid state flow-through detector. This consisted of an opaque PVC block having a 50 mm long 2 mm diameter central axial hole in the flow stream, with a high intensity green LED at one end, and light-to-frequency detector (combined silicon photodiode – CMOS amplifier) at the other. Data was generated at about 1 Hz, logged by Autosub, and post cruise, 10 s integrated values were used for data plots. Calibration of the system can either be through use of *in situ* selected blank and calibration solutions (using PEEK/PTFE valves operated by timers), or collection of up to 50 discreet samples using an Aquamonitor water sampler (WS Ocean Systems) and subsequent laboratory analysis and comparison to detector signals. For the analytical system to operate *in situ*, it needs to be physically and electrically isolated from the surrounding seawater, and to be able to function at ambient pressure. This was achieved by housing the manifold, pump and associated components in an acrylic tube (200 mm inside diameter) sealed with PVC end-plates and containing silicone oil (e.g. Gamo *et al.*, 1994). A flexible diaphragm at one end allowed for changes in volume of the oil with pressure and temperature.

This system was deployed on Autosub in Loch Etive in April 2000. Example data for dissolved manganese is shown in Figure 9.8(b), where Autosub was operating in terrain-following mode. The deeper basins have lower oxygen and higher manganese as can readily be seen. Additionally, there are smaller scale dissolved manganese features evident, which would not have been seen by conventional sampling and analysis methods.

Also deployed during the Loch Etive Autosub missions was a novel *in situ* methane sensor. The type of sensor used was the METS solid state sensor developed by GKSS in Germany for long-term monitoring of methane concentrations. The actual instrument deployed on Autosub was very generously loaned to us for this purpose by Drs Gary Klinkhammer and Bob Collier at the Oregon State University in the United States of America where it had previously been used to good effect in detecting methane-rich plume signals in the water column overlying cold-seep environments along the California–Oregon margin (Bussell *et al.*, 1999). In our study a general increase in methane concentrations was typically observed in the deeper, manganese-rich waters of Loch Etive but the approach was of limited effect in that the response time of the sensor was slow relative to the dive-and-surface cycle of Autosub on these missions and – importantly – we also lacked any mechanism/set-up for *in situ* and/or on-deck analytical calibration for dissolved methane, unlike the case for manganese.

9.4.3 Flow Cytometry from within an AUV

Flow cytometry measures scattering and fluorescence from individual particles that are entrained in a stream of fluid and passed through a tightly focused laser beam. The technique originated in the biomedical sciences, but it has been widely used in phytoplankton research (Yentsch, 1990). Flow cytometry is the only feasible method of rapidly determining the composition of mixed phytoplankton populations, and of counting and sizing phytoplankton in the presence of detrital particles. There is much current interest in combining flow cytometry with advanced data analysis techniques to achieve automatic identification of a wide range of microbes in the marine environment. However, commercial flow cytometers are laboratory instruments that require considerable bench

space and electrical power for their operation. The analysis of natural phytoplankton populations by flow cytometry traditionally involves the collection and
preservation of water samples, or the use of cytometers on board research
vessels.

As part of the Autosub Science Missions programme, we have designed and
constructed a submersible flow cytometer based on technology originally developed
for environmental monitoring from a moored buoy (Dubelaar *et al.*, 1999). The new
instrument is capable of operating to depths of 250 m and uses only 20 W of electrical power. Its distinctive features include solid state laser illumination at 675 nm,

(a)

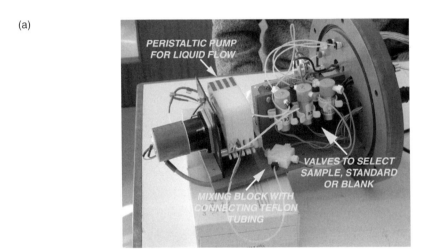

(b)

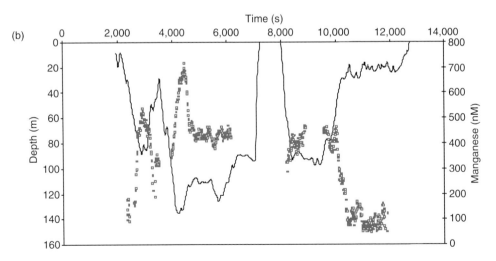

Figure 9.8 (a) The internal arrangement of the spectrophotometric dissolved manganese
sensor. (b) time series of dissolved manganese (squares) and vehicle depth as measured during
a mission in Loch Etive, Scotland in April 2000.

Figure 9.9 The Cytobuoy flow cytometer for use within an AUV. The leftmost section contains the system for sheath fluid injection and recycling; the lower right section contains the optics while the upper right houses the control electronics.

recycling sheath fluid, a very wide dynamic range (2–500 μm for spheroids) and the ability to measure optical pulse profiles generated by individual cells and colonies. It is housed in a pressure cylinder 415 mm diameter by 530 mm long, is able to operate autonomously, and has enough internal memory to log data from around 200 samples, Figure 9.9. The first deployment of the flow cytometer on Autosub took place in May 2001, when it was mounted in conjunction with an AC9 scattering and absorption meter and an Aquamonitor water sampler. The collection of water samples made it possible to validate the flow cytometer data by carrying out laboratory cytometry and microscopy on preserved samples, while the AC9 permitted direct correlations to be established between the particle content of the water column and its inherent optical properties. During this mission Autosub was used to map the distribution of intact cells and detached liths around the edge of a coccolithophore bloom in the western approaches to the English Channel and to measure the taxonomic composition of a mixed population of diatom chains and *Phaeocystis* colonies around the Isles of Scilly. At the same time, measurements of the underwater light field at discrete points along the Autosub cruise track were made using a Satlantic profiling radiometer. Airborne CASI radiometry was carried out by the UK Environment Agency, and satellite images of ocean colour were obtained through collaborative work with the NERC Remote Sensing Unit at Plymouth Marine Laboratory. The complete data set extends from distributions of size and morphology for individual suspended particles to large-scale variations in reflectance observable from satellites. The use of Autosub for optical instrument deployment allowed near-surface transects to be obtained over distances of tens of kilometres without the physical disturbance associated with towing an instrument package behind a research vessel. Preliminary analysis indicates that this spatially continuous sampling will allow us to assess the significance of sub-pixel variability in degrading the accuracy of sea-truth data for satellite imagery. We expect that the newly developed capability for *in situ* particle analysis will find applications in future autonomous vehicle science missions.

9.4.4 Measurements of Flow Over a Deep Sill: CTD and ADCP Data Quality

The aim of the Strait of Sicily project was to use Autosub in a combined undulating and terrain-following navigation mode to obtain high quality hydrographic and current measurements in deep water (down to 900 m) over the sill between the Eastern and Western Mediterranean Seas, Stansfield *et al.* (2001). This was first survey of its type for the vehicle and therefore an assessment of the quality of the data collected is important. For our work in the Strait, Autosub carried two pumped Seabird SBE-9+ CTD systems mounted about 0.5 m from the front of the vehicle, and two 300 kHz RDI ADCPs, one upward and one downward looking. The ADCPs collected data in 4 m bins at a range of up to 108 m from Autosub with one bottom track and two water track pings every 3 s. The two water track pings were averaged internally by the ADCP. For an average vehicle speed of $1.4 \, \mathrm{m \, s^{-1}}$ this sampling interval gave a horizontal resolution of 70.7 m at 100 m range from the ADCP transducer head, with adjacent samples overlapping by up to 62.3 m. The CTDs sampled at a rate of $24 \, \mathrm{s^{-1}}$ and then averaged into 3 s bins. This averaging reduced the horizontal resolution of the CTD data to 4.2 m at a mean vehicle speed of $1.4 \, \mathrm{m \, s^{-1}}$ and gave a vertical resolution of 0.48 m at a typical climb rate of $0.16 \, \mathrm{m \, s^{-1}}$.

The average error of the terrain following navigation after the longest mission, as measured by GPS surface fixes (at 0, 93 and 186 km), was 1.2% of the distance travelled.

The Autosub CTD data showed excellent agreement between the two temperature and conductivity sensors. However, as an independent check, the data were compared with data from four shipboard CTD stations in a region where the water mass characteristics were relatively constant. Although intended to lie close to, or along, the Autosub track in fact the shipboard CTD stations lay between 2.2 and 2.6 km north of the track, due to the 1.2% navigation error of the AUV. Shipboard CTD profiles reached to within 6 or 7 m of the bottom whereas Autosub was never closer than 30 m to the bottom. As a consequence of the northward offset and the deeper profiling depth, the shipboard CTD sensors reached slightly colder water. Otherwise both data sets generally showed the same $T-S$ ranges and variability, although the Autosub data also clearly shows mixing occurring between the two different types of profile recorded by the CTD (Stansfield *et al.*, 2001).

In particular, Autosub is a good platform for using acoustic current profilers. It can operate below the depth-range of conventional shipboard, hull-mounted systems

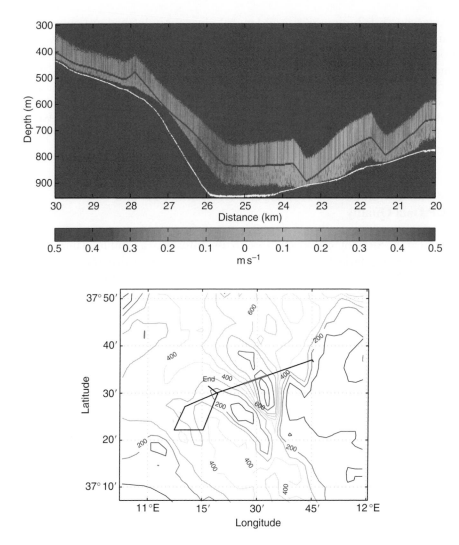

Figure 9.10 Autosub mission 239 in the Strait of Sicily used both upward- and downward-looking ADCPs to study the near-bottom velocity structure across the sill between the eastern and western Mediterranean. Upper panel shows the north component of velocity along the red section of the track shown in the bottom panel. *See* Color Plate 9.

(typically 300–400 m with a 150 kHz ADCP) and, due to its stable motion, less temporal/spatial averaging of the data is required resulting in greater horizontal resolution (Figure 9.10). A comparison of the "raw" 3-s velocity data from the upward and downward looking ADCPs with corresponding velocity data measured by the shipboard lowered ADCP (LADCP) shows that the two types of sensor agree to within $\pm 10°$ in direction and $\pm 10\,\mathrm{cm\,s^{-1}}$ in speed with the LADCP generally giving higher speeds than the Autosub ADCPs.

Acknowledgments

We acknowledge the sterling efforts of the Masters and crew of the ships used with Autosub who all gave us excellent support. This work was funded by the UK Natural Environment Research Council as part of the Autosub Science Missions programme under contract F3/G12/51/01, and grants GST/02/2144, GST/02/2145, GST/02/2151, GST/03/2153, GST/02/2155 and GST/02/2157, and ably managed by Mr Steve Hall.

10. DEFENCE APPLICATIONS FOR UNMANNED UNDERWATER VEHICLES

SIMON CORFIELD[a] and CHRISTOPHER HILLENBRAND[b]

[a] *QinetiQ, Winfrith Newburgh, Dorchester DT2 8XJ, UK and*
[b] *Naval Undersea Warfare Center, 1176 Howell Street, Newport,*
Rhode Island 02841, USA

10.1   INTRODUCTION

Military underwater systems have been in the vanguard of underwater technology development since the earliest times and the research and development of Unmanned Underwater Vehicles (UUVs) is no exception. While specific details of potential UUV missions and military UUV technology remain classified, there are a number of historical and generic aspects to the business of fighting at sea and from the sea that can be used to illustrate some of the current thinking on the way ahead for the development of military UUV missions and technology.

The intention of this chapter is to provide an overview of some historical aspects of naval underwater systems and warfighting relevant to current and future themes in naval UUV development, to present some of the generic ideas for UUV military roles and the associated functionality that is required to realise these roles and to explain some aspects of current naval UUV technology research.

It should be noted that the views expressed in this chapter are those of the authors and that these do not represent the views of any company or government organisation.

10.1.1   Reconnaissance from the Sea

In almost all Julius Caesar's Gallic campaigns, he had noted that British reinforcements had been sent to aid the Gaulish tribes (Caesar, 1982). However, he had not been able to learn anything about the Britons, the size of their island, their character, the strength of their tribes, their manner of fighting and customs or the harbours that would be capable of accommodating a large fleet of big ships. In late summer 55 BC, a Roman warship slipped out of harbour and headed for the coast of Britain. Its commander, Volusenus, had orders to reconnoitre the coast and report back as soon as possible while Caesar assembled an attack force on the northern coast of Gaul including about 98 infantry and cavalry transport craft and several warships. Volusenus completed his mission and returned to report to Caesar after four days.

It is evident that this early example of coastal reconnaissance as a precursor operation to an invasion was only partially successful. When Caesar arrived offshore with his first ships, he considered the initial landing point too hazardous due to the narrowness of the beach and the proximity of cliffs from which the Britons could hurl spears. When the rest of his ships arrived he moved 7 miles along the coast and ran his ships aground on an evenly sloping, obstacle free beach. However, the beach landing phase of amphibious warfare has never been simple and the value of good intelligence information about beach topography, sandbanks and shallows has always been highly valuable in planning offensive action. In 55 BC, the beach landing was made difficult by the lack of accurate intelligence information on water depths and the layout of the shallows. Due to the relatively deep draught of the transports and the complexity of the shallows, the Roman infantry found themselves having to jump into and wade through deep water that severely hampered their ability to fight their way onto the beach. However, using their local knowledge of the shallows and sandbanks, the Britons were able to attack the divided landing parties as they emerged from each ship. It was only Roman quick thinking, use of artillery from the warships and deployment of mobile reinforcement parties in the warships' boats that a foothold was gained and the Britons pushed back.

Over 2,000 years later, the importance of reconnaissance and the gathering of many types of intelligence information by naval forces have remained critical aspects of successful warfare. An on-going recognition of the military worth of such information has meant that the execution of reconnaissance missions and the gathering of tactical data have become the primary focus for current and near term UUV developments at the start of the twenty-first century.

10.1.2 Trade, Ships and Submarines

As Europe emerged from the last Ice Age, there appears to be strong evidence that the early peoples established a trade in stone tools, and presumably other desirable goods. After the flooding of the English Channel in about 5400 BC, migration of people to Britain continued, but now required the use of boats. The Celtic tribes that Caesar encountered appear to have arrived in Britain as a result of a westward, sea-borne migration of Belgic tribes. By the time of the Bronze Age, rivers formed well established trade routes and a cross-channel continental trade in arms was established by 1100 BC, as evidenced by the 350 Bronze Age swords, daggers, axes other goods found in a shipwreck off Dover in the 1970s. This continental trade, including long distance trade with the Mediterranean peoples, had become a way of life and the intimate link between trade, shipping and the wealth of nations well established.

The capability of ships to transport men and goods as a basis for travel and trade quickly became supplemented by the capability to provide groups with mobility for their raiding forces. The ability to conduct surprise attacks on rich coastal trading areas, to retreat by the time organised defence could be mustered and to move along the coast to further targets proved to be a long standing model for wealth creation from the earliest times to the nineteenth century.

The importance of specialised warships grew in parallel with the need to protect one's own trading routes and ships, the increasing financial benefits in being able to attack enemy trade ships and as a means of achieving superiority against enemy military forces. Tactics moved from the Mediterranean preference for ramming and hand to hand combat to the use of shipboard artillery, the firepower of which has continued to increase through to the present day.

In the struggle to achieve superior levels of maritime surprise and attack effectiveness, it is not surprising that the idea of war machines that could swim below the sea surface emerged. This idea is centuries old and accounts of inventions for manned underwater vessels exist from pre-Armada times. Significantly, all these inventions were for military systems and this trend has continued until the latter part of the twentieth century when, in addition to military importance, the economic and scientific importance of gaining information about the underwater environment has been recognised.

A major step towards the building of military underwater vessels occurred at the end of the eighteenth century when an American, David Bushnell, created a design for the *Turtle*, a small submarine intended to aid the Americans to break British naval blockades during the War of Independence. This tiny craft was equipped with an armament of a wood screw and a detachable explosive charge and conducted the first recorded submarine attack, albeit unsuccessful, on the British Flagship HMS *Eagle* in 1776.

Manned submarine development continued sporadically in Britain and Europe over the next century, with various countries seeking to develop blockade breaking devices. During the American Civil War, the Confederacy developed several submersibles for breaking the Union blockade of southern ports. In 1864 one of these, the CSS *Hunley*, was the first to sink an enemy ship when it detonated its spar torpedo charge against the USS *Housatonic*. Unfortunately the crude nature of its weapon led to its own destruction during the attack. In the 1870s Robert Whitehead developed and manufactured a more successful weapon type, the first free running torpedo. When Whitehead's torpedoes were used to arm Swedish Nordenfeldt submarines during the 1880s and French submersible designers introduced duel power systems, steam engines for surface running and batteries for submerged operation, a workable principle for effective warfighting submersibles was starting to emerge.

In 1900 John P. Holland won a US Navy procurement competition with a submersible design powered by a duel petrol/battery system. Shortly afterwards, the Royal Navy started negotiations with the Electric Boat Company for a licence to build five Holland submersibles at Barrow in the United Kingdom and the first of these dived for the first time in 1902. Over the next 12 years the numbers of submarines and submarine types increased significantly. By the start of the First World War, a number of countries possessed a credible underwater warfare capability that included the ability to lay mines and to attack enemy ships and submarines directly with gunfire and torpedoes.

In parallel with the development of large submarines, inventive minds were also developing small underwater vehicles designed to carry two men for carrying out covert harbour attack missions. In 1918 the Austrian flagship, the *Viribus Unitis*, was sunk by a mine attached to its hull by two intrepid Italian frogmen who had transited into Pola harbour riding on the back of the *Mignatta*, a converted 14 inch B-57 bronze torpedo. The success of this approach stimulated a new interest in special manned underwater vehicles and the development of a new generation of manned torpedoes, led again by the Italians in 1935 with their two man 6.7-m *Maiale* vehicles armed with a 300-kg charge in a detachable head. Successes in operating the *Maiales* against allied capital ships during the Second World War, such as the British battleships *Valiant* and *Queen Elizabeth*, prompted the British to develop their own two man systems, known as Chariots, which also achieved success, this time against Italian ships.

During the inter-war period conventional submarine capabilities and numbers had increased significantly and the effect of submarine attacks on merchant shipping during the Second World War was devastating. Submarines also scored many significant successes against major warships and found new roles in the deployment of agents and special forces to enemy coasts. In an effort to further enhance the ability to conduct clandestine operations, several countries embarked on the development of true miniature submarines that offered greater endurance than the Chariots and were able to carry more effective weapons. Between 1939 and 1945 a number of mini-submarines were built by the Italian, German, Japanese and British navies. Displacements were generally between 13 and 36 ton and the craft carried between two and five personnel. In addition to their torpedo or limpet mine attack roles against merchant ships and major naval battleships, these miniature submarines provided an alternative means to deliver agents and to carry out the familiar old missions of delivering raiding/sabotage parties and gathering intelligence.

10.2   TAKING THE MAN OUT OF THE WATER

While a certain amount of operational flexibility had been achieved with mini-submarines by the 1960s, the financial viability, functional limitations and safety of manned mini-submarines had led naval staffs and commercial management to start the evaluation of the potential of unmanned systems. Military research programmes had again taken the lead in the development of the new underwater technology in addressing the feasibility of unmanned Remotely Operated Vehicles (ROVs) for naval operations. The overall concept of robot use in land, air and sea warfare scenarios had long been considered to have merit in terms of:

- the removal of humans from areas of danger;
- the ability to implement new warfare capabilities that normally would not be feasible when human physical limitations would be exceeded;
- the potential for cost savings.

In underwater warfare, the ability to inspect and handle explosive ordnance at depth and to remove human operators from areas of danger acted as the initial focus for naval ROV development. The first ROV is generally considered to be the Cable-controlled Undersea Recovery vehicle (CURV) that was developed in the United States by the Pasadena Annex of the US Naval Ordnance Test Station. CURV was designed to recover test ordnance at depths down to 600 m and achieved fame in 1966 when it was used to recover a Hydrogen bomb off the coast of Spain.

Military research efforts into general ROV technology were soon overtaken by civil sector research and development as many companies raced to build and exploit ROVs within the emerging, lucrative subsea oil and gas production markets. As the subsea oil and gas production business grew, new safety and financial requirements emerged. The main themes of these requirements were to:

- remove human divers from the dangers of subsea production tasks;
- increase the depths at which these tasks could be carried out;

- reduce the cost of undersea operations by removing the high costs of diver training and diver support logistics.

A corollary of this expansion of civil underwater engineering and robotics was that military underwater robotics research became focussed on specific naval issues associated with mine countermeasure (MCM) tasks and underwater search rather than more general areas of ROV technology.

10.2.1   Taking the Man Out of the Mission Execution Loop

The footprint of ROV operations is dominated by the dynamics of the tether. The effect of this in MCM operations is to restrict hunting distance ahead of host platform and thus stand-off distance from a mine kill radius. For many years the tether has also been the most vulnerable element of an ROV system and often subject to damage during launch, vehicle operation underwater or recovery. Tether systems can significantly increase the complexity and cost of the vehicle handling system and dedicated tether management systems are often required to help ensure adequate reliability. While advanced tether designs and tether management systems can decrease the tether vulnerability and increase host platform stand-off distances, these are merely improvements rather than wholly satisfactory long term solutions.

The ultimate solution to many of the problems associated with tethered systems, and one that provides a UUV with the freedom of movement required to conduct a wider range of naval operations, is to remove the tether. There are, of course, a number of issues implied by the removal of the tether:

- the UUV will need to be pre-programmed with key mission plans and goals;
- the degree of subsequent interaction of the UUV with its human operators will be a function of communications capability and any sensitivity to prevailing operational situation constraints;
- the UUV will need to employ non-tether based means of communicating data back to its host platform, or to a third party, if on-board data storage with post-mission retrieval does not meet the requirements of the UUV's mission;
- the UUV will need an on-board navigation system that will meet the accuracy requirements of its mission including the relative positioning accuracy of objects detected by the UUV's sensors.

If the UUV will be required to react to prevailing mission conditions on-line, it will need to have:

- the sensing capability to be able to measure relevant characteristics of prevailing mission conditions;
- the processing capabilities to compile a situation assessment;
- the necessary 'intelligent' functionality incorporated into its mission management system to make decisions based on its goals and its situation assessment.

The technology available to provide cost effective autonomous capability within mobile robots has increased dramatically since the advent of advanced materials, power and microprocessors in the 1980s. These developments have meant that a wide range of potential UUV mission types are now considered to be feasible and that many of the military robot missions encountered previously only in science fiction are now the subject of serious research and development effort worldwide (e.g. East and Bagg, 1991; Steiger, 1992; Cancilliere, 1994; Iwanowski, 1994; Corfield, 1995; Anderson, 2000). This research and development has generally focussed on achieving advanced vehicle, vehicle subsystem and sensor performances where the performance of standard civil systems would not meet warfighting effectiveness and reliability requirements.

10.3 ROLES FOR UUVs IN MODERN WARFARE

In developing military UUVs, the first stage involves careful consideration of the roles that the UUVs will be required to play and their contribution to overall force effectiveness. Current concepts for the conduct of modern warfare are focussed strongly on the close integration of land, sea and air units and coordination of their actions for maximum effectiveness through the compilation and use of identical scenario pictures and situation assessments. These data compilation and assessment processes rely on the fusion of data derived from many types of sensor deployed by the different land, sea and air units distributed across the battlespace. They also rely on the timely dissemination of the fused data back to all units so that their individual and coordinated actions will maximise their combined effectiveness in carrying out the desired warfighting task. This of course is not a trivial exercise, but fundamentally it is not dissimilar to the task facing Caesar and all war commanders, albeit that the amount of data and numbers of units involved are greatly increased.

To realise this complex collation and dissemination of data across a modern battlespace requires an effective inter-unit communications network and this network and its robustness to battle conditions becomes a key factor in achieving warfighting success. In an abstract sense within this concept of warfare, UUVs represent entities that can act both as remote data gathering nodes and as agents that can execute specific actions, including actions against other entities. Hence, the military roles that UUVs can play can be either passive defensive, active defensive, passive offensive or active offensive, where passive roles are considered as those where the UUV does not specifically need to react to changes in the sensing scenario. Within these categories there are many hundreds of possible variations of potential naval UUV tasks and missions. Like other military assets, the timing of UUV deployment to conduct military operations within an actual or potential conflict will depend on a country's judgement as to the prevailing political and military positions including the level of perceived threat. An important aspect of military UUV research over the past 15 years has been the conduct of operational analysis studies on these mission variations to ascertain in which roles UUVs appear likely to contribute most to overall warfighting effectiveness.

10.3.1 Defensive Roles

Defensive roles for UUVs tend to require the UUVs to act as the host platforms for advanced underwater sensor systems and are generally associated with the covert determination of enemy force characteristics and disposition. Examples of defensive roles might be:

- the detection, location and trail of enemy submarines within Anti-submarine warfare (ASW) scenarios;
- the detection and identification of mines within own waters;
- the conduct of hydrographic and environmental operations in support of wider naval or joint force operations;
- underwater and abovewater gathering of enemy system data that may be exploitable by own forces.

Key UUV functionality required for the effective conduct of the ASW detection and trail role includes the ability to:

- carry, deploy/recover and operate sensors and sensor processing systems capable of detecting submarines at distance;
- host a high energy power system to allow both staying on station for an extended period and transit over long distances at significant speeds;
- maintain good positional data over an extended period to aid in the maintenance of an accurate target location;
- communicate enemy target position to own forces and planners in a timely fashion.

Achieving the required sensor functionality is not trivial since, to obtain various required performance characteristics, the sensors tend to have large physical apertures that need significant UUV body surface area and/or volume to house them. Passive acoustic sensors capable of long range detection are flank hydrophone arrays and towed arrays. Currently these systems are developed primarily for submarines, but where operational benefit could be gained from the use of single or multiple UUVs equipped with such sensors, it is likely that special UUV flank arrays and thin towed arrays could be developed to help maximise UUV detection and localisation of enemy submarines.

It is interesting to note that some of the first applications for military UUVs have been MCM applications (e.g. Trimble, 1996; Madsen and Bjerrum, 2000; Thornton and Weaver, 2000). One of the reasons for this is that the problems associated with the MCM roles are well understood and bounded, allowing appropriate UUV research and development goals to be well defined. Key to the MCM mission requirement is the ability to detect mines and mine-like objects with an extremely high probability of detection in environments that are fundamentally very challenging for any sensor system. Hence, sensor and processing capabilities dominate UUV MCM operations. Although MCM sensors operate at much higher frequencies than ASW sensors, the low target strengths of modern mines and the problems imposed by shallow water environments tend to mean that MCM sonar apertures need to be of sizes comparable to the size of the larger UUVs available at present. MCM sonar integration to UUVs is thus not a trivial exercise. The advent of wideband and synthetic aperture sonar technology will help to improve MCM sensor performance and potentially offer some advantages in the physical integration process to UUVs.

Like the ASW role above, the MCM role also requires the UUV to have an on-board energy system that will allow high mission endurance and good speed capabilities, both of which are required to support large area coverage and good speed of advance. MCM probably imposes the greatest requirement for UUV navigation accuracy since, in addition to basic detection, accurate positioning of mine or mine-like targets relative to the UUV is required to aid in subsequent minefield mapping, mine avoidance and mine relocation for disposal. In principle, UUV MCM operations in home waters in peacetime could be carried out using on-board data storage with subsequent data retrieval on the host platform following mission completion. However, in other MCM scenarios, the transmission of data back to a host platform or some other command centre will be required. It is thus likely that MCM UUVs will need a good long range communication capability to ensure that the criteria for timely battlespace data compilation and dissemination can be met to help achieve wider warfighting goals.

One of the obvious roles for military UUVs is to support naval hydrographic operations and this role is probably the closest to the majority of current civil UUV applications (Figure 10.1). In conducting this role, naval UUVs will need to be able to host standard hydrographic instrumentation and will ideally possess high endurance and good navigation capabilities. There are several UUVs in the world, some covered elsewhere in this book, which are already considered capable of carrying out this type of role.

Some of the more useful fundamental characteristics of UUVs for military operations are that they are relatively small, potentially hard to detect and potentially able to access areas that other craft would not be able to do so safely. The ability to maintain a good level of covertness relative to other types of vessel means that UUV deployment

Figure 10.1 US Naval Underwater Warfare Centre's Manta Test Vehicle fitted with a swath bathymetry sonar system for seabed mapping and use in non-traditional navigation techniques (an expanded view of the sensor is shown in the bottom left).

into enemy controlled areas to obtain data using both underwater and abovewater sensors could be an attractive addition to any intelligence, surveillance, targeting and reconnaissance (ISTAR) capability. Many types of acoustic, bathymetric, electric and electromagnetic spectrum sensor can be mounted on or within UUVs and the data from such sensors potentially provides useful indications and warnings of enemy capability and activity.

10.3.2   Offensive Roles

The role of UUVs as naval sensor platforms can be extended to include their use in offensive operations such as:

- dispensing of weapon systems;
- acting as targeting aids for weapon systems deployed from other own-force units;
- acting as active attack systems using their own targeting systems, weapon launch systems and weapons.

A typical example of the first type of offensive role is where a UUV might be used to place mines/charges at given seabed locations. In this case the weapons would be standalone systems and would not be activated directly or indirectly by the UUV. The role of the UUV would thus be confined to that of ensuring that the weapon is placed in the correct position. As such the UUV system requirements are not dissimilar to those of mobile mine systems, the key UUV functionalities being payload capacity and good navigational accuracy.

An example of the second type of offensive role might be the use of a UUV in a target detection, identification and illumination role (Figure 10.2). In this role the UUV would attempt to penetrate an enemy harbour, convoy or task force with the aim of detecting, classifying and localising a specific type of target of interest (e.g. a major naval ship) using its own sensor data. On achieving this, target information would

Figure 10.2   Surveillance information – infra-red images of a surface vessel.

Figure 10.3   Manta Test Vehicle data received at a submarine and remote control of payloads.

be sent to an own force unit capable of deploying missiles or other weapon systems into the target area (Figure 10.3). Once the weapon entered the target area, the UUV could be used to illuminate the correct target and thus potentially improve the effectiveness of the attack while reducing the risk of counterattack on the weapon launching unit.

The active attack role represents a huge challenge, but one whose feasibility is increasingly being assessed due to a perception that this may represent a way ahead for naval UUV systems in the future and a possible means of increasing attack capability in a cost effective manner. However, weapon designers have wrestled for over 50 years with the problems of ensuring that conventional weapons fitted with autonomous guidance systems can be both targeted correctly and homed onto targets accurately to maximise weapon effectiveness and to avoid collateral damage. The issues of technical functionality associated with the design of an autonomous weapon that has a human operator in the targeting loop are very significant. These issues are compounded when one considers the UUV on-board functionality that would be required before one would wish to let a UUV designate targets by itself and initiate a weapon attack. It is clear that there will be a strong case to maintain human operator interaction in the final attack and fire decision within any future attack UUV Rules of Engagement.

Key to the feasibility of future operational attack UUVs will be subsystems to aid in robust, fail-safe target identification, target motion analysis and weapon fire control. The required subsystems will include comprehensive underwater and abovewater sensor suites; advanced data processing, situation assessment processing and decision making systems; high accuracy navigation systems and robust communication systems. Additionally, it is likely that such UUVs will need high endurance capability for extended patrols and station keeping. At present and in the near term, most of these system requirements simply cannot be met with the degree of proven reliability that would allow both political and military planners to proceed with development and deployment of such systems. However, long term research into enabling technology for attack UUVs is viable and it is likely that significant progress in establishing a technical basis for attack UUVs will be made over the next 20 years.

10.4 CURRENT TRENDS IN NAVAL UUV TECHNOLOGY RESEARCH

Current trends in naval UUV research can be illustrated by the work of the US Department of Defence in developing technology to field multi-mission reconfigurable UUVs that can operate independently over long periods to support clandestine US operations. The US Navy's program plan identifies four basic mission areas for application of UUV technology: maritime reconnaissance, undersea search and survey, communication/navigation aids, submarine track and trail (Dunn, 2000).

10.4.1 Energy Systems for Military UUVs

Current UUV mission capabilities are often energy limited and advanced energy sources are a fundamental requirement for future high endurance military UUVs. This importance and the relatively high cost of achieving advanced, compact UUV energy systems have tended to mean that UUV energy system research has been led by military research programmes. A key aspect of this work has not only been the investigation of novel electro-chemical and thermo-chemical systems but also the efficient packaging of these systems to allow their integration with tactical size military UUVs. However, it should also be noted that where UUV size has not been a major design issue in recent civil and military UUV missions, a pragmatic approach tends to have been adopted whereby large numbers of standard COTS battery cells have been assembled to give the required level of UUV energy.

Military UUV energy sources must be affordable, safe, capable of a long shelf life, not prone to spontaneous chemical or electrochemical discharge, compatible with logistic support processes including safe refuelling, compatible with operation from naval vessels and ships of opportunity and environmentally sound.

Military research is addressing a number of candidate energy sources for future UUV applications (Carlin, 2000). A recent US Navy requirement has been for a UUV energy system capable of providing a minimum energy density of at least four to five times the energy density of silver–zinc rechargeable batteries (the batteries commonly used for exercise torpedoes). In trying to achieve high levels of specific energy (e.g. $>200\,\text{Wh kg}^{-1}$) and energy density, efforts have tended to focus on novel fuel cell and semi-fuel cell systems, novel thermal engines and advanced battery technology, often where lithium or aluminium form the primary fuel elements.

The thermo-chemical systems typically offer the highest available specific energies. However, there are significant system design tradeoffs in the development of such systems including prime mover complexity, relatively high levels of radiated noise and the high cost of high temperature, high velocity components. At this stage, the implications of these tradeoffs and their implied constraints tend to indicate that such systems will be longer term developments.

Electric systems typically involve fewer moving parts and possess inherent quietness at the expense of lower specific energy. Advanced fuel cell and semi-fuel cell systems are being developed in a number of naval laboratories with some prototypes having been fielded in larger UUVs. These systems are starting to offer attractive levels of energy content within compact packages and appear to offer exciting near-to-medium term energy options for military UUVs. At low powers (e.g. <1 kW), low temperature electric systems are of interest since, at these power levels, the energy overheads required to maintain the operating temperatures of high temperature systems begin to impact negatively on overall system efficiencies. Primary batteries tend to be attractive in comparison with secondary batteries since they are generally lower mass and volume subsystems that also have reduced maintenance requirements. Primary batteries offer very high reliability with no issue of charge capacity at time of use, but are often significantly more expensive than secondary batteries in terms of UUV life-cycle costs. Secondary batteries remain attractive from a cost standpoint under the conditions of frequent use and providing that a suitable operational recharging infrastructure will be available.

A number of near term solutions that meet energy targets similar to that of the US Navy UUV energy requirement noted above already exist, albeit that some remain at a prototype stage. However, the development of safe, compact energy systems that meet longer term naval UUV performance demands and naval logistic support requirements remains an on-going major challenge.

10.4.2 Propulsion Technology for Military UUVs

Military UUV propulsion system research is focussed on achieving high levels of propulsion system efficiency and reliability over a full UUV operational speed range while maximising manoeuvring potential and UUV controllability.

The need for UUV operation at low speeds (e.g. $<2\,\mathrm{m\,s}^{-1}$) poses significant technical challenges. The effectiveness of traditional control surface configurations decreases significantly at these low speeds while UUV inertia and buoyancy effects tend to dominate the UUV dynamics. To meet this challenge a variety of advanced main propulsor designs, auxiliary thruster designs and alternative control surface configurations have been developed and tested. The NUWC 21UUV Thrust Vectored Pump Jet (TVPJ) was designed and developed by Applied Research Laboratory/Pennsylvania State University (ARL/PSU). This type of propulsor maximizes control surface effectiveness at low speed by locating the control surfaces fins in the high-velocity propulsor exhaust flow. This positioning also allows significant increases in the effective angular range of the control surfaces before stalling occurs. The UK Marlin UUV uses a similar principle. This type of design can achieve propulsive efficiencies of up to 85% while also meeting cavitation inception targets and broadband noise goals, both of which are important for military UUV systems.

In parallel with the development of advanced propulsors has been the development of advanced electric motor technology. Design effort at NUWC has focussed on achieving a 50% size and mass reduction for the whole motor/propulsor system when compared with conventional motor/propulsor technology. An advanced radial field motor and motor controller have been developed during this project and these subsystems help achieve good low rotational speed control of the TVPJ. The motor uses a single rotation, outside radial field configuration and features a gimballed rim-driven propulsor rotor with a full afterbody and shroud assembly that is cantilevered from the aft stator vanes to provide directional propulsion force. This arrangement also provides attractive inflow to the rotor and the potential for vibration isolation, both of which are important in a military UUV system to reduce radiated noise.

By integrating the motor, rotor and propulsor into a single unit mounted external to the main 21UUV body, several system benefits have been obtained. These include the elimination of several primary sources of acoustic and electromagnetic energy signatures and a significant increase in reliability due to the elimination of mechanical couplings and seals linking the individual components. Again, both of these benefits are critical to the achievement of effective military UUV systems.

10.4.3   Control Systems

One of the main themes of current military UUV control research and development is the design of an environmentally adaptive autopilot that can maintain trajectory control and operate robustly in energetic shallow marine environments. This type of development is considered to be a key technology challenge for low-speed control of tactical scale UUVs in the littoral environment where simultaneous operation of propulsion system, control surfaces, auxiliary manoeuvring thrusters and variable buoyancy systems will be coordinated to allow effective mission task execution under difficult UUV operating regimes.

Over the last 15 years many UUV research groups have looked at advanced UUV control systems and algorithms including multi-actuator systems, non-linear adaptive control techniques, H-infinity robust control techniques and neuro-fuzzy intelligent control techniques. An example of military research in this area is the simulation based design and validation of a non-linear sliding mode autopilot combined with its hardware-in-the-loop and in-water testing on the NUWC 21UUV. Successful low speed trials of this system meant that further development of the controller was pursued to widen the scope of its applicability to other vehicles by the addition of adaptive neural network extensions to the sliding mode controller. The introduction of neuro-fuzzy techniques to UUV control systems is seen as an option that will provide either real time on-line modelling of and/or compensation for high-order vehicle dynamical characteristics that are poorly understood or unknown, *a priori*. This type of approach is considered valuable for military UUVs because of the potential advantages of on-line controller reconfiguration. A single baseline UUV autopilot would then be able to supply robust control functionality for numerous standard and degraded vehicle configurations. This approach will result in cost savings by eliminating the design and validation cycle for every UUV configuration and by providing an overall reduction in software development, testing and integration.

As with present UUV designs, future multi-mission reconfigurable UUVs will be volume and weight limited allowing little room for system redundancy. Military UUV research programmes are investigating the need for redundancy, subsystem designs that fundamentally increase system reliability and technologies to provide the UUV system with on-line analytical detection and in-stride compensation of faults. As part of a feasibility study, Fault Tolerant Control (FTC) algorithms were installed in the NUWC Large Diameter Unmanned Undersea Vehicle's (LDUUV) autonomous controller. These algorithms assumed control of the vehicle when simulated navigation and propulsion system faults were detected and the autonomous controller properly modified the mission plan in real time to compensate for the perceived system degradation. It is considered likely that greater use of analytical algorithms for fault detection and compensation will appear in military UUV systems to help achieve the required high levels mission reliability and to allow a degree of mission continuity while the UUV is in a degraded state.

10.4.4 Launch and Recovery

The current and near term limited endurance of UUVs means that most military UUV systems will need to be carried and deployed into their operational areas from surface ship or submarine host platforms. Options also exist for deployment from transport aircraft. In a number of military UUV roles it will be important that the launch and recovery processes remain covert.

While the launch and recovery of UUVs from surface ships can be hazardous and take a long time, a number of systems have already been developed to assist in these processes. For example, deployment and recovery of the UK Autosub UUV has been demonstrated in sea states up to Sea State 5. However, the deployment and recovery of UUVs from submarines is currently less well proven and, together with wider UUV/submarine integration issues, the development of robust launch and recovery techniques is considered to be another critical aspect of military UUV research (e.g. Chapuis *et al.*, 1996).

Current submarine launch and recovery research and development is focussed on launch and recovery from and to torpedo tubes and secondary UUV hangar structures. At present, the US Near-term Mine Reconnaissance System (Thornton and Weaver, 1998), US Long-term Mine Reconnaissance System (Emblem *et al.*, 2000) and UK Marlin UUV (Tonge, 2000) are 0.533 m diameter vehicles capable of tube launch and recovery. Other larger military UUVs will need to use alternative launch and recovery sites and facilities. Various restrictions on the number of weapons that can be carried in a weapon stowage compartment and the types of maintenance operation that can be carried out in magazine areas may also mean that smaller UUVs may ultimately be required to use alternative sites. It should be noted that, although torpedo tube recovery is a very challenging engineering problem, each of the alternative recovery

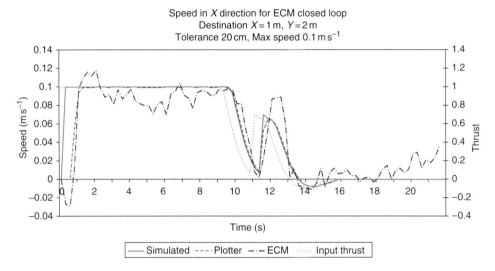

Color Plate 1   Speed in X direction during positioning – simulated, Cartesian robot, current meter and thrust. *See* Figure 1.6.

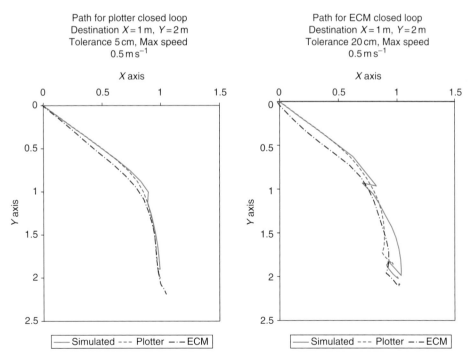

Color Plate 2   Varying UUV trajectory with varying time-slicing rollback error threshold. *See* Figure 1.7.

Color Plate 3 Photographs of both Slocum Battery (above) and Slocum Thermal (below). *See* Figure 3.6.

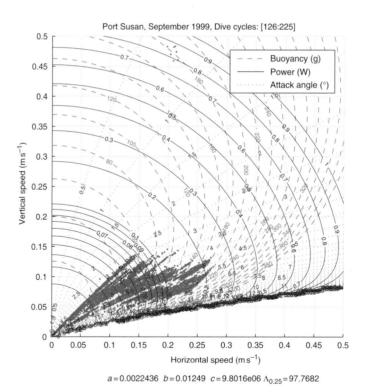

Color Plate 4 Glider performance diagram showing the buoyancy and power required to maintain a given speed and glide angle. This curve is for Seaglider but the behavior is similar for all designs. Note how at a given buoyancy, horizontal velocity U is maximized at a glide angle near $40°$, whereas at fixed power U is maximized nearer a $14°$ glide. Green marks show observed Seaglider operating points. *See* Figure 3.8.

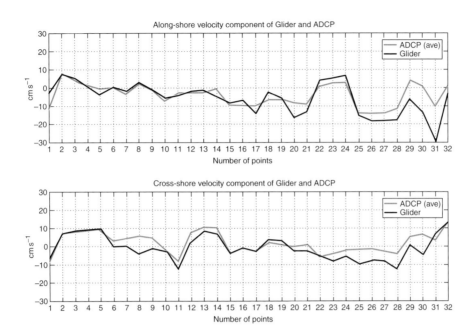

Color Plate 5 Time series of depth-average ocean velocity from Slocum (black), depth-average ADCP velocity (red) from the LEO-15 site during July 2000. *See* Figure 3.9.

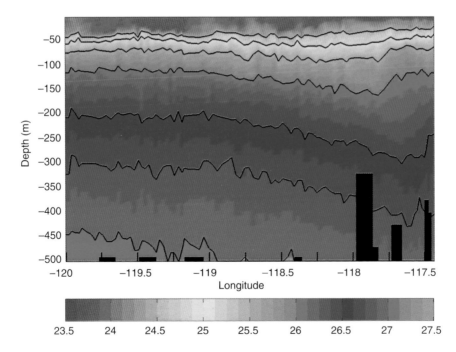

Color Plate 6 Density from the Spray section in Figure 3.10. Spacing of the temperature and conductivity profiles is about 3 km. The broad isopycnal slope downward to the east indicates the geotrophic shear of the California Current. The nearshore upward slope is associated with a near shore countercurrent. *See* Figure 3.11.

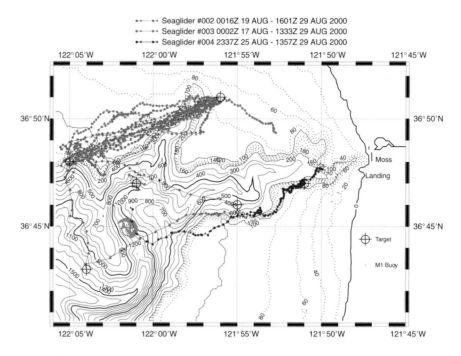

Seaglider #002 0016Z 19 AUG - 1601Z 29 AUG 2000
Seaglider #003 0002Z 17 AUG - 1333Z 29 AUG 2000
Seaglider #004 2337Z 25 AUG - 1357Z 29 AUG 2000

Color Plate 7 Tracks of three Seagliders in Monterey Bay (depth contours in meters). Two gliders made a total of 13 sections along the north rim of Monterey Canyon. At the end of the exercise, one of these (track in red) remained near a target about 2 miles north of a surface mooring (buoy positions shown in cyan). *See* Figure 3.12.

Color Plate 8 Prototype SAUV. *See* Figure 4.1.

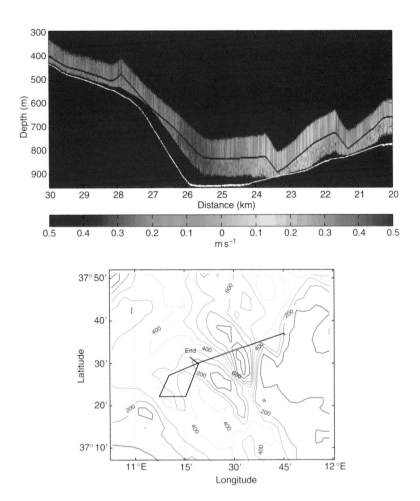

Color Plate 9 Autosub mission 239 in the Strait of Sicily used both upward- and down-ward-looking ADCPs to study the near-bottom velocity structure across the sill between the eastern and western Mediterranean. Upper panel shows the north component of velocity along the red section of the track shown in the bottom panel. *See* Figure 9.10.

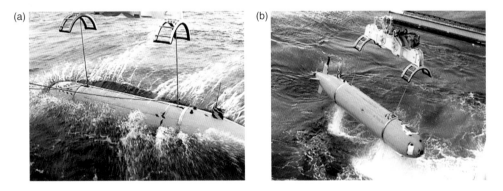

Color Plate 10 (a) Launch and (b) recovery of the Autosub vehicle from its purpose-built gantry as installed on the FRV *Scotia* in July 1999. *See* Figure 16.2.

Color Plate 11 Illustration of HUGIN 3000 operating scenario in UUV mode. *See* Figure 11.2.

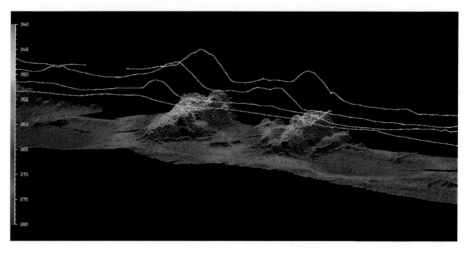

Color Plate 12 Bathymetry of an area with cold-water coral reefs off the West Coast of Norway. The vertical (depth) scale is exaggerated by a factor of 2. The white lines show the trajectory of HUGIN. *See* Figure 11.9.

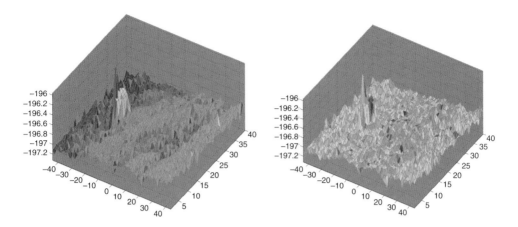

Color Plate 13 Mine countermeasures research EM3000 response from Manta mine dummy: Left, Depth colour coded. Right, Backscatter (echo strength) colour coded. *See* Figure 11.10.

Color Plate 14 Combined swath bathymetry and sub-seabed imagery, data from the Gulf of Mexico. *See* Figure 11.12.

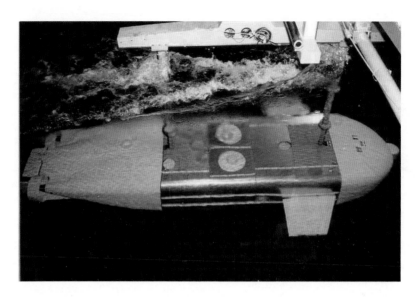

Color Plate 15 Hydrodynamic tank test of the 4.5 m long flat-shaped AUV 'MARIUS' in 1993. The test was carried out in a 240 m long towing tank at the Danish Maritime Institute. *See* Figure 12.1.

Color Plate 16 *Theseus* under the sea ice. *See* Figure 13.6.

site concepts have a variety of associated integration issues that mean that no single concept will provide an ideal recovery site solution.

Submarine torpedo tubes are configured in different numbers and geometries depending on the class of submarine. To produce through-life cost reduction and flexibility in usage, a design aim for tube recovery systems is that a given recovery system design can be reused on a number of different classes of submarine. This is not a trivial problem since differences in geometry exist, space and volume constraints are significant and severe limitations will be imposed by the proximity of other high priority submarine equipment in the vicinity of the tube exits.

A great deal of military research effort has focused on the development of accurate, physics-based, unsteady hydrodynamic models for the simulation and visualization of the hydrodynamic interactions between host platform, hangars/recovery systems and UUVs. This type of model allows assessment of the feasibility of UUV recovery with respect to host platform and recovery system geometries and likely hydrodynamic disturbances. As a result, improved UUV guidance and control systems can be designed together with improved homing and recovery system configurations. Recent UK and US studies on UUV recovery system hydrodynamics have resulted in the development of a vortex computational fluid dynamics code (VORTEL) capable of addressing the unsteady forces associated with complex flows in the vicinity of potential recovery sites and systems. Together with grid based methods, such as Reynolds Averaged Navier Stokes (RANS) codes and launcher system models, such as the UK Simulation of Submarine Launched Stores (SOSLS) model, VORTEL type techniques provide the basis for design and assessment of military UUV launch and recovery strategies.

10.4.5 Navigation

The importance of high accuracy, robust navigation for military UUVs has already been noted in the discussion on roles. Technically, the development of UUV navigation systems capable of meeting military mission requirements has involved the use of multiple, multi-mode, high-precision sensors together with advanced navigation processing algorithms that implement filtering and data fusion algorithms. During the early 1990s, high specification military and civil UUVs adopted a relatively standard navigation sensor arrangement comprising a ring laser gyro based Inertial Navigation System (INS) and a Doppler velocity log (e.g. Agoros, 1994). In this type of system the velocity log often provided the primary velocity input to a Kalman filter. The INS provided heading information into the Kalman filter and attitude data for lower level autopilot functions. A variety of data quality assurance algorithms were used to check for anomalies or errors in the input data streams. While this type of arrangement produced a navigation solution that degraded relatively slowly (e.g. typical performance in the order of 0.1–0.2% of distance travelled), it was clear from a very early stage that precise UUV navigation would require the addition of in-stride GPS updates to reset the Kalman filter and provide a low, bounded navigation error.

While GPS technology had been available at relatively low cost for a number of years, the issue of taking fixes from close to the air/sea interface, where GPS reception would be interrupted by wave washover, remained a problem. Two approaches were adopted by the military UUV community to solve the problem. The first approach was to raise a mast above the main level of wave influence and the second approach was support to the development of a GPS receiver where interrupted GPS ephemeris data could be reconstructed from successive partial receptions. Both approaches have achieved success and the use of GPS data by UUVs can now be considered a mainstream UUV navigation technology.

While generally not a problem for civil UUVs, the raising of a GPS mast or the surfacing of a UUV may compromise covertness in some military UUV missions. This has meant that military research has shifted towards the investigation of a range of non-traditional navigation (NTN) techniques and technology. Such techniques are generally aimed at achieving integrated correlation against geophysical data, either using prior knowledge of the operational area or using concurrent mapping techniques, such that UUV navigation error is bounded without the need for external navigation aids and while maintaining covertness. For close-in navigation and station-keeping vision-based navigation shows promise (Hillenbrand and Negahdaripour, 2000).

NTN sensor development is focused on the development of new generations of low cost geophysical sensors, such as those for accurately measuring the Earth's magnetic and gravitational fields, that can be made to work within the internal environment of a UUV. In terms of NTN processing a central problem in creating accurate geophysical parameter maps, such as for gravity, is to correctly register partially overlapping data patches. A number of different algorithms that use contours of constant field value and derivative data as the matching primitives for data registration have been developed. Although such algorithms need adequate feature content to achieve unambiguous matching and small areas of uncertainty, multi-modal approaches appear to offer great potential for high precision, bounded navigation provided that the relevant *a priori* maps are available. Typically, NTN processing algorithms comprise several modules including: data quality pre-processing, data gridding, parameter isocontour extraction and synthesis, search algorithms for finding corresponding pieces of isocontours in data patches, geometric transformations and calculation of multi-dimensional metrics for assessing quality of match.

However, despite the availability of proven NTN algorithms, if *a priori* map data is not available, this type of technique cannot be used. Hence, an alternative approach is necessary and recent military research efforts have investigated the feasibility of Concurrent Mapping and Localization (CML). The aim of this type of technique, discussed in Chapter 1, is to simultaneously build a map using UUV on-board sensors and navigate relative to the emerging map. The CML process can be formulated using a stochastic map representation. This representation consists of a single state vector that holds estimates of UUV location and the location of local features together with an associated covariance matrix. As the vehicle moves around its environment taking measurements of geophysical features, the stochastic map is updated using a Kalman filter (e.g. Ruiz *et al.*, 2001). The major technological challenges in performing CML are robustness, computational complexity and data association uncertainty. However, a variety of CML technical demonstrations have taken place and the technique appears to offer levels of performance and other advantages that will make CML a key navigation technique for future military UUVs.

10.4.6 Communications

The ability of military UUVs to operate independently and to integrate into the wider battlespace network is a key functionality (Fiebig, 1998). UUVs will need to communicate with host platforms, command centres, other UUVs and other mobile assets to maximise their effectiveness. Early UUV designs tended to store all data on-board and, particularly as data storage technology has improved, this technique remains valid unless other operational criteria dictate the need for data communication. The US Near-term Mine Reconnaissance System and UK Marlin system are capable of deploying fibre-optic cable to ensure that high bandwidth MCM sensor data can be communicated back to a host platform in real time for operational reasons. Although there are significant technical issues in assuring the robustness of this type of link, there are compelling reasons for its use if high bandwidth and real time are mission critical requirements. Where the role of a UUV does not require it to communicate high bandwidth data or communicate in real time, other technical options exist and these are being developed within military research programmes, for example using submersible acoustic networks (Welsh *et al.*, 2000).

The first technology that has been the subject of extensive research within military research programmes is acoustic communications technology. In these programmes a number of US and UK research teams are investigating the theoretical limits of such technology, the development of advanced algorithm building blocks for processing within acoustic communication systems and the development of advanced DSP hardware implementations. The algorithms include elements such as: interference suppression; waveform design; error corrector codes; data compression; adaptive Doppler compression; low probability of intercept/low probability of exploitation implementations and higher-order signal constellations. The military aims for this technology include the development of bi-directional medium bandwidth capabilities over relatively short distances (e.g. <10 km), and lower bandwidth capabilities over greater distances together with the achievement of various criteria associated with covertness and communication robustness when operating within shallow, acoustically challenging environments.

The second technology is satellite communications (SATCOM). Operationally this technology offers a UUV the ability to transmit sensor data over very long distances to command centre data consumers. However, the use of SATCOM will require a UUV to either surface and deploy an antenna, deploy a SATCOM buoy or deploy a mast. While this implies that communication will be on an intermittent basis, provided that the communication can be achieved in a timely fashion, SATCOM still offers an option that can be very attractive.

There is no doubt that many UUV sensors will generate large data sets and that transmission of data to third parties will often require on-board pre-processing within the UUV prior to transmission if acceptable transmission times are to be realised. A variety of data processing and data compression technologies will need to be optimized to support specific sensor generated data and these technologies will also need to offer adaptive levels of compression to match the bandwidth of the desired UUV communication link. For these reasons, military UUV research programmes are continuing to investigate data requirements and a number of standard and specialised data compression techniques.

10.5 CONCLUSIONS

In the twenty-first century, Naval Staffs are forced to continuously review defence capability requirements with respect to available budgets. While such deliberations may often result in cuts and refocusing of resource, there are also opportunities for the research and development communities to showcase emerging technologies that offer the potential for maintaining military capability at reduced cost or provide the means of filling capability gaps in a cost effective manner.

Aside from the continual force projection requirements of the submarine community, the tactical areas of remote mine countermeasures, ISTAR and tactical oceanography have been identified as important military roles where autonomous UUV operations can enhance force tactical capability. Because of this, a number of military UUV research programmes continue to research, develop and transition unique enabling technologies that will allow military UUV systems to act as cost effective force multipliers operating independently from various host platforms in a forcenet context. Present affordability issues are driving the development of these technologies in a manner that addresses a number of common design issues for a wide variety of UUV concepts and sizes while enabling the insertion of advanced mission-unique payloads.

Current military UUV research and development activities continue to provide a base of UUV technology that will not only often find duel-use and transition into civil UUV applications, but will also provide the basis for advanced military UUV systems deployed in joint force warfare into the twenty-second century.

11. DETAILED SEABED SURVEYS WITH AUVs

BJØRN JALVING[a], KARSTEIN VESTGÅRD[b] and NILS STORKERSEN[a]

[a] *Norwegian Defence Research Institute (FFI), PO Box 25, N2007 Kjeller, Norway*
[b] *Kongsberg Simrad AS, Strandpromenadan 50, N3191-Horten, Norway*

11.1 INTRODUCTION

During the last few years Autonomous Underwater Vehicle (AUV) technology has evolved from concept demonstrators developed by research institutes and universities to commercial products. In the short-term, the driving forces are the trend towards deeper waters for hydrocarbon exploitation, several emerging naval applications and the Internet-driven need for more intercontinental underwater communication cables. Other applications emerging for the future will be within environmental research and monitoring and deepwater exploration.

Deepwater developments beyond the continental shelf will require the same level of survey data quality and intervention access as established for shallow water. The potential for cost savings is two-fold. First, there is a reduction in the survey cost and second there is a considerable cost saving by avoiding over-design of subsea installations due to the lack of sufficient documentation of the area. In this scenario, there is an increasing understanding that underwater robotics and in particular AUVs will play an important role in future survey and subsea engineering work.

This chapter concentrates on the application of AUVs for pipeline, site/block hazard and cable surveying, among areas where AUVs have found their first true commercial applications. The chapter starts with a description of typical AUV instrumentation for detailed seabed surveying followed by an analysis of achievable depth and position accuracy. Finally, the HUGIN vehicles, which have been used in several commercial and scientific operations, are described and operational results presented.

The AUV, as a free swimming underwater survey sensor carrier, has several advantages over cable controlled Remotely Operated Vehicles (ROV) and deep-towed systems:

- *High data quality from payload sensors*: AUVs are ideal survey sensor carriers for obtaining high quality survey data due to low acoustic self-noise, hydrodynamic stability, effective control of optimal position and altitude and absence of cable-induced motions.
- *High positioning accuracy*: The operational freedom in relative position between the AUV and the survey vessel allows the survey vessel to be directly above the AUV. This will minimise the effects of ray-bending errors on the acoustic positioning accuracy.

179

In addition, minimal horizontal distance minimises the error contribution from survey vessel heading sensor and acoustic transducer installation misalignment.

- *High survey speed*: An AUV relieved from an umbilical can operate with a high speed independent of water depth, providing a significant increase in survey effectiveness compared to other survey tools.
- *Curvature changes and line turns*: Heading changes along a survey line and turns from one survey line onto the next can be accomplished within minutes as opposed to hours for deep-towed systems.
- *Portability*: An AUV system can be maintained in standard cargo containers and can be air freighted throughout the world.

A survey sensor suite in an AUV for site and pipeline route surveys will typically consist of:

- high resolution multibeam echosounder (MBE) for detailed bathymetry survey;
- dual frequency Side Scan Sonar (SSS) for hazard object detection;
- low frequency Sub-Bottom Profiler (SBP) for sub-bottom structure and geophysical mapping;
- CTD;
- Magnetometer.

If the AUV operates in an unmanned, untethered rather than autonomous mode a survey vessel follows it. The survey vessel tracks the AUV with a super-short baseline (SSBL) acoustic position system. SSBL systems are interchangeably referred to as ultra-short baseline (USBL) systems in the literature. The survey vessel is usually equipped with the following navigation instrumentation:

- SSBL;
- DGPS;
- heading, roll and pitch attitude sensors;
- acoustic communication links.

By combining DGPS with SSBL data compensated for attitude, range and bearing an AUV position estimate in global co-ordinates is obtained. This position estimate can be sent to the AUV using an acoustic communication link. The AUV navigation system serves three main purposes:

- provide the guidance and control system with real-time estimates of attitude, velocity and position;
- store the navigation solution and navigation sensor data for use in post-processing of survey sensor data;
- provide survey sensors with real-time attitude data for sensor stabilisation.

AUVs for detailed seabed mapping are equipped with an Aided Inertial Navigation System (AINS). The Inertial Navigation System (INS) calculates position, velocity and attitude of the vehicle using high frequency data from an Inertial Measurement Unit (IMU). An IMU consists of three accelerometers measuring specific force and

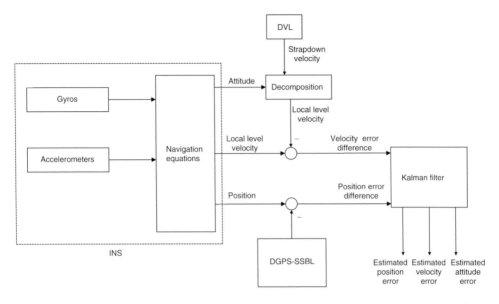

Figure 11.1 Kalman filter structure for an AINS. DGPS–SSBL position and DVL aiding is illustrated. Other aiding sensors, for instance a pressure sensor and a LBL acoustic navigation system are integrated similarly.

three gyros measuring angular rate. A Kalman filter will, in a mathematically optimal manner, utilise a wide variety of navigation sensors for aiding the INS. The Kalman filter is normally based on an error-state model and provides a much higher total navigation performance than is obtained from the independent navigation sensors. The structure of an AINS is shown in Figure 11.1.

The Kongsberg Simrad HUGIN 3000 AUV is equipped with the following navigation sensors:

- IMU;
- Fibre-optic Gyrocompass (FOG);
- pressure sensor;
- Doppler Velocity Log (DVL);
- DGPS receiver;
- SSBL transponder;
- optional long baseline (LBL) navigation transceiver.

Position updates are necessary to bound the position error drift in the navigation system. When the AUV is operating in UUV mode (refer to Section 2.1), the AINS will regularly receive combined DGPS–SSBL measurements from the survey vessel on the acoustic link. When the AUV is working in autonomous mode (refer to Section 2.2), position updates can be provided by the AUV DGPS receiver when at the surface, or from a LBL or SSBL system when submerged. For military AUVs, various terrain navigation systems may provide the AINS with position updates.

Table 11.1   INS classes.

Class	Traditional Application	Gyro Technology	Gyro Bias	Accelerometer Bias
1 nmi h^{-1}	Marine, air and land navigation	RLG, FOG[1]	≈0.01° h^{-1}	≈50 µg
>10 nmi h^{-1}	Tactical systems	RLG, FOG	≈1° h^{-1}	≈1 mg

Notes
1 RLG – Ring Laser Gyro; FOG – Fibre-Optic Gyro.

The DVL velocity bias is dependent upon the system's acoustic frequency, but will typically be of the order of 4 mm s^{-1}. This high accuracy makes the DVL a very important aiding sensor. When the AINS has no position aiding, but is aided by a DVL with bottom track, the AINS position error drift is limited to typically 0.05–0.1% of travelled distance (AUV speed 2 m s^{-1}) depending on the IMU and DVL quality and the survey pattern. If the DVL loses bottom track and velocity data becomes relative to the water column, the position drift rate will be dominated by the magnitude of the sea current.

Inertial navigation systems are usually classified by the standard deviation of the positional error growth of their free inertial (unaided) performance (see Table 11.1). A free inertial INS will, after a short period of time, have unacceptable position errors. IMUs for AUVs will probably be in the >10 or 1 nmi h^{-1} class.

11.2   AUV OPERATION

11.2.1   UUV Mode

An AUV with a real-time acoustic link connection to a survey vessel is defined to be in untethered underwater vehicle (UUV) mode. The acoustic link functions as an 'acoustic tether', which enables the operator to supervise operation and control the vehicle. Operator interaction can optimise the payload sensor parameters or can allow reprogramming of the mission plan. For example, should hazards be located along a pipeline route, the mission can be altered to enable closer inspection. The mission plan is a set of waypoints, altitude, depth and speed references as well as commands for controlling mission modes, payload sensors etc.

In the survey vessel, the AUV position is tracked by combining DGPS and SSBL data. Combined DGPS and SSBL position data is transmitted to the AUV AINS to bound the position drift. In Figure 11.2 the HUGIN 3000 AUV is shown in UUV mode.

11.2.2   AUV Mode

A vehicle operating independent of a survey vessel is defined to be in AUV mode. In this mode the operator does not have real-time supervision or control of the vehicle. The navigation system position drift is bounded by DGPS fixes at the surface at regular intervals (for shallow water operations) or operation within a pre-calibrated transponder array (LBL navigation system).

Figure 11.2   Illustration of HUGIN 3000 operating scenario in UUV mode (not to scale). *See* Color Plate 11.

For surveying a limited area (e.g. a site survey) in shallow water, the AINS can maintain good position accuracy with only DVL aiding. A prerequisite is that the dive phase is limited (when the DVL has no bottom track) and that the vehicle mission geometry is optimised to cancel out most of the systematic error growth. A favourable geometry is the 'lawnmower' pattern. Basically, the difference between UUV mode and AUV mode can be the presence of a survey vessel transmitting position updates.

11.2.3   Data Logging and Post-processing

During a mission, survey sensor and navigation sensor data is stored on the vehicle hard disk, while DGPS, SSBL, sea level recorder, CTD, sound velocity and atmospheric pressure data is stored on the survey vessel. All data is time-tagged to GPS UTC. Proper time tagging is ensured by synchronisation of the AUV with GPS UTC prior to a mission. A low-drift clock in the AUV maintains an accurate time reference for the whole mission.

In the post-processing, all data logged in the survey vessel is combined with the data logged in the AUV to produce the final Digital Terrain Model (DTM) and to give

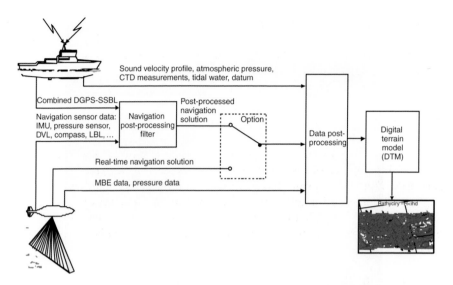

Figure 11.3   Illustration of the data logging, data flow and data processing required to produce a DTM. Depending on the DTM accuracy specification, the DTM can be based on either the real-time navigation solution or the post-processed navigation solution.

an accurate georeference for the survey sensor data. Data post-processing can take different routes and are normally under the control of a survey-processing specialist.

The navigation solution stored by the AINS can be used to produce the DTM. However, if maximum positioning accuracy is required, post-processing should be applied. The post-processing is a two-stage process. The first step is Kalman filtering using all AUV navigation sensor data and DGPS–SSBL position measurements. This is similar to the real-time AINS that runs in the AUV, except that the DGPS–SSBL position measurements are available at higher rate with no time delays and that all *a posteriori* estimates with their error covariance matrices are saved for a smoothing process. In the second step, an optimal recursive smoothing algorithm adjusts all the estimates, starting at the last estimate and running backwards in time, improving the accuracy further.

Figure 11.3 illustrates the data logging and post-processing process. Wild points in the data from the different survey sensors and navigation sensors are removed and the data is filtered, either automatically or manually. The pressure sensor measurements are compensated for tidal height, atmospheric pressure and water density to yield depth. Finally the processed data is gridded, a DTM constructed and contour plots etc. are produced.

11.3   RESOLUTION AND ACCURACY

The survey end product, for instance a DTM, has no better accuracy than the least accurate component in the total measurement chain, from surface navigation all the way down to the acoustic seabed footprint.

This section provides an analysis of achievable depth and horizontal position accuracy. The case investigated is seabed topography mapping with a high-resolution multibeam echosounder and the AUV operated in the UUV mode, that is, closely followed and tracked by a survey vessel. The basic principles of the case study also have relevance for the autonomous mode of operation.

11.3.1 Depth Accuracy

Establishing an accurate seabed depth estimate is a well known, but non-trivial challenge involving a range of instruments and parameters to be carefully handled and understood. The results presented in this section are based mainly on the work by Jalving (1999). The estimated depth of the seabed is the sum of the MBE depth measurement and the AUV depth, referred to a datum, for instance WGS-84. The error sources contributing to the depth error can accordingly be grouped into two categories:

- MBE depth accuracy;
- AUV depth accuracy.

The AUV depth category comprises the problem of relating the time varying sea surface height (tidal height) to a datum.

11.3.2 MBE Depth Accuracy

The depth accuracy of a MBE data is influenced by the following elements:

- echosounder measurement accuracy;
- attitude sensor accuracy;
- orientation sensor and echosounder transducer mounting axis misalignment;
- sound velocity profile.

The acoustic measurement accuracy of a MBE is dependent upon the signal-to-noise (S/N) ratio. Provided a S/N ratio above 10 dB exists at the detector, the typical depth measurement error is of the order of 0.1% of AUV altitude above the sea bottom.

Correct mechanical mounting of the echosounder transducer and the orientation sensor(s) to achieve axis alignment are critical and may be a major error source if not addressed properly. A reasonable specification on axis alignment is 0.06°, limiting the depth error contribution to 0.03 m (1σ) at 50 m altitude. For an AINS this mounting accuracy specification will apply to the IMU, for AUVs with less advanced navigation systems it will apply to the dedicated orientation sensors.

The typical AINS accuracy in roll and pitch is 0.06°, corresponding to a DTM depth uncertainty of 0.03 m (1σ) at 50 m altitude. DTM depth error due to AINS heading error is a function of multibeam echosounder beam angle and the slope angle of the seabed terrain.

The sound velocity profile can be determined either by computation of sound velocity from CTD data or direct measurement by a sound velocity probe. An accuracy of better than $0.5\,\mathrm{m\,s^{-1}}$ is obtainable from both methods, equivalent to $\sim 0.03\%$ of the AUV altitude. The dominating error source is probably the temporal and spatial variations in sound velocity and the corresponding measurement frequency.

11.3.3 AUV Depth Accuracy

The vehicle depth estimate is produced by combining AUV pressure sensor data with a density profile estimate, measurements of tidal height and atmospheric pressure. In order to estimate a density profile, CTD measurements of conductivity, temperature and pressure are needed. The relationship between these quantities and salinity are described by an international standard – the Practical Salinity Scale, 1978 (PSS-78). The density profile can be computed by inputting temperature, pressure and salinity to the International Equation of State of Seawater, 1980 (EOS-80). The depth is given by:

$$z = \int_{P_{\mathrm{ATM}}}^{P_{\mathrm{AUV}}} \frac{1}{\rho(S, t, p)\, g(L, p)}\, \mathrm{d}p, \tag{11.1}$$

where z is depth (m), p is pressure (Pa), S is salinity, t is temperature (°C), L is latitude, $g(L, p)$ is gravity $(\mathrm{m\,s^{-2}})$ and $\rho(S, t, p)$ is density $(\mathrm{kg\,m^{-3}})$. A numerical solution to the integral in Eq. (1) based on EOS-80 is derived in Fofnoff and Millard (1991).

Simplified, the main error sources for a pressure sensor are repeatability and hysteresis. Assuming a quality pressure sensor, these error sources should be less than 0.01% of full scale. Thus, for a 6 MPa unit, the depth accuracy is 0.06 m in water depths of 600 m.

The conventional way of measuring tidal height is to place a pressure sensor on the seabed. Assuming the same accuracy specifications as for the AUV pressure sensor, we have a depth error contribution in the order of 0.06 m. A standard atmospheric barometer accuracy of 1 hPa contributes to at DTM depth uncertainty of 0.01 m.

Referring to the above description of the steps involved in computing the density profile, the density profile accuracy is dependent upon:

- CTD temperature measurement accuracy;
- CTD pressure measurement accuracy;
- CTD conductivity measurement accuracy;
- absolute accuracy of PSS-78 and EOS-80;
- density profile variations in time and space versus the CTD measurement frequency.

An error analysis can be based on a simplified version of Equation (11.1):

$$z = \frac{p_{\mathrm{AUV}} - p_{\mathrm{ATM}}}{\bar{\rho} g}, \tag{11.2}$$

where p_{AUV} is vehicle pressure sensor reading (Pa), p_{ATM} is atmospheric pressure (Pa) and $\bar{\rho}$ is average density. $\bar{\rho}$ is given by:

$$\bar{\rho} = \frac{1}{p_{\mathrm{AUV}} - p_{\mathrm{ATM}}} \int_{P_{\mathrm{ATM}}}^{P_{\mathrm{AUV}}} \rho(S, t, p)\, \mathrm{d}p. \tag{11.3}$$

By modelling the error in estimated average density, $\Delta\bar{\rho}$, and the corresponding depth error, $\Delta z_{\Delta\bar{\rho}}$, we obtain:

$$z + \Delta z_{\Delta\bar{\rho}} = \frac{p_{\mathrm{AUV}} - p_{\mathrm{ATM}}}{(\bar{\rho} + \Delta\bar{\rho})g}.$$

Obviously, increased vehicle operation depth means stricter requirements on $\Delta\bar{\rho}$. Demanding a depth error contribution less than 0.05 m at 1000 m depth translates into an error in estimated average density of less than 0.05 kg m^{-3}, which is readily obtained with a good quality CTD. In the real ocean, the most important error source is likely to be the temporal and spatial variations in the density profile and the chosen measurement frequency to account for these changes. The survey vessel should have arrangements for continuous CTD measurements of the whole water column throughout the operation.

11.3.4 Total Depth Accuracy

Assuming statistically independent error sources, the total depth accuracy is calculated by root-square-summing the error contributions. In Jalving (1999) the DTM depth uncertainty was derived as shown in Figure 11.4.

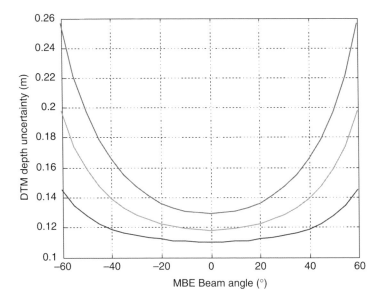

Figure 11.4 DTM depth uncertainty as a function of MBE beam angle. AUV altitude $h = 30$ m (lower), $h = 50$ m (middle), $h = 70$ m (upper), see Jalving (1999). Assumed sensor accuracy as stated in Sections 3.3 and 3.4.

11.3.5 Horizontal Position Accuracy Analysis

The AINS position aiding depends on whether the vehicle is operating in UUV or AUV mode. In the following analysis we have assumed that the AUV is running in UUV mode, that is the AINS receives regular DGPS–SSBL position updates.

In the survey vessel the AUV position in global co-ordinates is determined by combining DGPS and SSBL data compensated for attitude and lever arms. The following error sources affect the accuracy of the combined DGPS–SSBL estimate:

- DGPS accuracy;
- SSBL accuracy;
- survey vessel attitude accuracy;
- sound velocity profile accuracy;
- systems installation accuracy.

Typically, DGPS systems have an accuracy of 2 m (1σ), while an SSBL system produces its position estimates based on measured relative distance and direction between the survey vessel and the AUV and whose accuracy depends on the signal-to-noise ratio. In Table 11.2 the Kongsberg Simrad High Precision Acoustic Positioning system (HiPAP) accuracy for varying S/N ratios is shown.

A prerequisite for accurate SSBL position tracking is that the survey vessel's attitude is well known. GPS-based vessel reference systems have a typical angular accuracy of $0.06°$ (1σ) in roll and pitch and $0.12°$ (1σ) in heading (Table 11.3). Standard North seeking gyrocompasses are inferior to GPS-based heading measurements and have a typical accuracy of $0.7°\times$ secant (latitude) (1σ).

Accurate alignment of the SSBL transducer, attitude sensor and gyrocompass are taken care of by proper installation and post-survey calibration. The final error in alignment and offset should be better than $0.1°$ (Table 11.4).

For vertical sounding, the sound velocity profile has only a minor influence on the acoustic positioning accuracy, however for highly accurate positioning and for offsets from the vertical, the sound profile must be measured and compensated for.

Table 11.2   HiPAP accuracy for varying signal to noise ratios (S/N).

HiPAP	S/N (dB)		
	20	30	40
Direction accuracy (1σ)	0.3°	0.2°	0.1°
Range accuracy (m)	<0.2	<0.15	<0.1

Table 11.3   AUV position error due to survey vessel attitude error of 1 mrad (1σ) in roll and pitch and 2 mrad (1σ) in heading. The relative horizontal position between the AUV and the survey vessel is $x = 100$ m, $y = 100$ m.

AUV depth (m)	300	1,000	2,000	3,000
AUV position error (m at 1σ)	0.5	1.4	2.8	4.3

Table 11.4 AUV position error due to a 0.1° systems installation error in the survey vessel in all axes. The relative
horizontal position between the AUV and the survey vessel
is $x = 100\,\text{m}, y = 100\,\text{m}$.

AUV depth (m)	300	1,000	2,000	3,000
AUV position error (m at 1σ)	0.8	2.5	4.9	7.4

Table 11.5 DGPS–SSBL accuracy in metres (1σ),
not Kalman filtered, for varying depths and signal to noise ratios. The relative horizontal position
between the AUV and the survey vessel is $x =
100\,\text{m}, y = 100\,\text{m}$.

AUV depth (m)	S/N (dB)		
	20	30	40
300	3.0	2.5	2.1
1,000	7.7	5.3	3.2
2,000	15.0	10.1	5.3
3,000	22.3	15.0	7.7

It can be shown that use of a reasonably accurate sound velocity profiling sensor will
reduce the positioning error to a negligible level compared to other error sources
in the positioning chain. Again, it is important to select a sound velocity profiling
measurement frequency sufficient to compensate for temporal and spatial variations.

The combined DGPS–SSBL accuracy (not Kalman filtered) for varying depths are
shown in Table 11.5. In the calculations we have assumed DGPS and HiPAP accuracy as
stated above. Contributions from survey vessel attitude sensor accuracy, sound velocity
profile accuracy and systems installation accuracy are not included in the calculations.

The DGPS–SSBL position accuracy is improved by the AINS by means of a Kalman
filter that combines DGPS–SSBL position estimates, DVL data, pressure sensor data
and an INS in an optimal way. The Kalman filter is only capable of estimating zero
mean time-varying errors with faster dynamics than the AINS error drift. Hence, survey vessel attitude errors (assuming slower error dynamics than DVL-aided INS error
drift), sound velocity profile errors and survey vessel system installation errors are not
estimated for straight trajectory operations. Some of these errors can become observable by manoeuvring, but a conservative error budget for normal AUV operations
should include the following error sources:

- AINS position accuracy;
- survey vessel attitude accuracy;
- sound velocity profile accuracy (negligible);
- survey vessel systems installation accuracy.

The AINS position, velocity and attitude accuracy is strongly related to the accuracy of
the core INS system, see Table 11.1 and the SSBL direction accuracy, see Table 11.2. In
Table 11.6, the AINS accuracy for a 1 nmi h^{-1} IMU and a >10 nmi h^{-1} IMU and HiPAP

Table 11.6 AINS heading and position accuracy estimates (real-time) for different IMU and SSBL accuracy. Not shown in the table is roll and pitch accuracy which is in the order of 0.06° (1σ) for the >10 nmi h^{-1} IMU and in the order of 0.003° (1σ) for the >1 nmi h^{-1} IMU.

AUV Depth (m)	Parameter (1σ)	HiPAP: 0.3° IMU: >10 nmi h^{-1}	HiPAP: 0.1° IMU: >10 nmi h^{-1}	HiPAP: 0.3° IMU: >1 nmi h^{-1}	HiPAP: 0.1° IMU: >1 nmi h^{-1}
300	Heading (°)	0.15	0.14	0.030	0.030
	Position (m)	2.0	1.7	1.7	1.5
1,000	Heading (°)	0.18	0.15	0.03	0.03
	Position (m)	3.6	2.1	2.9	1.8
2,000	Heading (°)	0.22	0.17	0.032	0.031
	Position (m)	5.7	2.8	4.4	2.3
3,000	Heading (°)	0.25	0.18	0.033	0.031
	Position (m)	7.6	3.6	5.6	2.9

Table 11.7 Horizontal position accuracy for EM 3000 and EM 2000.

Pointing angle (°)	0	45	60	70	75
Position accuracy (% of AUV altitude)	0.8	0.9	1.3	2.0	2.8

direction accuracy of 0.1° (1σ) and 0.3° (1σ) has been calculated. A comparison of Tables 11.5 and 11.6 illustrates the improvement in position accuracy that can be expected from an AINS.

The real-time AINS accuracy estimates have been calculated using an extensive navigation system simulation tool developed by the Norwegian Defence Research Establishment called NavLab (Gade, 1997). In the simulations we have assumed an AUV speed of 2 m s^{-1}, constant heading and depth and DGPS–SSBL position updates every 30 s. Simulated AUV aiding sensors have been a magnetic compass (3° (1σ) accuracy), pressure sensor and DVL. Post-processing with smoothing and higher DGPS–SSBL measurement frequency will improve the accuracy further. Note that the accuracy estimate is theoretical and an indication of what is obtainable with an AINS. The results are dependent on the error models used and the simulated AUV trajectory. However, the accuracy figures have been confirmed in 1,300 and 2,200 m water depths in the Gulf of Mexico using HUGIN 3000, see Section 4.4.

The MBE horizontal footprint position accuracy is mainly determined by the along-track beamwidth of the MBE. Kongsberg Simrad EM 3000 and EM 2000 echosounders both have an along-track beamwidth of 1.5°. The position accuracy as a percentage of altitudes for various beam-pointing angles is shown in Table 11.7.

The MBE footprint position relative the AUV must be compensated for AUV attitude. AINS heading accuracy is shown in Table 11.6. Typical roll and pitch accuracy is in the order of 0.06° (1σ) for the >10 nmi h^{-1} IMU and in the order of 0.003° (1σ) for the >1 nmi h^{-1} IMU. Relative mounting axis misalignment between the IMU and the MBE transducer should be less than 0.06° after proper mechanical design, careful assembling and calibration. The contributions of these individual error sources to the total DTM horizontal position uncertainty are summarised in Table 11.8.

Table 11.8 Contributing error sources and resulting DTM horizontal position uncertainty.

Error Source	Resulting DTM Position uncertainty contribution
AUV AINS position accuracy	See Table 11.6
Survey vessel attitude accuracy	See Table 11.3
Survey vessel systems installation accuracy	See Table 11.4
MBE measurement accuracy	See Table 11.7
AUV AINS attitude accuracy	0.3 m (1σ) for >10 nmi h^{-1} IMU at 50 m altitude and 60° MBE beam angle
AUV systems installation accuracy	0.3 m (1σ) at 50 m altitude and 60° MBE beam angle
AUV clock drift	0.3 m (1σ) assuming crystal oscillator with a drift specification better than 1 ppm (part per million)

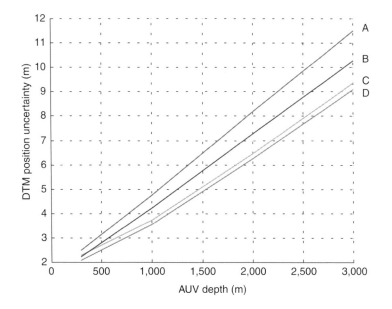

Figure 11.5 Resulting DTM position uncertainty for four different combinations of IMU and SSBL accuracy as a function of AUV depth. A: 10 nmi h^{-1} IMU and HiPAP S/N 20 dB. B: 1 nmi h^{-1} IMU and HiPAP S/N 20 dB. C: 10 nmi h^{-1} IMU and HiPAP S/N 40 dB. D: 1 nmi h^{-1} IMU and HiPAP S/N 40 dB. The relative horizontal position between the AUV and the survey vessel is $x = 100$ m, $y = 100$ m. AUV altitude above the seabed is 50 m. MBE beam angle is 60°.

Under the assumption of statistical independent errors, the total DTM position uncertainty can be calculated by square-root-summing the different error contributions. In Figure 11.5, the total DTM position uncertainty for four different combinations of IMU and SSBL accuracy as a function of AUV depth is shown. For a navigation system with SSBL position aiding, the accuracy of the SSBL system is clearly more important than the accuracy of the IMU. Unless the survey vessel system installation accuracy is better than 0.1° (1σ) the benefits of high quality navigation sensors on the AUV are significantly reduced.

In order to ensure this high DTM position accuracy in real missions, the following items must be addressed properly:

- highly accurate alignment of SSBL transducer and survey vessel attitude sensors;
- selection of high quality navigation sensors for the AUV navigation system;
- installation of the SSBL system in the survey vessel in an optimal location with respect to acoustic noise;
- calibration survey to measure and compensate for any misalignments.

Further reduction of the DTM error can be achieved by proper smoothing of all data in the post-processing and the DTM production phases. Modern post-processing systems also contain facilities for cancelling fixed position errors observed in the data.

11.4   THE HUGIN AUVS

The HUGIN vehicles are examples of AUVs designed primarily for detailed seabed surveying. The HUGIN I and II vehicles have been used in a wide range of seabed surveying operations down to 1,200 m, both for tests and commercial work. HUGIN I is used as a research and concept demonstration vehicle and is continuously upgraded with the latest technology developments. HUGIN II was completed in 1998 and is owned and operated by Norwegian Underwater Intervention (NUI) providing survey services in the North Sea on a commercial basis. HUGIN 3000 is a further development towards deeper water, extended payload suite and improved navigation and autonomy. HUGIN 3000 was delivered to the US survey company C&C Technologies (Lafayette, Louisiana) in July 2000, and is currently being used for seabed mapping, imaging and sub-bottom profiling in the Gulf of Mexico (Chance *et al.*, 2000; Vestgård *et al.*, 2001). In summer 2001 HUGIN II finished an update program including a 1.2 m lengthening of the vehicle and upgrade of electronics, control system and navigation system to bring it to the HUGIN 3000 level. In summer 2002 a new Hugin 3000 vehicle was delivered to the Norwegian survey company Geoconsult.

11.4.1   The HUGIN I and II Vehicles

The HUGIN I and II vehicles have a hull shape designed for low-drag, they weigh approximately 700 kg in air and have an overall length of 4.8 m and a displacement of $1.2\,m^3$. The HUGIN II power source was an 18 kWh aluminium–oxygen semi-fuel cell, sufficient for 36 h of operation. A single propeller driven by a DC-electric motor gave the vehicle a nominal speed of $2\,m\,s^{-1}$, while four control surfaces provided heading and pitch control.

The HUGIN I and II vehicles are equipped with three independent acoustic communication links (Kristensen and Vestgård, 1999). The command link is a robust, medium speed, bi-directional acoustic link used for vehicle control and event monitoring. It has a bit rate of 55 bit/s based on Frequency Shift Keying (FSK) modulation.

The data link is a high-speed unidirectional acoustic link used to transmit real-time sensor data to the surface. It has a bit rate of 2000 bit/s based on Multiple Frequency Shift Keying (MFSK) modulation.

Command link functionality has also been integrated in the HiPAP acoustic position tracking system and provides an independent emergency command link. This link has a separate housing, electronics and back-up power, providing a totally independent emergency system.

The software architecture is based on a hierarchical structure (Jalving *et al.*, 1998). The Mission Manager constitutes the highest level, executing the mission plan. The mission plan is entered prior to vehicle launch, but the operator may alter single values or complete mission steps at any point during the operation using the acoustic command link. An error detection system continuously supervises the vehicle status, including deviations of vehicle internal or operational parameters. The Error Handler reacts by sending a message to the Operator Station and initiating appropriate action (i.e. switch operation mode). If the error is critical, the vehicle is brought to the surface by activating the emergency ascent function, releasing a drop weight and inflating two air bladders. The operator may also activate the emergency function through the acoustic links.

Ahead of deployment, the mission is programmed on the HUGIN Operator Station and downloaded to the vehicle Control Processor. Normally, the vehicle is in altitude mode (constant altitude), producing multibeam echosounder data of constant swath width. During the survey, subsets of the payload and the vehicle data are compressed and transmitted to the surface and displayed in real-time, providing on-line quality control of the data as it is being collected.

The vehicle is deployed from the launch and recovery system located at the stern of the survey vessel. When initiated by the mission plan or commanded by the operator, the vehicle ascends to the surface using its propulsion and rudder system. Upon reaching the surface, the vehicle drops its nose cone and releases a retrieval line. The line is used to pull the vehicle onto the launch and recovery ramp that projects into the water from the aft deck of the survey vessel.

HUGIN I and II are equipped with the Kongsberg Simrad EM 3000 MBE. The EM 3000 has 127 beams spaced at 0.9°. The EM 3000 depth data together with vehicle navigation sensor data are stored on a hard disk. After a completed mission the data is off-loaded to the post-processing system.

In the post-processing, data logged on the survey vessel (vehicle position, sound velocities, barometric pressure etc.) are combined with data logged in the vehicle (multibeam and vehicle navigation sensor data). Prior to launch a high precision clock (<1 ppm time drift) in the vehicle Control Processor is synchronised to GPS UTC time. Data logged at the surface and subsea is thus time-tagged to the same reference.

11.4.2 The HUGIN 3000 Vehicles

HUGIN 3000 is characterised by a high degree of flexibility, high quality sensors and an accurate AINS. The following describes extensions compared to the first versions of HUGIN I and II.

The HUGIN 3000 vehicles have twice the volume of HUGIN I. The distributed systems approach is further extended, especially in the payload section. The weight is approximately 1,450 kg in air, the overall length is 5.3 m and the displacement is 2.5 m³. The power source is the aluminium–oxygen semi-fuel cell, extended to 50 kWh, sufficient for 50–60 h of operation, depending on payload configuration and vehicle speed. A new high torque synchronous motor allows direct drive of a single large three-bladed propeller, providing a significant improvement in electrical to hydrodynamic efficiency.

HUGIN 3000 has a depth rating of 3,000 m. The pressure vessels are made from titanium or glass. Glass- and carbon-fibre laminate material and high performance syntactic foam are used in the vehicle design. The complete propulsion system is installed in an oil-filled pressure compensated section.

The distributed sub-systems are placed in separate pressure containers for easy maintenance and replacement. Mechanical flexibility is obtained by dividing the vehicle into separate sections with an open internal structure, allowing payload reconfiguration either by replacing individual electronic containers and transducers or complete electronic containers and transducer bays, see Figure 11.7.

The HUGIN 3000 payload system consists of the Payload Processor, the Payload Network, the Payload Power Distribution and the different payload sensors (Figure 11.6). The payload sensor interface software is written as separate software modules for each type of payload sensor, implementing all sensor specific functions. The communication between a surface Payload Operator Station and its subsea payload sensor is "generic", that is the data transfer is transparent and common for all sensor types. This ensures that new payload sensors can be added without the need to change the basic system software. Signal interfacing of payload sensors is either Ethernet 10/100 MBit or serial line RS232 or RS422. Payload sensor power on/off is controlled from the Payload Processor. Connectors for distributing signal and power from the Payload System Container to the individual payloads are standardised. Subsets of the

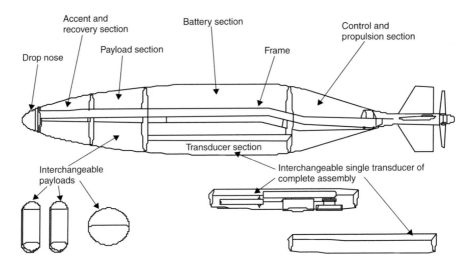

Figure 11.6   HUGIN 3000 modular concept.

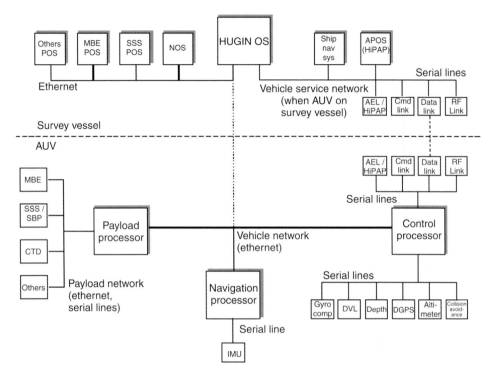

Figure 11.7 Schematic illustration of the HUGIN 3000 hardware architecture. Abbreviations: POS – Payload Operator Station, NOS – Navigation Operator Station, AEL – Acoustic Emergency Link, Command link – Acoustic Command Link, RF link – Radio Link.

payload sensor data and/or QC parameters are transferred in real-time to the surface and displayed on the Payload Operator Station.

HUGIN 3000 can be equipped with a range of payload sensors depending on application and user needs, including:

- MBE;
- side scan sonar;
- sub-bottom profiler;
- CTD sensor;
- magnetometer;
- fishery research echosounder.

11.4.3 HUGIN I and II Field Experience

Over the past years the HUGIN I and II vehicles have been used for several commercial and scientific applications. The summary below illustrates the variety of experience obtained with the vehicles.

In 1997, a full-scale test survey was carried out with HUGIN I on the West Coast of Norway to qualify the system for commercial survey operations. An area of approximately 200 m by 2,000 m with water depths between 320 and 350 m was surveyed over two dives. The first dive at 20 m altitude was run with 15 m line spacing. The second dive at 50 m altitude was run with 30 m line spacing. A Stolt Comex Seaway SOLO ROV was run at 20 m altitude with 15 m line spacing surveying an overlapping area of $200 \times 1,000 \, m^2$. The results from these trials documented the ability of the HUGIN I system to conduct commercial survey operations, both operationally and in terms of data quality. The maps produced by data from HUGIN I and the ROV showed no significant differences. The trials also showed that HUGIN I was a very stable and low noise sensor platform that allowed the survey to take place at 50 m altitude.

The first commercial operation was undertaken on the Åsgard gas transport pipeline route in autumn 1997. The main characteristic of the survey were:

- detailed seabed topography survey;
- 200 km corridor route length;
- 100 m minimum corridor width;
- depth ranging from 100 to 370 m;
- line surveys 30 m to either side of the route centre line at 40–60 m altitude;
- 460-line-km surveyed;
- 140 h diving time.

The main characteristics of the mapping exercise were to provide a DTM with a resolution of 1 m by 1 m with 0.2 m depth contouring and 160 maps of the seabed at a scale of 1 : 2000.

The results from these operations documented the following:

- In general, HUGIN showed an offshore operational performance and reliability well beyond most expectations.
- Even in shallow waters the vehicle collected data with a significantly greater efficiency (four times) compared with what was achieved by the ROV.
- A number of seafloor areas were mapped using both HUGIN I and the ROV. No significant differences in depth between the two independent survey systems were observed.
- The prototype handling system proved to be an adequately efficient and safe method for launch and recovery of the vehicle throughout the weather conditions of this operation (even when operated in Sea State 5).

Figure 11.8 shows a sample map generated during this survey. The bathymetric contours are 0.2 m, while the scale is 1 : 2000.

A particular challenging survey was carried out with HUGIN in October 1998 in a deep fjord on the West Coast of Norway. A subsea condensate pipeline was planned to be deployed between the oil and gas processing plants at Sture and Mongstad. The detailed pre-engineering survey had to cover a seabed area of rapidly varying topography, with water depth in the range of 300–550 m, including narrow passages where HUGIN had to operate very close to the almost vertical sides of a rocky fjord. This survey operation fully confirmed the capability of HUGIN to operate even in this

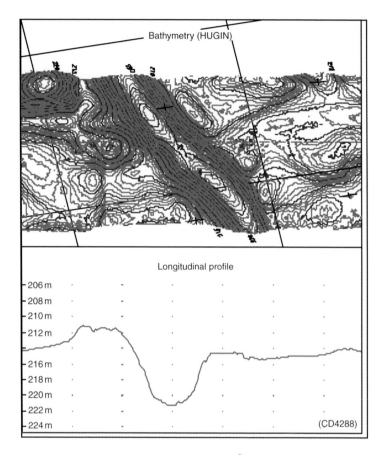

Figure 11.8   Sample map from the Åsgard survey 1997.

type of terrain, both with respect to navigation, positioning, communication, vehicle operation, as well as the performance of the survey sensors.

In June 1998, the Institute of Marine Research (IMR, Norway's national centre for research on coastal and ocean life and the marine environment) used HUGIN II to map an area that was presumed to be scattered with cold-water coral reefs. In a single 6-h mission, an area of 7,000 m by 600 m (4.2 km^2) was mapped. An extract of the data recorded is displayed in Figure 11.9.

The main objectives of this mission were to gain experience with the possibilities and data quality offered by the AUV and to create documentation on the number of coral reefs and their locations, and partly to assess the amount of damage done to the reefs. The vehicle may be used more extensively on similar missions in the future.

A general problem with the towed net and acoustic observation methods currently used in fisheries research is that the platforms from which the sensors are operated may actually scare away the fish, causing biased and uncertain estimates. In January 1999 initial tests were performed by IMR. For these missions, the EM 3000 MBE was replaced with a 38 kHz Simrad fishery echosounder. Apart from realistic testing of the

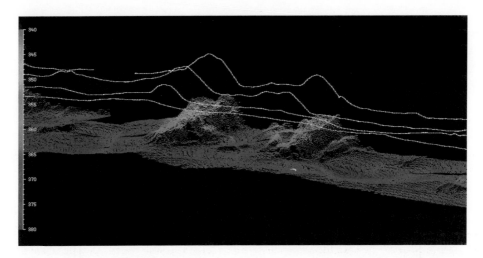

Figure 11.9   Bathymetry of an area with cold-water coral reefs off the West Coast of Norway. The vertical (depth) scale is exaggerated by a factor of 2. The white lines show the trajectory of HUGIN. *See* Color Plate 12.

system configuration, the objective of the operation was to determine the radiated noise levels attainable with the AUV. Excellent quality echograms were recorded, and noise measurements verified that HUGIN has a very low radiated noise level. Based on these successful results, new programmes for fishery research have been launched (Godø *et al.*, 1998).

In 1998 HUGIN, equipped with the EM 3000 MBE (both bathymetry and imaging function in one unit), undertook a series of mine hunting research operations. The tests were conducted by the Norwegian Defence Research Establishment (Hagen *et al.*, 1999). A number of dummy mines were lowered to the seabed at depths ranging 80–200 m, with HiPAP mini-beacons attached for accurate positioning. HUGIN was run over the dummy minefield at altitudes ranging 9–30 m. An example of the data collected with the EM 3000 from low altitude is shown in Figure 11.10. The images display the data recorded from one pass (from bottom left to top right) over a Manta dummy mine. The response from the mine is clearly seen in the left half of the image. From the backscatter data (right-hand image) even the recovery chain running from the Manta is seen to give significantly higher echo strength than the surrounding seafloor. Furthermore, the peak at one end of the depth response can be attributed to the HiPAP beacon and flotation device, which was floating a few dm above the dummy mine.

Figure 11.11 shows a 3D view from the Ormen Lange field. The Ormen Lange Field is a gas province located off the North West coast of Norway, in an area with very rough and steep terrain at water depths exceeding 800 m. The data is from autumn 2000, surveyed with the HUGIN II vehicle equipped with an EM 3000 multibeam echosounder. This figure shows an area of 4 km by 4 km with peaks of 30–40 m in height. The detailed information unveiled by the survey will contribute significantly to

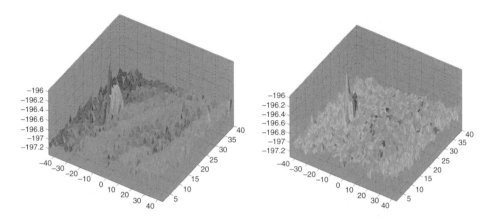

Figure 11.10 Mine countermeasures research EM3000 response from Manta mine dummy:
Left, Depth colour coded. Right, Backscatter (echo strength) colour coded. *See* Color Plate 13.

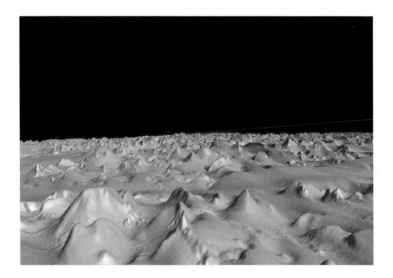

Figure 11.11 The Ormen Lange field survey. Data processed from EM 3000 MBE on HUGIN II.
Approximately 850 m water depth, 4 km by 4 km view, peaks are 30–40 m high.

the work of planning and selecting the most optimal site and track for the production
and pipeline installations.

11.4.4 HUGIN 3000 Field Experience

The first HUGIN 3000 was delivered to C&C Technologies in autumn 2000. The
vehicle was subjected to extensive sea trials and a customer acceptance test program

was concluded. All subsystems, vehicle control and navigation accuracy were tested and verified in water depths down to 2,300 m.

The vehicle has since January 2001 been in routine use for deepwater pipeline route surveying in the Gulf of Mexico, in the Mediterranean and off the west coast of Africa. The vehicle control systems have performed exceptionally well in very rough bottom terrain and the navigation accuracy is better than predicted. All payload systems (MBE, dual frequency SSS and chirp sub-bottom profiler) have worked to full expectation, delivering excellent data. Figure 11.12 shows an example of combined swath bathymetry and sub-seabed imagery.

In Table 11.9 results from two accuracy tests in 1300 m water depth are shown. "Ref difference" is the difference between the mean value of the DTM observation and the pre-surveyed position of the well head. The well head was deployed from the BP drill platform *Ocean America* in 1999 and was positioned by determining the surface position of the drill string using DGPS. The drill string had a heavy well guide attached to the end. This made the drill string almost vertical but uncertainty of the effect of water current on the drill string was the largest source of uncertainty in position.

The AUV position accuracy was improved by post-processing the data with a forward/backward filter. The magnitude of improvement can be seen by comparing the "Kalman Filter σ" of the real-time position data with the post-processed results shown in Table 11.9. The filter produces an optimal estimate based on all logged measurements, both historical and future. Furthermore, the full set of position measurements stored on the survey vessel is utilised. The Kalman filter standard deviations are well in accordance with the standard deviation of the well head observations (population standard deviation), although the March 2001 post-processed results are a little high. Given the uncertainty of the verticality of the drill string, the "Ref difference" results are consistent.

Figure 11.12 Combined swath bathymetry and sub-seabed imagery, data from the Gulf of Mexico. *See* Color Plate 14.

Table 11.9   Position accuracy results in 1,300 m water depth.

Mission	No. of Meas.	Ref Difference (m)		Population σ (m)		Kalman Filter σ (m)	
		North	East	North	East	North	East
March 2001 real-time	6	1.2	6.1	2.4	4.1	4.4	4.5
March 2001 post-processed	6	0.5	3.1	3.7	4.2	1.7	1.7
October 2000 post-processed	11	2.7	4.0	1.2	1.7	1.7	1.7

11.5 CONCLUSIONS

This chapter has described AUVs intended for detailed, high quality seabed surveying in shallow and deep water. In recent years AUV technology has matured with respect to robustness, battery endurance, depth rating and navigation accuracy and for some years AUVs have carried out commercial surveys for the offshore market.

AUVs for detailed seabed mapping or autonomous operations for longer periods of time are now equipped with aided inertial navigation systems that combine an inertial navigation system with other navigation sensors and systems in an optimal manner.

For a good DTM, the required horizontal position accuracy can be achieved at all depths by periodic position updates to the AINS from an accurate SSBL system. The DTM position accuracy budget analysis predicted the benefits of Kalman filtering on the combined DGPS–SSBL position measurements. Results from HUGIN 3000 surveys in the Gulf of Mexico have shown that the predicted DTM accuracy has been achieved in commercial surveys. The analysis showed that survey vessel systems' installation and calibration and SSBL accuracy are the two most important factors to the DTM position accuracy. A complete DTM depth error budget has been derived from individual error sources, revealing that a DTM depth accuracy of 0.13 m (1σ) can be achieved at 300 m AUV depth, 50 m altitude and 30° MBE beam angle.

Acknowledgements

The authors are grateful to Kenneth Gade (FFI) for performing the AINS accuracy simulations and to Rolf Arne Klepaker (Kongsberg Simrad) for contributing with his unique knowledge in pioneering AUV technology in Norway.

12. AUVs FOR SURVEYS IN COMPLEX ENVIRONMENTS

ANDERS BJERRUM

Maridan AS, Agern Alle 3, 2970 Horsholm, Denmark

12.1 INTRODUCTION

When dead lobsters from Kattegat, Denmark were shown on television in 1986 it generated an immediate increase in the political awareness for the marine environment. The Danish Parliament reacted by adopting a \$17 million Marine Environment Plan and ordered farmers to reduce fertiliser run-off. The Marine Environment Plan called for frequent monitoring of the marine environment. In the same year, the government initiated marine research projects to improve the understanding of eutrophication processes (oxygen depletion) in Danish waters which, again, called for underwater surveys. Interest in the marine environment was also increasing in other European countries and, under the EUREKA initiative, EUROMAR was adopted in 1986 as an umbrella programme for European marine research and technology development. In 1989, the EC prepared the research programme for Marine Science and Technology (MAST) as a component of its framework programme of research, technology development and demonstration.

The first European survey AUV called 'MARIUS', was developed under the MAST programme from 1991 to 1994 by a consortium of Danish, Portuguese and French partners (Ayela *et al.*, 1995). The prototype AUV was designed for 600 m water depth. The Danish partner, COWIconsult, was responsible for the hydrodynamic design and manufacturing of the hull and propulsion system for 'MARIUS' and for meeting the requirement for a low-drag hull that could provide sufficient space for sensors and down-looking acoustic instruments. A flat streamlined low-drag shape was developed that had the additional advantage of a flat bottom for easy transportation and handling. Electronic equipment was housed in cylindrical pressure vessels inside a wet hull construction that allowed water to run in and out. The mechanical design of the central steel section that houses the batteries was a welded multideck construction, well known from shipbuilding. The pressure vessels were aluminium/stainless steel cylinders with aluminium end-covers sealed by double piston O-rings.

Based on the computer drawings, wooden models were constructed by an experienced boat builder. He was not completely satisfied with the shape and, before fibreglass moulding, he finished the models with a few intuitive strokes of his plane. This experience has undoubtedly paid off in contributing to the very low drag. The drag coefficient measured by tank tests was 0.14 (Egeskov *et al.*, 1994). A comprehensive tank test program was carried out to measure the hydrodynamic parameters. The tank tests were carried out in cooperation with the Danish Maritime Institute

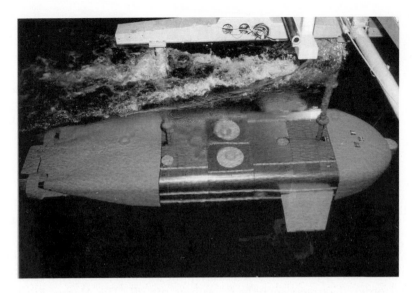

Figure 12.1 Hydrodynamic tank test of the 4.5 m long flat-shaped AUV 'MARIUS' in 1993. The test was carried out in a 240 m long towing tank at the Danish Maritime Institute. *See* Color Plate 15.

and the Department of Offshore Engineering at the Technical University of Denmark (Aage and Smitt, 1994). Non-dimensional coefficients of the hydrodynamic forces and moments of the AUV were determined by full-scale tests, using a towing tank equipped with a Planar Motion Mechanism (PMM). The test results enabled the designers to refine the AUV dynamic performance by optimising mass, inertia, damping, lift and drag coefficients of the vehicle and its control surfaces and balancing resistance and propulsion characteristics. The PMM was installed on the towing carriage and via struts and force dynamometers connected to the submerged body. During the tests, the towing carriage was running at constant speed (Figure 12.1). Except for the towing tests the propellers were operated at the 'self propulsion point' giving zero net longitudinal force in steady motion with all control surfaces at zero angle and no drift or pitch angle. The PMM was used to fix the vessel at a constant yaw or pitch angle or in the oscillatory mode, to generate oscillatory motions in sway, yaw, heave or pitch. The results in terms of the hydrodynamic properties were used in the algorithms for the autopilot. Furthermore, the results were used to set up a hydrodynamic computer simulation of the vehicle.

One of the future applications for the AUV was autonomous pipeline inspection (Egeskov *et al.*, 1995). Therefore, forward wings were required to enable the vehicle to lift fast to avoid obstacles when flying at low altitude. The wings gave good vertical manoeuvrability but their forward location made the vehicle unstable in pitch and further tests were carried out on 'MARIUS' to develop the control system solution. The hull was equipped with two main thrusters and four lateral thrusters. The lateral thrusters were installed for hovering and docking purposes but as explained later they have never been in use because hovering and docking could be controlled more accurately by using the control surfaces (rudders and elevator at the aft and ailerons at the forward wings). In the early stages of development the lack of an efficient autopilot made it difficult to follow a straight path but the problem was solved by the use of adaptive autopilots originally developed for ships (Bech, 1983).

An obstacle detection system based on a Reson Seabat multibeam echosounder MBE was implemented for tests (Henriksen, 1994). The system provided promising image information, which could be useful in a further development program but, for priority reasons, a simpler solution based on a single echosounder was adopted.

12.2   CONTROL AND COMPUTER SYSTEM

After completion of the AUV project in 1994, COWIconsult concluded that substantial work remained before AUVs would be able to perform any practical surveys and the company decided to withdraw from the project. Two of the engineers involved continued to have a strong belief in AUVs and founded the company Maridan with a three-year business plan to complete the AUV development based on results from the European MAST project. The company was financed by the founders' own resources together with funding from national and European development contracts.

A second prototype AUV called 'MARTIN' was constructed. The hull dimensions were the same as 'MARIUS' but 'MARTIN' was designed for 150 m water depth. The pressure vessels were glass fibre reinforced plastic (GRP) cylinders with POM end-covers sealed by double piston O-rings.

A distributed control system was designed in 1996 with the following control levels defined for manoeuvring:

(1)   Mission Management System (MMS);
(2)   survey execution;
(3)   path control;
(4)   autopilots;
(5)   actuator and thruster control.

Adequate 'intelligence' of the control system was distributed to the lowest possible level. The low-level control systems had direct access to vehicle data (speed, position, altitude etc.) organised in a table known as a 'Blackboard' (Englemore, 1989). The Blackboard solution serves as a common dynamic database for information exchange between connected computers (Smith, 1994).

The MMS is the highest control level, the 'Captain' onboard the vehicle, and is able to execute surveys without communication with the operator. The function of the survey execution level is to carry out surveys as defined in a pre-programmed survey plan. A survey plan includes a set of sequential instructions corresponding to the planning carried out by a surveyor before starting a survey. To support this a 'Basic-like' survey plan definition language MARPRO was developed for preparing survey plans. The routes to be followed by 'MARTIN' are defined by waypoints and speed, depth and the switching on or off of payload sensors are defined by SET commands in the survey plan.

The following example shows a survey plan for video recording a site area of 1,000 m by 1,000 m with 50 m line spacing (the code is explained in lines starting with REM):

```
REM Survey plan for a lawnmover site survey, 1000 m, 1000 m (20 lines spaced 50 m)
REM All units are metric
VAR I;
REM Go to first Waypoint defined as Lat-Lon coordinates
SET NextWaypoint=LATLON(DMS(55:41.533:0), DMS(12:04.980:0));
SET Depth_Cmd=1.0;
SET Speed_Cmd=1.0;
SET FollowMode=fmLoose;
SET NextWP_Dir=330;
SET WPdev=2;
SET FollowDev=1;
SET DepthMode=dmDepth;
SET NextWP_Speed=Speed_Cmd;
WAITUNTIL (WaypointHit);

REM Turn on video and follow first survey line: 1000 m, 330°
SET I=1;
SET Video=1; SET Videolights=1;
LABEL LoopAgain;
SET NextWaypoint=NextWaypoint+POL(1000.0,330);
SET FollowMode=fmExponential;
WAITUNTIL (WaypointHit);

REM Turn portside and head for next survey line
SET NextWaypoint=NextWaypoint+POL(50.0,150);
SET FollowMode=fmLoose;
SET NextWP_Dir=60;
WAITUNTIL (WaypointHit);

REM Follow survey line: 1000 m, 60°
SET NextWaypoint=NextWaypoint+POL(1000.0,60);
SET FollowMode=fmExponential;
WAITUNTIL (WaypointHit);

REM Turn starboard and head for next survey line
SET NextWaypoint=NextWaypoint+POL(50.0,150);
SET FollowMode=fmLoose;
SET NextWP_Dir=330;
WAITUNTIL (WaypointHit);

REM repeat the loop until last survey line
SET I=I+1;
IF (I<10) GOTO LoopAgain;

REM Follow last survey line: 50 m, 330°
SET NextWaypoint=NextWaypoint+POL(1000.0,330);
SET FollowMode=fmExponential;
WAITUNTIL (WaypointHit);
STOP;
```

The path controller guides the vehicle along the specified path. When the vehicle encounters a crosscurrent the path follower is able to adjust the heading to compensate for the current. The path controller reports to the MMS in the case of heavy crosscurrent or countercurrent. Three independent autopilots were implemented, one for speed control, one for pitch control and one for course control. The Autopilot uses a model of the vehicle for maintaining control bandwidth and at the same time ignoring wave oscillations.

Fail-safe features were built in at all levels, ensuring that a module, not receiving commands from the layer above, would execute a set of pre-defined safe mode procedures. A number of built-in self-tests were defined in all modules for monitoring proper functionality. If a module failed, it would be restarted on another computer or alternatively, control would be transferred to a higher level. The fail-safe features include releasing a drop-weight for emergency surfacing. All modules include a watchdog for generating local diagnostic messages. A diagnostic system for 'MARTIN' was developed in cooperation with the National Laboratory Risø (Christensen *et al.*, 1997; Madsen, 1997). A risk analysis was carried out based on a functional model of 'MARTIN'. A Goal Tree/Success Tree study was used to define the structure for the diagnosis system.

Early in the project, it became clear that the central computer architecture used for 'MARIUS' had serious limitations and the distributed control system called for a computer network that could support high-speed communication between the subsystems. To minimise the amount of data to be transmitted between subsystems, raw sensor data gathered in the nodes should be pre-processed at local node level before broadcasting the information over the network. Therefore, the node processors should have sufficient computing power in addition to that required for data communication. The solution was found in 1995: A computer network based on a serial bus, the Controller Area Network (CAN). The CAN controller, originally developed by Bosch for use in automobiles, is increasingly being used in industrial automation and underwater applications.

Many communication systems for networking 'intelligent' I/O devices as well as sensors and actuators are commonly available. However, most of them are not well suited for real-time applications. When developing 'MARTIN' different solutions were considered:

- Point to point RS232 and Ethernet. This solution was not appropriate due to the limited bandwidth and/or poor real-time capability.
- I^2C (a standard protocol widely used in electronic equipment, such as TV sets). This solution did not provide sufficient noise immunity for use in the underwater environment with electromagnetic interference from motors and power cables.
- The LONTalk network (most commonly used with the Neuron chip). This solution has been selected for the Autosub AUV (McPhail and Pebody, 1998) and for the Ocean Explorer (Smith, 1994). However, Maridan found the LONTalk processor lacked in true real-time performance because of too few priority-levels.

The CAN solution was selected for 'MARTIN' due to its suitable specifications (Table 12.1).

The development of a decentralised computer network was a major technological step towards autonomy. As part of this a single board computer called Communication Interface and Power (CIP) was developed (Nielsen, 1997). The CIP was designed for high-speed data communications and power control for subsystems. It supports internal data communication between subsystems onboard the AUV and external

Table 12.1 Features and advantages of the CAN bus for onboard network communications.

Feature	Advantage
Multi-master protocol	Broadcasting does not stop in the case where a single node stops responding
The existence of compliant silicon components	Suitable for micro-controllers
Low overhead	Accurate time tagging and synchronisation
Arbitration on bit-level	Broadcasting does not stop in the case of bottleneck problems
'Party line' broadcasting	Automatic error-correction
	All nodes get the same information
Thermal and ESD protected	High noise immunity
Clock frequency 250 kHz	Accurate synchronisation (10 ms)

data communication between the AUV and the topside control system. The CIP data communication system originally developed for 'MARTIN' has performed very well since the vehicle's first sea trial in 1996. Some of the useful features include:

- versatile I/O facilities (digital and analogue, power switch);
- remote programming of microprocessor boards (FLASH memory);
- monitoring of performance and physical conditions (power consumption, temperature, humidity).

All sensors, actuators and thrusters are connected via the CIP computer. To make the CIP board fit into small diameter tubes it was designed as a small 10 cm diameter circular board. The CIP system and Blackboard development added two years to Maridan's development program, but five years on, the advantage of robustness and modularity in the computer system have proven to be a major benefit.

Surface communication was implemented by the acoustic modem developed for 'MARIUS' by the French partner Orca Instrumentation. In addition, a radio modem was used for sea tests near the surface (Bjerrum, 1997). Subsequent sea tests demonstrated and proved the efficiency of the control system (Aage, 1997).

12.3 ACCURATE NAVIGATION

The first successful sea trials with the 'MARTIN' prototype in 1996 encouraged the Maridan team to start another important development, the positioning and guidance system. A high accuracy underwater Inertial Navigation System (INS), in combination with a Doppler Velocity Log (DVL) would be the best solution. The INS would be a state-of-the-art strapdown system with high performance gyros and accelerometers. Together, the DVL and the inertial sensors would give accurate heading and speed. By processing the sensor data in a Kalman filter, the systematic errors would be removed

Figure 12.2 Doppler-inertial navigation instrument (MARPOS®) including the RDI DVL fixed to a pressure compartment housing the Kearfott optical gyro and the navigation computer. Diameter 37 cm, length 67 cm including the DVL.

and the processed speed and heading data would be used for dead reckoning (e.g. as discussed by Jalving and Vestgård in Chapter 11).

The cost of accurate gyros based on ring laser technology was high, but Maridan decided to go for the best solution. So began a fruitful cooperation with Kearfott Guidance & Navigation Corporation (New Jersey, US) for the development of what became the most accurate underwater positioning system of its type, the MARPOS® Doppler-inertial navigation system (Figure 12.2). A sea test program was carried out in 1998 in cooperation with Kearfott with the purpose of determining the accuracy and to optimise the initialisation and calibration procedure. The tests showed remarkable positioning accuracies for lawn mower site surveys of 0.02% of distance travelled and for line surveys an accuracy of 0.1% of distance travelled (Bjerrum *et al.*, 1999).

MARPOS® was designed to provide positioning and attitude information (roll, pitch and heading) for 'MARTIN'. This was achieved with ring-laser-gyro (RLG) technology and the fusion of Inertial (INS), Doppler (DVL) and Differential GPS data in a Kalman filter architecture. MARPOS® is a self-contained solution but one that can use external sources (e.g. USBL) to further improve positioning accuracy. Positions relative to a known starting point are continuously calculated by dead reckoning using the Kalman filter to process the INS and DVL measurements. Compass course (heading) is determined by MARPOS® in an alignment process wherein the Kalman filter uses the gyros and accelerometers to sense the centre of gravity and the rotation of the Earth. Alignment can be performed either statically or on a moving base using DGPS and DVL aiding.

Based on the high performance of MARPOS®, the Maridan AUV was able to navigate without the need of a support vessel following the AUV. This observation radically changed the perspective of what was achievable in terms of autonomous operations

(Bjerrum and Krogh, 1998). In 1998, after seven years development, Maridan could confidently predict commercial opportunities as light at the end of a much-shortened tunnel.

12.4   THE SUCCESSFUL 'DIAMOND-SURVEY'

In March 1999, Maridan carried out its first demonstration of precise underwater navigation under strict commercial conditions. A 10-day site-survey was carried out by the 'MARTIN' AUV on board the survey vessel *Zealous* owned by De Beers Marine in Cape Town (Figure 12.3). The location was selected by the client for exploitation of diamonds from the seabed in 60–120 m water depth. Their interest was to examine the use of 'MARTIN' for autonomous surveys and thereby reduce significantly the cost and time required for exploration of potential diamond fields. Equipped with side scan sonar 'MARTIN' was required to perform pre-programmed autonomous surveys and to log the data for further analysis at the surface.

'MARTIN' was launched every day during the campaign and its computer system and instrumentation worked non-stop during the 10 consecutive survey days. The most successful part of the mission was a 16 km autonomous survey carried out on 31 March 1999 and known appropriately as the 'diamond-survey'. The path included a lawn-mower pattern site survey and a pre-programmed 2 km long diamond-shaped path performed at the end of the main survey. The survey was carried out autonomously without contact between 'MARTIN' and the support vessel. This was not planned but,

Figure 12.3   'MARTIN' during preparation for launch offshore Namibia 1999. The vehicle was equipped with a Tritech side scan sonar (Photo by courtesy of De Beers Marine Pty).

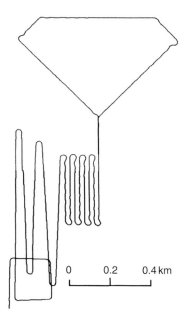

Figure 12.4   Trajectory from one day of the 'diamond-survey'.

due to a failure in the acoustic modem onboard the support vessel, the Maridan AUV crew was forced to pre-program the complete survey plan: 'MARTIN' was converted from a 'supervised' to an 'autonomous' vehicle overnight.

Navigation for the 'diamond-survey' was carried out using MARPOS® as a stand-alone instrument with GPS input before diving and Doppler aiding during the survey. The pre-programmed survey plan shown in Figure 12.4 included four parts:

(1)   alignment of MARPOS by making a square near the seabed (10 min);
(2)   performing a trajectory of survey lines defined by the AUV heading;
(3)   performing a trajectory of survey lines defined in global co-ordinates;
(4)   performing a diamond-shaped trajectory to demonstrate the accuracy of MARPOS®.

During the survey, 'MARTIN' was tracked by an acoustic tracking system (USBL) but no communication was possible. After launch, 'MARTIN' dived as planned and started the alignment of MARPOS. The first survey lines were defined by time and heading. It appears that 'MARTIN' was influenced by current. The following survey lines were defined in global co-ordinates and now 'MARTIN' compensated autonomously for the current. To add to the challenge the AUV crew had thought they had programmed in six lines per grid but had in fact put in six pairs of lines. As a result, the AUV did not return at the hour the team felt had been programmed and they grew concerned. When it was finally recognised that the AUV had in fact correctly travelled along the prescribed trajectory it had been given, the team drew a collective sigh of relief and, from that day, the concept of autonomy got a new meaning for the team. The

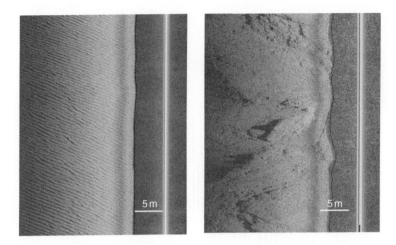

Figure 12.5 Sand ripples (left) and rocky seabed (right) recorded by a 675 kHz Tritech side-scan sonar during the 'diamond-survey".

last part of the survey was a positive surprise. 'MARTIN' was programmed to follow the diamond path. After travelling some 2 km, 'MARTIN' returned to the tip of the diamond. The offset was only 0.6 m, which was verified by the side scan sonar since the seabed pattern recorded at the tip of the diamond could be recognised on return. All side scan sonar data were logged together with time-stamped navigation data. Thereby, features on the side scan sonar data could be located in global co-ordinates (Figure 12.5).

Next day a more relaxed autonomous operation was carried out. The survey plans were coded in MARPRO and reviewed over and over again. After the AUV crew launched the vehicle they went for lunch. Why worry when there's nothing you can do? All subsequent surveys with 'MARTIN' have been carried out autonomously.

After the successful diamond survey in 1999 more demonstrations were made to prove the high navigation accuracy provided by MARPOS®. On a survey for the Naval Materiel Command (NMC), Denmark, the vehicle demonstrated the applicability of an AUV for military operations. The demonstration included a mine reconnaissance survey in Storebælt, Denmark. Five dummy mines were fixed at the seabed at some 20 m water depth. The positions of the dummy mines were quickly located from the side scan sonar recorded by 'MARTIN' and, from several repeated passes, the position of each was determined to within 0.3 m. 'MARTIN' repeated the success from Namibia by making a 2 km turn before returning to the mines. On its return, 'MARTIN' had an offset of 0.3 m, which was verified on the side scan sonar (Madsen and Bjerrum, 2000), Figure 12.6.

12.5 ARCHAEOLOGICAL SURVEYS AND SEARCH OPERATIONS

The first survey contract was for a marine archaeological search for the Danish National Museum, Centre for Maritime Archaeology (NMU). The job was carried

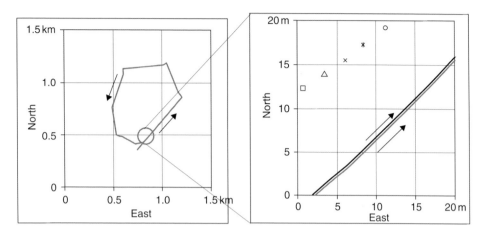

Figure 12.6   Repeated passes of targets shown in UTM coordinates. After making a 2 km turn, 'MARTIN' returned to the targets with only 0.3 m offset.

out from February to June 2000 in different locations off the coast of Denmark and Sweden.

Underwater reconnaissance and site mapping for marine archaeology requires major data gathering and refined evaluation and analysis. Facing a growing political awareness of the underwater cultural heritage, along with a number of scheduled or notified large exploitation projects in the sea surrounding Denmark, NMU initiated the project. NMU has the responsibility for most of the underwater rescue-archaeological work carried out in Denmark, including long-term, scheduled research excavations that are especially suited for methodological development work. The aim of the project was to demonstrate the feasibility of AUV technology in:

- reconnaissance for wrecks, underwater constructions and submerged settlement or refuse areas;
- stratigraphic mapping of archaeological underwater sites;
- savings and improving the data quality.

Maridan fulfilled the goals and delivered high quality data using side scan sonar and sub-bottom profiler payloads. As a bonus, a 12-m oak cod from the twelfth century was spotted below 1 m of mud. The recorded side scan sonar images (Figure 12.7) clearly indicated the contour of the well-protected cod that was originally discovered in 1940 but had no precise record or indication of its location.

For this type of survey, the AUV and its 3-man crew could cover an average of 30 km per day. With the new Maridan M600 AUV, the daily production was increased by larger battery tubes and the capability of hot-swap between two sets of batteries.

A search operation was carried out for the Danish Police in October 2000 in Arresø, a shallow water lake in Denmark. The purpose of the search was to find a helicopter that has crashed in the lake. 'MARTIN' equipped with a side scan sonar was used to locate the target. After few hours mobilisation the search started in the Southern

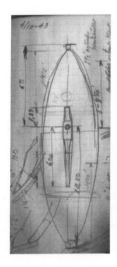

Figure 12.7   Left: Reconstruction of the Kolding Cog from the twelfth century discovered in 1943. Right: The Cog was mapped by 'MARTIN' during a marine archaeological site survey (Kolding Fjord, Denmark, 2000). The instrument used was a Tritech 675 kHz side scan sonar.

part of the lake. After a seven-day search the helicopter was found in the centre of the 40 km^2 lake. The Arresø survey was the first search operation carried out with an AUV. Search had never been possible in the shallow water lake with low visibility prior to using an AUV, the client stated after the survey.

In 1999, a new type of AUV for 600 m water depth was developed based on the same flat-shape design as 'MARTIN'. The new model was called Maridan M600 and 'MARTIN' designed for 150 m water depth was accordingly renamed Maridan M150. The first Maridan M600 vehicle was launched in 2000 and instrumented with acoustic instruments for seabed mapping: side scan sonar, sub-bottom profiler and multibeam echosounder (MBE). The second Maridan M600 was launched in 2001 equipped with the same type of instruments. Both vehicles are in commercial operation providing high-resolution 3D terrain models and sediment profiles.

12.6   NAVIGATION IN COMPLEX ENVIRONMENTS

Autonomous navigation in unknown environments has long been the most difficult challenge in AUV research. Maridan is actively participating in research into advanced navigation tasks including deep-sea navigation, under-ice surveys, docking, landing and terrain navigation.

With MARPOS® new standards were set for Doppler-inertial navigation accuracy in shallow water. Synthetic Long Baseline® (SLBL) is a new concept that extends such high-accuracy navigation technology to deep-sea applications (Larsen, 2000). Using a single acoustic transponder as a fixed reference point on the seabed MARPOS® can make ongoing range measurements while the AUV moves along a straight path.

Thereby simultaneous range measurements from multiple transponders are substituted by a time series of range measurements from only a single transponder, that is, *synthetic baselines* are provided. The SLBL concept provides adequate accuracy and redundancy to eliminate the need for a USBL equipped survey ship or a full LBL transponder network, even for long duration deep-sea surveys. Experimental results with MARPOS® have demonstrated that SLBL will provide sub-meter positioning accuracy for 1×1 km site surveys at $3,000+$ m water depth using a single transponder as reference.

The concept of docking AUVs that can be supplied with power and communication from Subsea Production Systems opens up new deep-sea applications of AUV technology, in particular within the offshore oil and gas industry. The idea is that a 'Work Class AUV' can be launched from a surface vessel (e.g. FPSO or supply vessel), swim autonomously to a subsea installation, dock to the installation for connection to the subsea system's umbilical, receive power and control from the surface during intervention, and finally un-dock and free swim back to the surface. A homing and docking system based on underwater acoustic telemetry (USBL) is currently being sea tested with 'MARTIN' (as discussed in Chapter 6).

An AUV for pipeline inspections is being developed under a European Union funded project called AUTOTRACKER based on the Maridan M600 AUV which will be equipped with tracking systems and an obstacle avoidance system.

Autonomous landing on the bottom for sediment sampling is another important task that will require inspection of the landing area for any obstacles. Image analysis of data from multibeam sonars seems to provide promising results. To perform advanced autonomous navigation tasks, such as navigation in complex environments, and landing on the seabed etc. it is important to extract information from the acoustic sensors to model the environment. Image analysis and fast model updating will be an important challenge for development of the next generation AUVs.

Acknowledgements

I want to thank my colleagues for their valuable contributions: Mogens Bech, Leif Wagner Smitt, Mads Kjær and Christian Aage for design and hydrodynamic test of the 'MARIUS' prototype, Per Egeskov, Anders Ishøy Rasmussen, Henrik Østergaard Madsen and Mikael B. Larsen for their pioneering contributions to AUV guidance and navigation systems, Per Fogt Nielsen and Bo Krogh for their operational support.

13. CARGO CARRYING AUVs

JAMES S. FERGUSON

*International Submarine Engineering Ltd, 1734 Broadway Street, Port Coquitlam, BC,
Canada V3C 3M8*

13.1 INTRODUCTION

One of the future missions for AUVs will be to deliver payloads or cargoes to places
that manned ships or submarines cannot operate cost-effectively or safely. Some potential cargoes include sonar arrays, underwater cables, scientific instrumentation and
smaller AUVs. Areas where this requirement could arise in the near term include the
continental shelf and margin, ice covered regions and sites where accurate positioning
of the cargo on the seabed is required.

The potential to use an AUV for cable laying was explored and successfully demonstrated by the Canadian Defence Research Establishment and the US Office of Naval
Research under the *Spinnaker* programme between 1991 and 1996. As part of this programme an AUV (Figure 13.1) was developed by ISE Ltd, for a cable laying mission in
the Arctic in the spring of 1996 (Ferguson and Pope, 1995; Thorleifson *et al.*, 1997).
During this mission, two armoured fibre optic cables were laid under the ice over
distances up to 175 km. These were connected to a mission device at the terminus.

Figure 13.1 The *Theseus* cable laying AUV.

In this chapter, the rationale for using an AUV to lay cable and the accompanying design parameters, are discussed. The features of the *Theseus* AUV developed for this mission are described and an overview of the cable laying mission is presented.

13.2   CABLE LAYING RATIONALE

The AUV is a good platform for laying cable, particularly in deep water. It can operate close to the seabed and it can accurately follow the bottom contour. This results in two important benefits. First, it is far more likely that the AUV will lay the cable continuously on the seabed. Cables laid from the surface run the risk of being 'clothes-lined' from peak to peak on the sea-bottom and when this happens, the cable does not usually last for long.

Second, in laying cable close to the seabed, the vehicle is not affected by sea state. This permits the use of cables that are much lighter and cheaper than those cables laid by surface ships through reduction in snap loading and because of the shorter catenary.

Finally, the AUV is capable of laying cable in areas where traditional cable laying platforms become impractical. This was one of the driving factors in the selection of the AUV for cable laying work under multi-year Arctic sea ice as part of the *Spinnaker* programme.

13.3   ENVIRONMENTAL AND DESIGN CONSIDERATIONS

Both the environment and the complexity of the *Spinnaker* mission imposed con-straints on the vehicle design. In the operating area, north of Ellesmere Island, Canada, the ocean is completely ice covered, mostly by multi-year ice 3.5–10 m thick. As the launch site was some distance from the airfield at Alert (see Figure 13.3) and no roads existed, it was necessary to transport the AUV to the site by helicopter or by small aircraft. To facilitate air transport to the launch site, a modular design was required with each completed section weighing less than 1,400 kg.

Currents varied from $0.25\,\text{ms}^{-1}$ near the launch site to $0.1\,\text{ms}^{-1}$ at the array site. This factor, along with the ice cover, dictated reliance on a relatively accurate inertial navigation system, to ensure that the vehicle could reach its terminus and second, to ensure that cable was not wasted. As a means of recovering the cable at the terminus, a 200 m wide cable recovery loop was suspended from two holes in the ice. The concept was that the vehicle would 'home in' on a beacon in the middle of the recovery loop and after it had passed through the loop, the cable could be recovered to the surface. In adopting the concept of an inertial navigation system with terminal homing, it was determined that an overall navigation accuracy of 1% of distance travelled would be needed to bring the vehicle through a 200 m wide cable recovery loop at the cable delivery site.

Cable can be laid passively or with an active payout mechanism. The *Theseus* team chose the passive option in the interest of improving reliability by keeping things simple. In opting for this approach, and to preclude the possibility of the cable 'self-dispensing' under its own weight, the AUV was required to follow the bottom at an altitude of 20–50 m. This decision was also prompted by the fact that ice keels can extend to depths of 30 m within 10 km from the launch site, and to 50 m further out.

As cable is dispensed, the vehicle will become lighter. In the case of *Theseus* with a 220 km cable, the potential loss in displacement over the mission was 660 kg, and thus a requirement existed for a buoyancy compensation system capable of correcting for this as cable was laid. Again, because of the environment and the possibility of losing communication with the vehicle, a passive approach to buoyancy compensation was selected.

At the site, the air temperature varies from −40 to −20 °C; and water temperatures vary from −1 °C in the shallow waters near the launch site to 4 °C near the bottom at a depth of 600 m. Near the bottom, pockets of fresh water are also found. To avoid a drastic change in temperature as the vehicle was launched it was kept in a shelter with a temperature around 10 °C. A greater concern was the possibility of fresh water near the bottom, and to preclude the possibility of freezing in the compensation or ballast tanks, anti-freeze was used.

In the 1996 demonstration, an optical fibre cable was the cargo. This raised the possibility of controlling the AUV during the cable laying portion of the transit, and considering the alternative of recovering the vehicle at the terminus, the possibility of developing the vehicle more along ROV lines was investigated. The deciding factor, however, was that the cable laying system had to be capable of returning to its launch point in the event of a cable break, and to do this, it would have to have a full AUV capability. Accordingly, it was decided to operate the vehicle as an AUV on both the outward cable laying and return transit portions of the mission, but to monitor the vehicle systems through the optical fibre on the outbound leg.

Cable laying operations, as well as those conducted under ice, have an impact on the mission plan and on the failsafe strategy. As examples only, the primary obstacle avoidance strategy was to climb over (or dive under) obstructions rather than to turn around them. In this way, less cable would be wasted. Because of the ice cover and the fact that it would be easier to locate the vehicle on the bottom than floating against the underside of the ice, the vehicle was programmed to 'park' on the bottom in the event of a major failure, contrary to what would usually be done in open water operations.

Concern over the possibility of collision with ice keels or bottom objects led to the development of a sophisticated Obstacle Avoidance System (OAS). In retrospect, this was a mistake and future missions will likely be conducted with a simple OAS, similar to that used in the ISE *ARCS* AUV.

13.4   VEHICLE DEVELOPMENT

The cable laying vehicle was named *Theseus*, after the mythical Greek hero who slayed the Minotaur and escaped through the labyrinth of Knossos by following a ball of thread given to him by Ariadne and laid on the outward leg of his trip. The analogy is quite apt. During one of the cable laying missions, it was decided to temporarily park the vehicle on the bottom near the end of the transit. Subsequently, it was determined that *Theseus* parked directly on top of the cable laid on the outward leg.

Vehicle development was carried out in several phases. Between 1989 and 1993, the existing *ARCS* AUV was used as a half-scale model to predict the hydrodynamic performance of the *Theseus* and to evaluate various equipment that would be used in the *Theseus* AUV. The availability of *ARCS* was particularly helpful in determining the final navigation strategy, demonstrating the feasibility of cable laying and selecting (through trial) the best cable for the lay. The vehicle was also used to test software developments relevant to the *Theseus*.

In January 1993, design and development of the *Theseus* began in earnest. The vehicle was completed in early 1994 with local trials commencing in the summer. A mini-deployment to the Arctic was conducted in the spring of 1995 and final developments were undertaken in the following year prior to the 1996 deployment.

13.5   VEHICLE DESCRIPTION

A cross-section drawing of the *Theseus* AUV is shown in Figure 13.2. Note the exit tube that keeps the cable clear of the propeller. The principal characteristics of the vehicle are listed in Table 13.1.

13.5.1   Cable Dispensing and Compensating system

The cable was stored on a series of spools that were stacked longitudinally along the vehicle axis. The ends of each spool were spliced to adjacent spool ends prior to launch. The cable wound off the spools from the inside out, and exited through a tube in the stern. A deployment peel tension of 1 or 2 kg was maintained through the use of a special adhesive, which was applied to the cable jacket during the spool winding process. To keep the system simple and reliable, no active tensioning or dispensing devices were used.

As the cable left the vehicle, weight was lost. To prevent this from affecting vehicle trim, the loss in cable weight was counteracted by an automatic buoyancy compensation system. Surrounding each cable spool was a toroidal hard ballast tank, which was filled with water as the cable was dispensed from its companion spool. This kept the net buoyancy of each spool/tank assembly near neutral. Metallic tabs at the end of each cable spool signalled the vehicle control computer as each pack was emptied.

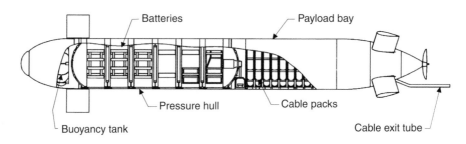

Figure 13.2   Cross-sectional view of *Theseus*.

Table 13.1   Principle characteristics and features of *Theseus*.

Length	10.7 m (35 feet)
Diameter	127 cm (50 inches)
Displacement	8,600 kg (19,000 lbs)
Speed	2 ms^{-1} (4 knots)
Range	900 km (380 nm)
Maximum depth	1,000 m (3,280 feet)
Cable capacity	220 km
Navigational accuracy	Specified as 0.5% of distance travelled
Propulsion	6 hp brushless DC motor and gearbox/single 61 cm diameter propeller
Power	360 kWh Yardney silver–zinc battery consisting of 280 individual cells
Variable ballast	±95 kg (250 lbs) in each of 2 toroidal tanks, located fore and aft
Controller	Proprietary real-time kernel running on MC68030 microprocessors.
Transit navigation	Honeywell H-726 MAP inertial navigation unit
System	EDO 3050 Doppler sonar (bottom tracking)
Terminal homing	Datasonics AUC-206 acoustic homing system with ORE 6701 transponders (8–14 kHz)
Acoustic telemetry	Datasonics Model ATM-851 with MFSK – 15–20 kHz
Surface tracking	ORE 4336B Transponder mounted on *Theseus'* top side
	ORE LXT ultra short baseline tracking system operating at 19/22 kHz
Emergency beacon	ORE 6702 acoustic transponder located in the tail section. Interrogated with ORE LXT Ultra-short baseline acoustic tracking system operating at 11 kHz
Obstacle avoidance	Sonatech STA-013-1 forward-looking sonar, 5 by 4 beams – 200/230 kHz
Pressure hull	2-inch thick Aluminium (7075), 4.5 m by 1.27 m in 5 sections plus end domes
Payload bay	Free-flooding fibreglass shell with syntactic foam lining, top half removable. Inner diameter 114 cm, length 228 cm. Payload up to 1,960 kg dry, 320 kg in water
Current payload	11 packs of 20 km cable, each weighing 60 kg in water, 11 torodial compensation tanks fill as cable is payed out. Tank inner diameter 76 cm
Transportability	Modular construction in sections under 1,400 kg each

13.6   TESTING AND TRIALS

The first Arctic trials of *Theseus* were carried out in April 1995, in shallow ice-covered waters near Ellesmere Island. The objectives of the trials were to verify launch and recovery procedures, test all vehicle systems in an under-ice environment and refine techniques for delivering the fibre optic cable and bringing it up onto the ice.

Four under-ice dives were carried out. The longest individual dive covered a total distance of 5 km, and the cumulative distance for all dives was 13 km. The acoustic telemetry performed rather poorly initially, but an investigation revealed that the propulsion motor controller was a strong source of interference. When the inter-ference was reduced, the telemetry range increased to 3 km, which is considered reasonable for shallow water.

The cable deployment system worked well, just as it had done during earlier trials and the on-ice cable recovery system was successfully tested. A 9 km length of fibre optic cable was left on the ocean bottom until April 1996 to provide some information on long-term survivability.

Following this first Arctic trial, the vehicle controller software was completed between June 1995 and January 1996. The Obstacle Avoidance Control System algo-rithms were not fully tested during this time; hence, it was decided to carry out the subsequent Arctic missions without an avoidance capability. The silver–zinc battery banks were commissioned and tested in November 1995.

The next extensive trials were carried out at the Canadian Forces Maritime Experi-mental Test Range in Georgia Strait during January 1996. Here, a full mission-length trial (360 km) was carried out in order to verify the endurance of the vehicle, to test the projected power budget and to test the navigational accuracy. The Tracking Range is 20 km long, and the endurance trial consisted of eight round trips plus transit to and from the range from the launch point at Winchelsea Island. *Theseus* ran a total distance of 360 km in a time of 51 h without stopping. An extra 18 km on previous trials brought the total distance that was travelled on the battery to 378 km, which used a total of 145 kWh. Since approximately 360 kWh was available from one charge, the energy safety margin was more than adequate.

Theseus' navigation was checked by comparing its true position, as determined by the range, with its own calculated position. On the first leg the vehicle's cross-track error was 0.45% of the distance travelled (a bearing error of 0.26°) and its along-track error was about 0.5%. This was much better than had been expected. At the start of the next leg, a heading error correction was sent to the vehicle over the acoustic telemetry link. This correction had the effect of rotating (in software) the heading of the Doppler sonar relative to the INU. After this correction was applied, the navigational accuracy was exceptional; the average cross-track error was of the order 0.05% of the distance travelled (bearing error of 0.03°).

The vehicle's ability to home to a beacon and do an automatic position update were tested on the final five round trips. The homing system worked very well out to a range of 5 km, the maximum range attempted during the trial, and the position update procedure worked flawlessly.

13.7 MISSION DEPLOYMENT TO THE ARCTIC

On 17 April 1996, a fibre optic cable was laid from Jolliffe Bay, just west of CFS Alert, to Ice Camp Knossos, a one way distance of 175 km (as shown in Figure 13.3). A more detailed map of locations in the Alert area is shown in Figure 13.4. At that time of year, the Arctic Ocean was completely ice-covered. *Theseus* was deployed through ice that was 1.7 m thick, and the fibre optic cable was delivered to an ice camp where the ice was 2.7 m thick (Ferguson, 1998). The launch and recovery procedures and the technique of catching the fibre optic cable at the ice camp are sufficiently new and different that they are discussed separately.

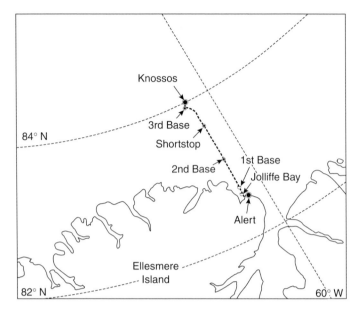

Figure 13.3 Map showing the cable route and homing bases.

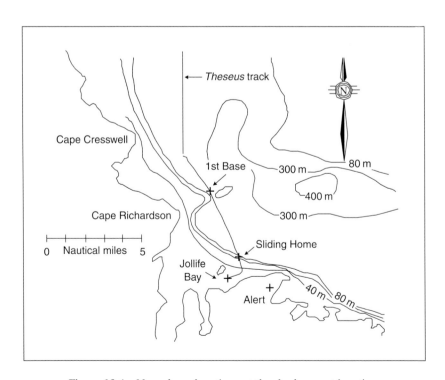

Figure 13.4 Near-shore locations at the deployment location.

Theseus was launched through a large (2 × 13 m) ice hole made by cutting 0.9 × 1.5 × 1.7 m thick ice blocks with a hot-water slot cutter. The ice hole was located at Jolliffe Bay (Figure 13.4) in a large (11 × 20 m) heated tent that housed the vehicle and served as a maintenance workshop. Two travelling gantries with 5,450 kg rated hoists were erected inside the tent to handle *Theseus* (8,600 kg) and to hoist the ice blocks (2,000 kg) out of the water. When *Theseus* returned after a mission it parked itself either on the bottom or up under the ice within tethered vehicle range of the large hole. A Phantom DHD2+2 remotely operated vehicle (ROV) was used to attach a recovery line to *Theseus*, which was then pulled back to the launch/recovery hole.

When *Theseus* arrived at Knossos it delivered the fibre optic cable by flying through a loop suspended from the ice surface. This loop was in the shape of an equilateral triangle, 200 m on the side, and it consisted basically of two ropes and a saddle-shaped weight. After the vehicle passed through the triangle, the cable sank slowly into the saddle at the bottom of the triangle. (The saddle kept the fibre optic cable from kinking over the small-diameter rope.) The saddle was then pulled to the surface, and the cable was recovered. However, before the cable could be cut and spliced to the array cable, it was necessary to allow for possible ice motion during the splicing period, which was estimated to be 2 h. To compensate for cable that might be pulled away by ice motion, one extra kilometre of slack was pulled up onto the ice.

The Knossos mission consisted of navigating from Jolliffe Bay to the array site via 35 waypoints. Acoustic beacons were located at six locations as shown in Figures 13.3 and 13.4: Sliding Home, First Base, Second Base, Shortstop, Third Base and Knossos. First Base and Second Base were manned in order to make acoustic telemetry contact in case the vehicle encountered problems. Figure 13.5 shows the depth profile along this track. The solid line is the bottom profile, and the dotted line is *Theseus'* depth.

The mission began at 00:22 PST on 17th April. *Theseus* passed First Base at 02:20 and Second Base at 11:12, with successful homing and acoustic telemetry contact at both. Shortstop was passed at 19:00 without successful homing but continued on to Third Base as programmed. (Later investigation revealed that the Shortstop beacon was not functioning.) At 01:18, 18th April, *Theseus* arrived at Third Base, homed to the transponder there and continued on to Catcher, located about 1.6 km away. Homing was good at first, but deteriorated as *Theseus* approached the catchment loop. As a result, the decision was made to fly *Theseus* through the loop under shore-based pilot control through the cable telemetry system. A surface-based tracking system at Knossos provided position information via voice radio to the pilot, who was located in the *Theseus* Control Room at Alert. Under human pilotage, *Theseus* successfully flew through the loop and parked under the ice some 600 m away. The cable settled down into the saddle at the bottom of the catchment loop and was recovered.

Theseus was given a final position update through the cable, and, after the ballast was adjusted, the cable was cut and *Theseus* was sent home. *Theseus* returned to Jolliffe via homing beacons at Shortstop (a new beacon was not in place), Second Base, First Base and Sliding Home. At First Base, the homing step failed to complete, possibly due to poor acoustic conditions. In this situation, the Fault Manager had been programmed to have *Theseus* stop and park under the ice to await further instructions. Acoustic telemetry and surface tracking were established, and the vehicle's health was checked. The vehicle's ballast was adjusted, and it was sent on its way. At 11:40 (19th April)

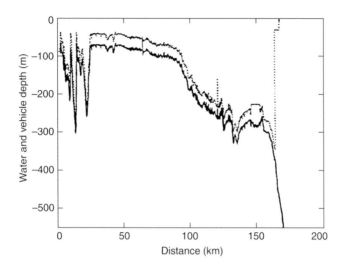

Figure 13.5 Depth profile along cable route: solid line is bottom depth; dashed line is *Theseus'* depth.

Figure 13.6 *Theseus* under the sea ice. *See* Color Plate 16.

Theseus came to a stop under the ice at the launch hole, Figure 13.6. Lines were attached, and it was recovered safely. The energy used on this mission was 149 kWh, less than half of the estimated battery capacity.

A second mission to lay a backup cable was carried out approximately one week later.

13.8   PRESENT STATUS

Theseus completed an impressive cable laying mission under the permanent polar ice cap. The vehicle successfully deployed 175 km of fibre optic cable and returned to the launch site for an overall mission length of 350 km. This constituted the longest cable laying and under ice mission undertaken by an AUV. Even at the time of writing, five years later, it continues to be the longest mission completed by an AUV.

Following this successful demonstration, the *Theseus* AUV was returned to Vancouver and put in a long-term storage condition. The Canadian Defence Research Establishment plans to use the vehicle for future cable laying activities.

13.9   FUTURE MISSIONS

The *Spinnaker* cable laying mission proved the viability and the advantages of laying cable with an AUV. With this, the feasibility of laying larger diameter or longer length cables becomes simply a matter of engineering.

Future cargo missions will involve not only the deployment of simple cables but also the rapid deployment of underwater surveillance and scientific instrumentation packages. The use of AUVs to tow underwater sensor arrays is also a possibility that only requires a small winch in addition to the equipment demonstrated in the *Theseus* AUV. With the continuing development of precise navigation technology that is not reliant on GPS, as well as techniques for homing and autonomous docking of AUVs, the ability of the AUV to host a number of smaller AUVs on data gathering missions will become feasible.

14. SEMI-SUBMERSIBLE AUVs

JAMES S. FERGUSON

*International Submarine Engineering Ltd, 1734 Broadway Street, Port Coquitlam, BC,
Canada V3C 3M8*

14.1 INTRODUCTION

The development of small, unmanned, semi-submersibles started at International
Submarine Engineering (ISE) in 1980. Since then, ISE has built 12 vehicles for military
and civilian applications. In 1996, Lockheed Martin commenced development of
a minehunting semi-submersible. These two activities comprise the main body of
experience in unmanned semi-submersible vehicles.

Small, research semi-submersibles have also been developed and tested in coastal
waters. For example, the *SASS* – from an UK consortium led by Seaspeed Engineering is a 5 m long semi-submersible design with a planned range of 600–1,000 km at
6–7 m s^{-1} with a 200 kg payload space. A one-third scale prototype completed its trials
in 1999 (Young and Phillips, 2000).

This chapter describes the operating principles of the semi-submersible and their
advantages over surface drones and over fully submerged autonomous underwater
vehicles. The factors affecting the basic semi-submersible design, and those that bear
on the use of the vehicle as a platform for towing minehunting sensors, are discussed,
and the chronological development of the ISE semi-submersible is described.

14.1.1 Development History

Development of the semi-submersible commenced in 1981 when the Canadian Hydrographic Service envisaged a requirement for the capability in surveying offshore waters
in Eastern Canada. Between 1981 and 1986, a total of four vehicles were built by ISE
and integrated with hydrographic sensors (Thomas, 1986). These vehicles were 7.6 m
in length and displaced approximately 3.4 ton. In the following few years, offshore
testing was carried out and a launch and recovery system was developed and the
Simrad EM1000 multibeam echo sounder was fitted. The capability was transferred to
the private sector in 1990. An elimination of offshore survey budgets in 1993 curtailed
further activity and development in this area.

Two vehicles, originally known as Sealions, were developed by ISE for the US Naval
Research Laboratory (NRL) in 1985 and successfully tested with above-water warfare
sensors, Ferguson and Bane (1995). Following these tests, they were used for the development of AUV concepts and in 1992, were overhauled and fitted with oceanographic
sensors including the Simrad EM950 multibeam echo sounder (Kalcic and Kaminsky,

Figure 14.1   The ISE semi-submersible on a US Navy operation – 1995.

1993; Scott, 1995). The vehicles, now renamed ORCAs (Figure 14.1), remained in active service with the Stennis Division of the NRL until late 1999, Bourgeois and Harris (1995).

Two further vehicles were delivered to the US Navy by ISE in 1988. These were fitted with a basic minehunting sensor suite and operated by the Coastal Systems Station at Panama City, Florida. In 1993, they were overhauled and fitted with the AN/AQS 14 minehunting sonar as well as a Reson 8101 Seabat forward looking sonar. Reclassified as Remote Minehunting System Version 2 (RMS-V2), the vehicles were still in use in late 1999.

In 1996, Lockheed Martin won a contract to develop an advanced version of the US RMS system. The first prototype (RMS-V3) has been on trial since mid 2000. At the same time, the Canadian Defence Research Establishment commenced development of a prototype minehunting semi-submersible vehicle. Initially, the hydrographic design was 'stretched' to allow more room for fuel and a larger engine. This prototype was used to test a number of concepts such as the contra-rotating propulsor, a deep towing winch, and navigation equipment appropriate to the minehunting application. In 1998, the lessons learned from this design were incorporated into a larger diameter hull with a displacement of 5,900 kg, similar to that of the Lockheed Martin prototype.

The Canadian vehicle, known as Dorado (Table 14.1), has operated towfish to depths of 200 m and is being used as the basis for further system development by the Canadian Defence Research Establishment. In 1999, it was mobilised to France and demonstrated to the French Naval establishment with successful results. A partnership

Table 14.1 Basic features of the Dorado semi-submersible AUV.

Hull diameter	1.17 m
Length	7.5 m
Diameter	1.15 m
Weight	5,900 kg
Speed	$9\,m\,s^{-1}$ – no tow; $6\,m\,s^{-1}$ – 200 m tow
Endurance	14 h @ 180 m depth with tow, without tow – 28 h
Propulsion	Caterpillar 3,116–310 kW, 6 cylinder, after-cooled, turbocharged, marine diesel engine
Command and control	Manual, autopilot, or autonomous DGPS based line following between geographic waypoints. 400–465 MHz 9,600 baud full duplex communication radio with five asynchronous RS 232 ports
Payload	ISER Aurora Towfish, Klein 5500 side scan sonar, 2.4 GHz Dataradio. Other sonar options available

between ISE, MacDonald Dettwiler Associates (MDA) of Canada and DCN International of France, has been concluded to develop and sell the vehicle as part of a RMS.

14.2 PRINCIPLE OF OPERATION

To date, the semi-submersible has been designed as a mobile platform for carrying hydrographic, above-water warfare, oceanographic, and mine-countermeasure sensors (Figure 14.2). With the exception of the above-water warfare application, these sensors have been sonars.

The vehicle is designed to operate like a snorkelling submarine and consequently, is limited to operations near the sea surface. Motive power is supplied by a marine diesel engine that draws air through two induction valves located at the top and bottom of the semi-submersible mast. The engine exhausts through two spring-loaded exhaust valves and out of an exhaust pipe located in the stabiliser assembly on the tailcone of the vehicle that exhausts below sea surface.

The engine is coupled to the propeller shaft through a reduction gearbox. For low power applications, a single propeller is fitted. For higher power applications, both a contra-rotating gearbox and a double propeller set are provided.

In addition to providing motive power, the engine drives an alternator to provide electric power to the vehicle control systems and sensors through a battery, as well as driving two hydraulic pumps that power the hydroplanes. For applications where towing is involved, a hydraulically powered winch is provided. This winch is located in the keel of the vehicle.

Fuel for the engine is stored in flexible fuel bags located under the dry electronics compartment forward of the mast. As fuel is used, the bags compress under the external water pressure, minimising the change of weight in the vehicle.

When the vehicle is launched, air in both the forward and after ballast tanks keeps it on the surface. Vent valves on the top of these tanks are opened by remote command allowing the tanks to flood. When the tanks are fully flooded, the vehicle retains a very small amount of positive buoyancy and is almost submerged. To 'resurface' the vehicle, air is blown into the tanks from bottles that are located in the after ballast tank.

Figure 14.2   A semi-submersible underway at $9\,\mathrm{m\,s^{-1}}$.

The semi-submersible is provided with both passive and active stabilisation. Passive stabilisation is provided by lead weights and by the weight of the winch located in the keel. The effect of this weight is to place the vertical centre of gravity below the vertical centre of buoyancy, and to provide longitudinal and horizontal righting moments which stabilise the vehicle in roll and pitch. Williams *et al.* (2000) have described extensive tank tests with a model of the semi-submersible to explore its stability, especially during turns (Figure 14.3).

Active stabilisation is provided by the hydroplanes and the rudder. The two forward hydroplanes act independently to control roll and in unison to control pitch. Independent operation of the afterplanes also contributes to roll control, and their combined operation controls the pitch of the vehicle. The rudder stabilises the yaw or heading of the vehicle. Corrections to the hydroplane settings are determined by the vehicle control computer, which compares the setpoint in any particular axis with the current position and velocity obtained from an inertial navigation unit (INU). The loop frequency of the control computer, or the rate at which corrections are sent to the hydroplanes, needs to be on the order of 50 Hz.

The control computer also varies the engine speed in response to commands from the surface console and monitors conditions on the semi-submersible. In the event of adverse or abnormal conditions, the computer is programmed to take either corrective or failsafe action.

The operation of the vehicle is supervised from a surface control console over a command and control radio link. In ISE's semi-submersibles, this radio link operates in the UHF band, at a rate of 9,600 baud. This full duplex (bi-directional) link continually

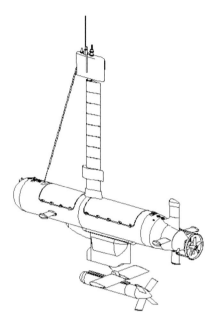

Figure 14.3   A 7.4 m semi-submersible with a towfish slung beneath.

transmits setpoint commands and other variables from the console to the vehicle and both operating and navigation data from the vehicle to the surface console. The radio operates over line of sight distances, which means the range is primarily determined by the height of the transmitting and receiving antennae. One-way radio links that operate at higher bandwidths are provided as necessary for sensor data.

Positioning of the vehicle is provided by a Differential GPS (DGPS) – Inertial Navigation System. This vehicle position updates the vehicle positioning computer to make corrections to bring the vehicle back to its pre-programmed track.

While the vehicle generally operates in an automated mode (autoheading and pre-programmed track), its operation can be immediately over-ridden by moving the rudder joystick or the throttle at the surface console.

14.3   ADVANTAGES OVER OTHER PLATFORMS

The semi-submersible is an alternative to surface drones and in some applications to the fully submerged AUV. Generally speaking, its advantages over the fully submersible AUV are in its ability to transmit large volumes of sensor data over high bandwidth radio links, its ability to use D-GPS continuously for highly accurate navigation, and finally, its inherent higher speed and range. While the semi-submersible is able to tow at reasonably high speeds, these speeds cannot be maintained at deeper depths due to drag on the towcable. Thus, for seabed surveys the semi-submersible is a good choice in coastal waters. In deeper water, the fully submerged AUV is a better choice.

Essentially, the semi-submersible is used for the same applications as the surface drone. As the control and guidance system of the semi-submersible is considerably more complex than that of a ship, one might ask why is it necessary to take all the trouble of developing a semi-submersible when the sensing equipment could just as easily be fitted to a small ship.

There are five principle advantages of the semi-submersible over an unmanned surface ship or drone. The advantages are:

- improved sensor stability;
- reduced transducer aeration;
- improved vehicle speed and endurance;
- reduced vehicle size and weight; and
- reduced signatures.

14.3.1 Sensor Stability

Surface ships are buoyant vessels that have a large and varying waterplane area (the area determined by the breadth and length of the vehicle at the waterline).

Because of this large waterplane, they are easily perturbed in roll, pitch, yaw and depth by sea state. Because of the large waterplane, the surface ship is also very difficult to stabilise with active hydroplanes. The task of actively stabilising narrow beam sonar transducers and directional radio antennae is even harder, and, therefore, more costly.

In contrast to a small surface ship, the waterplane area of the semi-submersible is several orders of magnitude smaller. It is described by the small, almost unvarying cross-sectional area of the mast, which is not perturbed by sea state to any significant degree. The result of this is a vehicle hull that is easily and precisely stabilised with active control surfaces and provides a stable platform for sonars, other sensors and radio communications.

14.3.2 Transducer Aeration

Small surface ships tend to have a relatively shallow draft and because of this, the risk of transducer aeration occurs in even relatively low sea states. Transducer aeration will render a sonar survey useless and it was for this reason that the Canadian Hydrographic Service abandoned the use of surface drones in favour of semi-submersibles in 1978.

14.3.3 Speed and Endurance

A vessel moving through water is subject to various forces that tend to slow it down. The two dominant forces are skin-friction resistance and wave-making resistance. In a body of revolution, such as a semi-submersible, wave-making resistance is negligible when the body is submerged three diameters below the sea surface and consequently the overall resistance or drag on the semi-submersible will be much less than that for the surface ship. Drag is proportional to the square of velocity, and the power needed to overcome the drag is proportional to the velocity cubed. Accordingly, the

Table 14.2 Comparison of surface drone
and semi-submersible hulls.

Vehicle	Surface Drone	Dolphin
Length (m)	18.3	8.3
Weight (ton)	29	6.4
Top speed ($m s^{-1}$)	6	9

powering requirement for a semi-submersible will be considerably smaller than that
for a surface ship.

The benefit of the semi-submersible is further enhanced by the fact that it operates in a nearly neutrally buoyant condition and does not require large amounts of
reserve buoyancy as does the surface ship. The inclusion of this additional buoyancy
in the surface ship design tends to increase the wetted surface area of the ship, again
increasing the overall drag and powering requirement.

The result is that the semi-submersible requires less motive power at any given speed
and thus can operate at higher speeds or for longer periods of time than the surface
ship, given the same amount of fuel.

14.3.4 Vehicle Weight and Size

The inevitable finding is that for any particular set of operating parameters, the surface
drone will be substantially larger and heavier than the semi-submersible. A surface
drone built with operating parameters almost identical to the ISE semi-submersible,
compares as shown in the Table 14.2.

This smaller size and weight provides the semi-submersible with the very distinct
advantage that it can be transported in civilian and military aircraft. This permits its
rapid mobilisation to support collective security or expeditionary operations abroad.
The semi-submersible is also more readily transportable by road or by rail than is the
surface drone and it is possible to load it onto ships which are too small to handle the
larger surface drone.

14.4 VEHICLE DESIGN CONSIDERATIONS

Two factors, which strongly impact the design of the semi-submersible vehicle, are its
size (weight and displacement) and the amount of power (engine and fuel) that is
to be installed. These directly affect the hull configuration, its design, the vehicle's
stability and buoyancy and finally, the requirements of the control system.

The hull configuration and design is also driven by the requirements for sensor
payload, field maintenance, and by the number of vehicles which are expected to be
required.

14.4.1 Hull Configuration

The hull must house the engineering plant, fuel, control system electronics, buoyancy
tanks and finally, the sensor payload. The volume and weight of these components will
determine the minimum length, diameter and displacement of the semi-submersible.

The diameter of the vehicle is generally driven by the diameter of the off-the-shelf diesel engine because engines tend to come in distinct sizes for various power levels. The length and displacement of the vehicle are then determined by the volume and weight of the remaining components.

Generally speaking, underwater bodies with a length to diameter (L/D) ratio of between 6 and 8 are optimised for drag and are easiest to dynamically stabilise. Longer L/D ratios are difficult to stabilise in pitch and shorter ratios are inherently unstable in yaw. While configuring the vehicle to fall within these design parameters is not a hard and fast requirement (torpedoes have an L/D $\cong$ 12 because of the need to maximise the amount of explosive carried and fuel in a fixed diameter), designers try to situate the design within these parameters to allow some room for growth through lengthening the vehicle.

14.4.2 Hull Design

Two hull design approaches can be selected – the monocoque hull or the modular 'AUV style' hull.

The modular AUV style hull consists of a series of cast and machined cylinders with a hemispherical end cap at either end. Streamlined nose and tail cones, generally made from GRP, are added at the bow and stern. The monocoque hull, on the other hand, is a single piece, all welded structure.

The principal advantage of the modular AUV hull is that its individual cylinder pieces are interchangeable between vehicles. This feature will reduce the assembly costs in production runs and will ensure that damaged hull sections can be replaced with nominal effort. The disadvantages to the approach are the high cost of design and the fact that unlike the AUV, disassembly of the cylindrical sections for maintenance requires removal of the exhaust piping, propeller shafting, the keel and possibly the engine. For this reason, access to the vehicle for routine maintenance is through small, O-ring sealed ports. This access strategy demands a complete and thorough understanding of the maintenance requirements and is not generally practical in prototype systems or a configuration that is likely to change.

The main advantage of the monocoque hull is that its design allows the use of large access hatches greatly simplifying the tasks of routine and unscheduled maintenance, as well as the interchange of equipment. Weighing against this and its comparatively low design cost, are the facts that sections of the hull cannot easily be replaced and the assembly costs are higher.

For prototype work, in which the design or the configuration of the vehicle is likely to change, a monocoque hull represents the most practical approach. When the design has been fixed, and the maintenance and access requirements are fully understood, the cost and benefits of each approach should be re-examined to select the one, which is most practical for the production run at hand.

14.4.3 Static and Dynamic Stability

The static stability of the semi-submersible is dependant on the hydrostatic characteristics of the vehicle, particularly the relationship between the centre of gravity (CG) and

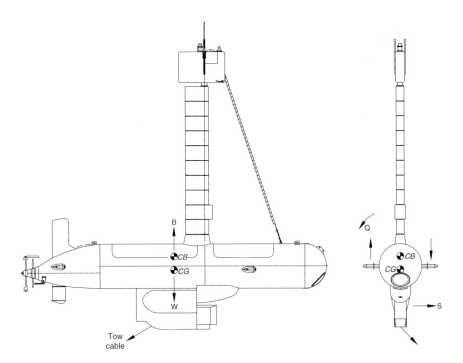

Figure 14.4 Submerged vehicle stability.

the centre of buoyancy (*CB*). This combines with the hydrodynamic forces developed by the hydroplanes to offset the perturbing effect of sea state.

The hydrostatic characteristics of a submerged vehicle are different from that of a ship (Figure 14.4). In a ship, *G* is always above *B* and at small angles of heel, a metacentre (*M*) above *G* is described by the intersection of lines of force acting through *B*. Righting moments are determined by the product of $gW\,GM\sin\theta$ where *W* is the displacement of the ship, *M* the metacentric height or distance between the centre of gravity and the metacentre and θ is the angle of heel.

In a submerged vehicle, there is no waterplane and therefore no waterplane inertia.

Because of this, the metacentre is co-incident with *B*. Furthermore, because the underwater form of the vehicle does not change as it heels, the position of *B* is fixed. In order to provide static righting moments, the *CG* must be located below the *CB*. Hence, the righting moment (*GZ*) is described as:

$$GZ = g\,W\,BG\sin\theta,$$

where θ is the heeling angle, *W* is now the buoyancy of the vehicle and *g* the acceleration due to gravity.

If the *BG* (distance between *CB* and *CG*) is maintained on the order of 15–20 cm or more, sufficient righting moments exist for control of the vehicle at low speeds.

A major consideration in fitting the vehicle with payload equipment (antennae, cameras, etc.) is to keep the weight that is added at the masthead to a minimum.

Because of its distance from the *CG*, even small additions of weight here can quickly reduce the *BG* to unacceptable levels.

At higher speeds, the vehicle also requires active control from the hydroplanes to maintain attitude, to correct for wave disturbances and to offset propeller torque (Q). While the effect of wave disturbances on the vehicle is generally constant, for any particular aspect (beam sea, head sea etc.) propeller torque creates a rolling moment about the vehicle centreline that is proportional to rpm or engine speed. Moreover, when turning, the flow about the hull becomes asymmetric and the resulting imbalance of side-forces causes a rolling moment (S). The size of this moment is proportional to the rudder angle and the resulting turning circle of the vehicle. The fully dynamic characteristics of the design have been explored by Hopkin *et al.* (1999).

In very early tests of ISE semi-submersibles, an induction mast was tested with various fixed fairings rather than the freely rotating fairings used in the existing design. Although more efficient, it acted like a wing generating a large amount of lift that overturned the vehicle in a turn.

The four hydroplanes and the rudder all develop lift in proportion to the speed of the vehicle. The two foreplanes use this lift to maintain depth while the aft planes use the lift to maintain pitch and roll angles. All four hydroplanes can be used to maintain the zero roll setpoint. Hydroplane corrections are determined by the vehicle control computer, which compares the setpoint in any particular axis with the current position and velocity obtained from an INU and the vehicle depth sensor. To avoid backlash, the loop frequency of the control computer, or the rate at which corrections are sent to the hydroplanes, needs to be on the order of 50 Hz.

A single rudder is sufficient to maintain vehicle heading on a straight course. Depending on the rudder angle the turning diameter of the vehicle will be between 70 and 115 m at all speeds with a 70 m diameter occurring with a rudder angle of 25°. At higher angles of attack, the rudder tends to stall. These turning characteristics also apply to towing at shallow depths. When towing at depths over 50 m, the turning diameter tends to be 2.5–3 times the towing depth.

Operations of the vehicle with this control system have been carried out in a variety of sea conditions up to Sea State 6.

14.4.4 Buoyancy

The semi-submersible vehicle is ballasted such that at all times its weight is less than its displacement. The reason for this is that if the vehicle stops, it will float on the surface. A minimum differential of 1.5–2.0% of the vehicle weight is adequate. The adjustment to achieve this trim is generally achieved by adding or removing lead weight from the keel of the vehicle. Lead weight can also be placed in the nose and stern sections of the vehicle, to adjust fore and aft trim.

Because diesel fuel is more buoyant than salt water, provision for this must be made in the operation of the vehicle. A vehicle fully fuelled with 500 kg of diesel fuel will be trimmed such that it is 175 kg positively buoyant at the start of operations. As the fuel is consumed, this positive buoyancy reduces to the point where the vehicle is 100 kg buoyant in the empty fuel condition.

The semi-submersible requires additional buoyancy for operations on the surface and in the event of flooding. In the basic vehicle, buoyancy is provided by two ballast tanks at the bow and stern of the vehicle. A vent and blow valve is provided on top of these tanks and free flooding holes are found in the bottom. To blow the tank, the vent valve is shut and the blow valve actuated from the surface console. The reverse is followed to flood the tanks on diving. In normal dived operations, the tanks are fully flooded with the vent valves shut.

During operation of the vehicle, a small amount of water or moisture is ingested down the air intake mast as a result of its proximity to the sea surface. Over time, this will build up in the bilge and if no action is taken, the weight of water taken on could affect the buoyancy of the vehicle. For this reason, a bilge pump is fitted. This pump is directly connected to the vehicle's battery and is operated by a float switch.

14.4.5 Power

In the basic semi-submersible, there are two requirements for power. These are the power needed to operate the auxiliary machinery such as hydraulics, the vehicle control system and radios, and to operate the payload or sensor equipment in the vehicle. Leaving the payload equipment aside, the auxiliary power requirement is approximately 7.5 kW in the current Dolphin configuration.

The second requirement is motive or propulsive power. Because the effects of Froude or wave-making drag are negligible at operational depths, the drag (F) can be expressed by the following formula:

$$F = \tfrac{1}{2}\rho A v^2 C_{\mathrm{d}}.$$

Here ρ is the mass density of salt water, A is the area of the body (usually wetted surface area), v is velocity and C_{d} is the co-efficient of drag of the body.

The equation for the propulsion power required (P) in kilowatts is as follows:

$$P = F\,v/3600.$$

As can be seen from these relationships, power varies as the cube of speed and doubling the speed will increase the power requirement by a factor of 8. Applying these formulae to the underwater vehicle will provide a measure of the power that must be delivered by the propeller to drive the vehicle at the desired speed. To determine the real or effective power needed to drive the vehicle at any given speed, the efficiency of the propeller as well as torsion, friction and gearing losses must be accounted for. By far the largest of these losses is in the propeller. For the basic vehicle, the propeller efficiency is likely to be on the order of 65–70%. For more demanding applications such as towing, propeller efficiency can drop to 50%.

Because of this and the cubic relation between speed and power, the power requirements at higher speeds become dramatic. For instance, the increase in power requirement for a vehicle from 15–16 knots is over 21%. The consolation is that a reduction in power usually results in a much smaller reduction of speed than would generally be expected.

14.4.6 Control and Navigation

The vehicle is controlled by a VME based computer. The vehicle control computer (VCC) performs a number of tasks that include:

- guidance and attitude control;
- positioning and navigation;
- operation of the engineering plant;
- operation of payload instrumentation;
- monitoring of critical functions or actions and fail-safe.

Commands from the surface control computer (SCC) are transmitted over a 9,600 baud data link operating in the commercial UHF band. These commands are received by the VCC and output to the various actuators within the vehicle. These include the engine on–off control, the engine speed control and transmission, and the vent and blow valves. In the case of the minehunting vehicle, winch commands are also transmitted over the link. Attitude setpoints (roll, pitch, depth, and heading) are also transmitted. The VCC also sends critical engineering data up the link for display at the console. This includes engine, hydraulic and air system data, hydroplane positions, vehicle attitude and cable and towbody data. Finally, the command and control radio link can be used to operate controls on payload instrumentation – such as the gain control on a sonar. Generally speaking, the operation of the vehicle requires less than a quarter of the 9,600 baud bandwidth. The remainder can be made available to support data transmission. This has worked well with low bandwidth equipment such as multibeam echo sounders (Simrad EM 950/1000) but is incapable of meeting the requirements for multibeam side scan sonars.

Attitude control is accomplished using an inertial navigation system. The inertial measurement unit determines vehicle position, velocity and acceleration in the pitch, roll, and yaw axes (position in the heave axis is determined by a strain gauge depth transducer) and passes this information to the VCC. The VCC compares this data with the setpoints in each axes and output corrections to the hydroplanes.

Positioning is accomplished by the inertial navigator, which is closely coupled to a D-GPS receiver. The position of the vehicle is computed and passed to the VCC where it is compared with the desired position. A measurement of the off-track distance is made and course corrections are output to the VCC. In the event that GPS coverage is lost temporarily, the inertial navigation capability of the INU will enable the vehicle to navigate accurately for several minutes.

Fail-safe activation is provided for the following conditions:

- overdepth;
- flooding;
- loss of control radio communications;
- excessive distance off-track;
- engine operating parameters exceeded.

The general strategy in the event that fail-safe action is required is to place the vehicle on the surface, stopped and with the air ballast tanks blown.

14.5   TOWING

The dynamics of towing place heavy demands on the near neutrally buoyant semi-submersible. These demands can be described under the following categories:

- cable tension and increased power requirements;
- downforce and buoyancy control;
- torque and roll control;
- pitch control.

14.5.1   Cable Tension

Towcable tension is caused by the hydrodynamic forces acting on the cable as it is pulled through the water. Cross-flow, which lifts the cable, is the dominating hydrodynamic force. Length-wise flow along the cable also contributes to tension; however, this amounts to less than 5% of the total axial tension. Because the cable is totally submerged, strumming is not a major factor and is not considered here.

The cable hydrodynamic force is primarily determined by the cable diameter and by the coefficients of drag (C_d) and lift (C_L) on the cable. The cable can be considered as a long cylinder, and as such, will have a fairly high C_d on the order of 1.5, based on cross-sectional area. Again, this high C_d and the cubic relationship of power to speed contribute to very large towing loads at higher speeds. These loads are extra drag, which must be offset with additional power. In the case of the semi-submersible, the power requirement to support a tow cable to 200 m depth at a speed of $5\,\mathrm{m\,s^{-1}}$ is between 2.5 and 4 times than needed to power the vehicle itself.

Because of the catenary of the cable, tension is highest at the towpoint and lowest at the end of the cable – usually the attachment point for a towed body. This towed body determines the actual shape or catenary of the cable as it provides the hydrodynamic force, or the weight, to offset the lifting forces and pull the cable down. Cable scope and the layback are thus determined by the actual downforce applied by the towfish (Figure 14.5).

14.5.2   Downforce

Downforce is required to operate the towbody at its desired depth. It is created by the towbody and to a lesser degree, by the towcable. The towing downforce is largest at the towbody or bottom end of the cable and smallest at the towpoint. However, its effect on a neutrally buoyant structure, such as a semi-submersible, cannot be disregarded as it can with a surface ship. In the case of the semi-submersible, even small downforces of a few hundred kilograms at the vehicle are sufficient to disrupt operations if they are not compensated.

Downforce can be compensated in several ways. The two obvious ways are with upward (lifting) hydrodynamic force from the control surfaces and second, with controllable ballast. Use of the control surfaces is attractive from the point of simplicity,

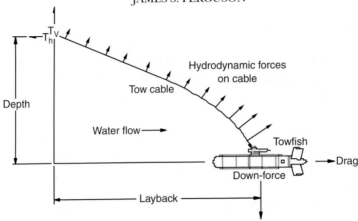

Figure 14.5   Towing dynamics of a towfish from a semi-submersible.

but it may overload their capability and increase drag, necessitating the need for larger surfaces. The use of controllable or variable ballast in the vehicle avoids this problem but adds to the complexity of the system. If variable ballast is added for this reason, the best strategy is to use the hydroplanes for small or fast changes in the towcable downforce, and the variable ballast for longer term or steady state changes. In the current ISE semi-submersible configuration, approximately 910 kg of variable ballast is provided.

14.5.3 Torque

The problem of torque developed by a single propeller has been mentioned earlier. For the Dolphin vehicles, the engine transmits power (P, here in kW) to the propeller in the form of torque (Q) and shaft revolutions (n, here in rpm) according to the following relationship:

$$Q = 9550\, P/n.$$

The resulting torque creates a static roll angle θ (degree) according to the formula:

$$\theta = 180Q/(g\, W\,(BG)\,\pi).$$

In the case of a semi-submersible travelling at $7.5\,\mathrm{m\,s^{-1}}$ with no tow, the propeller-generated torque is on the order of 173 Nm. This creates a roll angle of 14°. This angle is corrected by lift generated from the foreplanes.

When the same vehicle is required to tow a body at $5\,\mathrm{m\,s^{-1}}$ at a 200 m depth, the torque will be 397 N assuming the same propeller rpm. This creates a rolling moment of 32°. Again, this could be corrected by the hydroplanes; however, the vehicle has slowed from 15 to 10 knots. Hydrodynamic lift (L) is calculated by the formula:

$$L = \tfrac{1}{2}\rho A v^2 C_{\mathrm{L}},$$

where C_{L} is the coefficient of lift.

Because the speed of the control surfaces has slowed from 7.5–5 m s^{-1}, the maximum amount of lift that can be developed by the planes is reduced by 56%. Put otherwise, the lifting force available for roll control is less than half of what it was, however, the power and torque have more than doubled. Using the planes to correct this condition is possible but not practical, as they would very often be in an overburdened state and unable to meet other requirements.

There are several other approaches to mitigating the problem. These include a twin shaft vehicle with opposing propellers, and the use of compound propulsors such as contra-rotating, pre- and post-swirl propeller sets. The twin shaft vehicle approach was studied thoroughly in the mid-1980s and rejected because of cost, additional weight and complexity.

In the development process, a variety of compound propulsors were considered. Pre- and post-swirl propulsors are double propeller sets mounted on a single shaft. A static propeller is mounted ahead (pre-swirl) or aft (post-swirl) of the second moving propeller. This static set of propeller vanes creates offsetting torque, which partially or fully offsets the torque being produced by the moving propeller. Pre- and post-swirl propulsors can generally achieve torque balance at one particular shaft speed, but not across the regime of operating speeds. They are best suited to applications where the power and speed of the vessel is usually constant (such as in towing booms of logs).

Contra-rotating propulsors have two propellers moving in opposite directions on a concentric shaftline. They are more complex to design and build, and problems have been encountered with sealing the shafts. They do, however, provide torque balance over the entire operating range of the engine.

Contra-rotating propellers and gearbox sized for power levels up to 298 kW was developed and successfully tested in 1996. The arrangement provided torque balance to within ±11.2 N over the entire engine operating range. In contrast, the Lockheed Martin semi-submersible uses a pre-swirl propulsor to mitigate torque.

14.5.4 Pitch Control

If the tow point is vertically displaced from the centre of gravity, a pitching moment will be generated. Because the after planes that control pitch are less burdened, they can generally offset this pitching moment. The moment can also be offset with the variable ballast system if desired.

14.5.5 Design Considerations

Taken together, these constraints force a careful review of the operating requirements in the design stage. One of the most critical aspects is to determine what layback and cable scope is acceptable. From an operational point of view, layback directly affects the positional accuracy of the target detected from the towbody sonar and it will often determine how close to the bottom the towfish can be operated. This latter factor will in turn have a direct bearing on coverage or search rate.

From a vehicle design point of view, towing configurations with a long layback (1.3 : 1 cable to depth ratio in Figure 14.6) are characterised by moderately low cable tension and downforce. This results in lower power requirements and possibly eliminates the

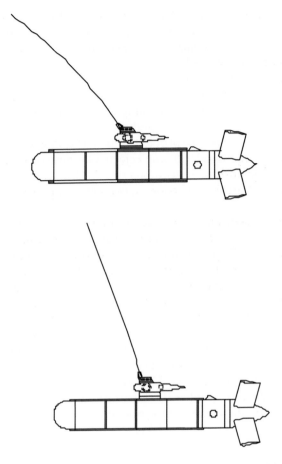

Figure 14.6 Towfish layback options: (upper) long layback (3.0 : 1 cable length to depth); (lower) short layback (1.3 : 1 cable length to depth).

need for variable buoyancy. The cable is longer, however, and weighs more. It also creates the need for a substantially larger cable drum in the vehicle.

Towing configurations with a short layback (Figure 14.6) offer much shorter cables and therefore, lighter and more compact handling systems. On the other hand, they do require much more power and ability to handle downforce.

The current semi-submersible design supports the power and buoyancy requirements for laybacks as small as 1.75 : 1 and there is sufficient length of cable for laybacks of up to 3 : 1. The system has successfully been tested in Sea State 5 and has towed a variety of sonar bodies to depths down to 200 m at 5 m s^{-1}.

14.6   TARGET POSITIONING

The position of the sonar target depends on the accuracy or resolution of the sonar that is used to detect the target and on the accuracy to which it is possible to locate the towed sonar body in geographical co-ordinates.

As the accuracy of the sonar is outside the parameters of this discussion, only towfish location will be considered. This breaks down into two components; first, the geographical position of the Dolphin vehicle and second, the position of the towed body relative to the Dolphin vehicle.

With the advent of GPS and a differential capability in the mid-1990s, obtaining an accurate position of the semi-submersible vehicle in most parts of the world is straightforward. In coastal waters, an accuracy of between 1 and 3 m should be expected.

The challenge lies in obtaining an accurate position of the towfish relative to the semi-submersible. Two approaches have been considered. These are the use of ultra short baseline acoustic positioning equipment (USBL) either in the normal or the inverted mode, and the calculation of the towfish position with cable modelling programmes.

USBL positioning systems have been around for some time and their use is commonplace with ROVs in the offshore oil industry. They provide range and bearing to a transponder, normally placed on the vehicle which is being positioned. In the inverted configuration, the transponder is placed on the towing vehicle, which the towbody interrogates. This equipment was developed for use with ROVs, which operate at speeds of $0.5–1 \, \text{m s}^{-1}$. Trials in the 1990s with the system at speeds in the range which Dolphin generally operates showed a higher than acceptable standard deviation and a proliferation of outlier points. However, improvements are being made to enable USBL operation at higher speeds and the possibility of using this technology in the near future cannot be overlooked.

The second approach involves modelling the position of the towfish with simulation programmes. The ability to predict the longitudinal or along track position of the towfish is well developed and easy to implement. Predicting the horizontal or cross track position involves measuring the cable tow-off angle. Several ways of doing this have been developed and a number are easily implemented on small vehicles such as a semi-submersible.

15. AUV SENSORS FOR MARINE RESEARCH

LAWRENCE C. LANGEBRAKE

*University of South Florida, 140 Seventh Avenue South, St Petersburg,
FL 33701-5016, USA*

15.1 INTRODUCTION

It has taken decades to develop the nearly one hundred practical Autonomous
Underwater Vehicles (AUVs) in use today (Wernli, 1999). The need for economical, free-swimming, and versatile marine research platforms has motivated AUV
development. Designs, which vary greatly, include small creature-mimicking robots,
automatically undulating probes and large submarine-like vessels capable of travelling
hundreds of kilometres.

AUV developers envision marine researchers plying their craft through a personal
computer: just a few simple 'point and click' commands send an AUV to the depths,
returning hours, days, weeks or even months later with its internal memory full of valuable data. This vision is rapidly becoming reality. However, an AUV requires a broad
suite of compatible sensors to be truly useful to the researcher. Without sensors, an
AUV is not much more than a water-mixing device.

AUVs have limited stored energy and internal volume, and most need to maintain
nearly neutral buoyancy. A plethora of commercially available marine sensors can be
adapted to AUVs but significant efforts have also focused on developing sensors to be
used specifically on these vehicles.

AUV sensors can be classified into two groups:

- sensors needed to provide situational information to an AUV's navigation and
  control system;
- sensors used for underwater research or experimentation.

There are several sensors that are common to both groups for example an acoustic
Doppler current profiler (ADCP) can be used to study water column currents as well
as provide positional information to an AUV's navigation system. This chapter focuses
primarily on sensors used for scientific research only. Sensors used to study biological, chemical, geological, and physical processes are covered. The state of current
research, design approaches, packaging, performance, and related topics are discussed. Miniaturization is also examined for its benefits in AUV sensor design. New
approaches to autonomous underwater systems that focus on small, low power or biomimetic AUVs (Ayers, 2000) are becoming more prevalent. Microelectromechanical
systems (MEMS) technology is taking its place in marine sensor development and
is providing opportunities to increase sensor capability while reducing energy and

volume requirements. A section on MEMS technology as it applies to underwater sensor systems is included at the end of the chapter.

15.2 GENERAL CONSIDERATIONS IN THE DESIGN OF AN AUV SENSOR

Designing AUV sensors requires attention to a host of interrelated design variables and their constraints. The functional requirement is the most important variable, but several general design issues specific to AUVs should be considered. They include:

- form factor and occupied volume;
- sampling system;
- energy requirement;
- data storage, control, and communication method.

Each factor has varying importance depending on the AUV deployment scenario. It is most important to consider the variables together rather than independently. For example, AUVs are energy-limited. Designing for maximum energy efficiency can impact the sampling capability of the instrument and thus affect the research being attempted.

Building sensors that are compatible with more than one type of AUV can add complications because of the constraints imposed by each of the intended AUVs. For example, if a sensor is needed for a payload space-constrained AUV as well as an AUV with highly limited payload energy, it must be designed with appropriate power, size, and form factor to be compatible with both vehicles.

15.2.1 Form Factor and Occupied Volume

Most AUVs do not have spare, watertight spaces available to host sensors, so the designer must house the sensor in an appropriate package. This package is typically a pressure vessel. In extreme cases, the designer may completely encapsulate the sensor in a polymer. AUVs of many shapes and sizes have been deployed in marine environments. It is impossible to design a sensor package that is compatible with every AUV. Nevertheless following a few basic guidelines will ensure a reasonable design.

The generally accepted rule is to design AUV sensor packages as small and as close to neutrally buoyant as possible. Most AUVs are designed with low drag, energy-conserving hulls, thus interior spaces similar to the AUV's shape are typically available. Packages with an overall form factor that maintain reasonable symmetry in two axes and allows the third axis dimension to extend (or contract) as necessary work well in AUVs, that is, cylindrical or hemispherical designs. In cases where the sensor cannot be designed as neutrally buoyant, adequate space within the AUV should be left so that weight (usually not needed) or buoyancy (e.g. syntactic foam) can be added. Syntactic foam occupies considerable volume for small returns in buoyancy, for example, packing the free spaces between the sensor's housing and inner AUV hull can be difficult (see Chapter 5). The designer should consider using buoyant foam if early calculations show that the instrument will be heavier than water. As long as there is enough space, it should not matter what the actual design is; however, the designer must choose the appropriate material for the vessel.

To reduce potential corrosion effects, some AUVs have a common power supply connection in electrical contact with surrounding water. This means the AUV sensor package must either have complete electrical isolation, or the only electrical connection to the housing should be the same (common) power supply connection as the AUV.

One solution to the problem is to use insulating materials such as PVC, polycarbonate, polyurethane, or other suitable insulating material for the instrument housing. Materials such as nylon are poor choices because they deteriorate over time in seawater.

To reduce the material cost or to reduce stress on sensitive components, some designers choose a pressure-compensated (also known as pressure-balanced) design. In this case, the designer should be especially cognizant of buoyancy. Most pressure compensation systems involve a bladder or flexible membrane to allow for trapped air compression. If a sealed vessel of this type has trapped air of even 500 ml the mass can increase approximately 0.5 kg when the sensor is at depth. For medium to large sized AUVs (2 m or longer in length) this is not really an issue, however, it can become a problem for small AUVs. Feature-rich internal structures such as stacked electronic circuit boards or complex mechanical systems can trap air in a distributed fashion. Electronic components such as electrolytic capacitors, crystal 'cans', windowed integrated circuits, etc., can also represent a potential crush risk when using pressure compensation. As an option, pre-evacuation of a pressure compensated package can help to eliminate trapped air before filling with compensation fluid. The designer should consider the deployment scenario and intended AUVs of use when choosing pressure compensation packaging. Other factors such as bio-fouling, corrosion in long-term deployments and maximum operating depth should also be considered in the design.

It is not always possible, or practical, to house an instrument within the space constraints of an AUV. In these instances it is convenient to take advantage of two features of most AUVs: flooded hull design and control system stability. The flooded hull design allows for penetrations or hull modifications without adverse impacts on other internal AUV systems. Since critical systems within most AUVs are housed within individual pressure vessels, hull penetrations are possible without the need to consider water integrity. AUV control systems are typically robust and can accommodate perturbations while maintaining controlled speed, heading, manoeuvrability, etc. Different AUVs have different stability bands within which they can successfully operate. The sensor designer must consider the limitations of the AUV's control system if major hull modifications are anticipated. In most cases minor changes in hull hydrodynamics can be tolerated without significantly impacting performance, Figure 15.1.

15.2.2 Sampling System

Just as AUV control systems can handle hull penetrations for sensor housings, penetrations for sampling systems can also be accommodated. There are several factors to

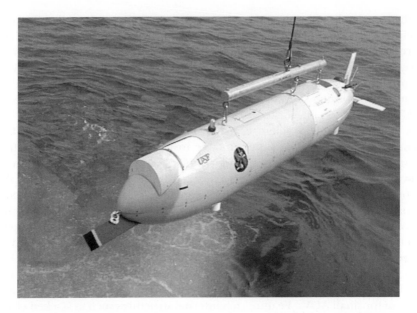

Figure 15.1 Hull penetration can often be made in the AUV without negatively impacting navigational performance; USF's SIPPER onboard and Ocean Explorer AUV.

consider when designing a sampling system for an AUV sensor. In general, sampling systems should:

- sample forward of the vehicle;
- be protected against impact;
- be designed so as to not effect the measured properties of the sampled water, that is, use inert materials, avoid turbulent mixing or undesired heating and cooling effects, etc.;
- use minimal energy; and
- not *significantly* alter the hydrodynamics of the host AUV.

Most non-acoustic sensors require access to water that has not been disturbed by the AUV. The analysis may consist of reactions between discrete samples and reagents; induced optical fluorescence or measurement of optical absorption; vaporization and ionization; exposure to intense electric or magnetic fields; or exposure to radiation. In each case there may be different requirements for sample size, sample rate, discrete vs. continuous sampling, pre-treatment or filtering, etc. The designer must consider the sensor's sample input requirements and the sampling strategy that is dictated by the research being attempted.

Sampling strategy is another significant aspect in the overall design of an AUV sensor. Sampling an aquatic environment – whether it is an ocean, lake, or river – requires careful evaluation of the parameter to be measured. Is the intent to sample a periodic event, a random event, or an ambient condition? What are the spatial and temporal scales of the targeted parameter? What is the sensor's required dynamic range, and is the AUV adequate in terms of deployment duration capability, depth limit, and manoeuvrability? Is the measured parameter a contaminant? Will it be a relative or absolute measurement? How precise and accurate do the measurements need to be? The designer must consider each factor and ensure the sampling system design is adequate for the research being attempted. Although a sensor might be capable of measuring the desired parameter with appropriate resolution, rate, dynamic range, etc., the sampling system must be designed to have minimal or no impact on the sensor's measurement capability. Since the sampling system and sensor are arranged in a logical AND configuration, one can consider that the sensor is going to measure the desired parameter and whatever relevant impact the sampling system has on the sample. If the sensor makes a measurement of a parameter that is affected by physical factors (e.g. temperature, light, chemical reduction, salinity, etc.) then the sampling system should be designed to not affect those parameters in any way, that is, negatively or positively. Examine closely the impact of integration of the sampling system into vehicle. Other instruments, control systems, and cooling or heating systems within the vehicle can adversely affect the sampling system. A good rule of thumb is for the design to be as simple as possible.

15.2.3 Energy Requirement

Energy consumption is another important factor in AUV sensor design. AUVs have a pre-determined energy budget. Typically, one-third to one-half of the available stored energy is used for propulsion, about one-fourth is used for navigation and control (hotel load) and the remainder is available for the payload. Extended deployment becomes possible if the payload energy requirement is less than that allocated by the AUV designers. Likewise, if the payload requires more energy than is allocated, the deployment duration will be reduced.

Most AUVs use some form of electrochemical energy storage. Energy is usually available as DC electricity via fuel cells, batteries, or some other system. To improve electrical efficiency, designers typically use higher voltages to reduce resistive losses in wiring and connectors (higher voltages allow use of smaller wiring and smaller connectors). Sensors to be connected to an AUV's main electrical supply must be able to directly use the available voltage. DC to DC converters provide a convenient and fairly efficient (as high as 90% in some cases) means by which to create the necessary voltages required by the sensor's internal systems. Most DC to DC converters accommodate a wide range of input voltages and they provide electrical isolation. Isolation becomes significant if the AUV uses a power system that is in electrical contact with the surrounding water.

15.2.4 Electrical Considerations

Fuses are typically used to protect system wiring. Although they are helpful in preventing excessive damage if an instrument experiences an internal electrical failure, fuses should generally not be used within the sensor's housing. Fuses fitted within the sensor housing can require significant effort to replace. Furthermore, opening a pressure vessel in the field carries the risk of reduced water integrity when reassembled; it also exposes sensitive components to potentially corrosive agents. Not having ready access to a fuse can also create a conundrum regarding sensor health, for example, is the sensor not functioning because of a software bug, loose connection or tripped fuse? Fuses should only be used in sensor subsystems that require significant electrical energy. Fusing only the high-energy portion of a sensor can allow uninterrupted communication with the sensor's controller or data outputs, thus increasing the likelihood of determining the cause of a failure without opening the sensor's housing.

Following are some guidelines for designing the electrical portion of an AUV sensor:

- Make the sensor's power supply connector pin male. DC voltages in excess of 40 V can be lethal. A pin male connector on the sensor will ensure that the potentially energized mating female connector will not have exposed 'hot' conductors.
- Keep current-carrying conductors that are outside the sensor's housing, (conductors connecting multiple housings) as short as possible. Conductors that carry 1 A or more and that are close to the AUV's magnetic compass can affect compass operation. Constant or switched high currents can also affect other sensors used for AUV navigation and control as well.
- Keep conductors carrying small signals shielded and short in length. The AUV's computer, actuators, navigation sensors, or propulsor motor(s) can all produce noise.
- Consider using an independent power supply if space allows. More energy will be available for AUV propulsion and hotel load allowing mission duration to be increased.
- Use fuses to protect system wiring, not instruments. Design the sensor's electronic systems to be fault tolerant and protect the high-energy portion of the sensor with a resettable link or thermal breaker. Also consider using breakers that automatically reset after a pre-determined time period or after a fault is cleared.

Although several battery systems are available with higher energy density (lithium ion, nickel metal hydride, silver zinc, etc.), gel cell lead acid batteries can be a reasonable choice for AUV sensors. They are inexpensive, rechargeable, and reliable. They are also not made of ferromagnetic materials that can affect an AUV's navigational systems.

15.2.5 Data Storage, Control, and Communication

The sensor designer should plan a sound approach for data storage, sensor–AUV communication, and internal control. Common computer network methodologies are often applicable in AUVs. Networked subsystems on an AUV reduce wiring and connector requirements, permit easy expansion, improve reliability, and decrease software programming. Consider a non-networked central control computer and the associated 'star' interconnect topology, Figure 15.2(a). This system is complex, requires several potentially expensive connections, and is not a reliable configuration because of the large number of housing feedthroughs, connectors, and cables. However, a looped or terminated network greatly simplifies internal wiring as shown in Figure 15.2(b).

(a)

(b)

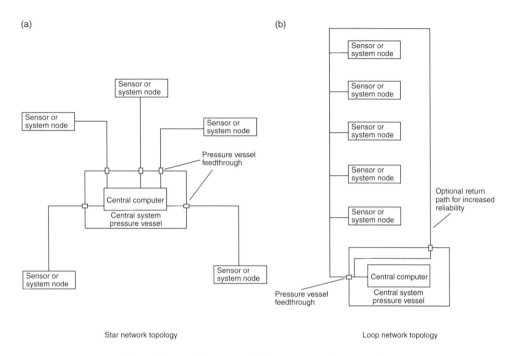

Star network topology

Loop network topology

Figure 15.2 (a) Star vs. (b) loop connection topology.

Several AUVs use an internal network for control and communication. A system such as Ethernet, EIA standard RS-485 or RS-422 bussed serial communication, or one of several commercially available distributed control systems, is typically used. In each case subsystems within the AUV are simply addressable nodes on a network, which allows a sensor package to be easily integrated into an AUV. The sensor is assigned a network address and is then simply connected electrically, optically or through another appropriate electronic communication medium. In the networked configuration, software can be written such that the sensor can access AUV data and/or transmit data to the AUV's main computer.

Consider an AUV that is hosting a sensor that needs conductivity information to calculate a derived parameter such as pH. (Most AUVs are fitted with the most basic of oceanographic instruments, the conductivity, temperature, depth probe or CTD.) A networked CTD continuously communicates information to the AUV's internal network. Any network-connected sensor can inquire the network for the CTD's data and then use it for internal calculations. In fact, any data that is transmitted onto the network by any AUV subsystem can be shared with any other subsystem. This approach has several advantages:

- data acquired by several sensors and AUV subsystems can be time-synchronized, thus easing post deployment data analysis;
- sensors can 'listen' to network variables transmitted by AUV navigation and control instruments, thus allowing intelligent turn on/turn off;

- the AUV's main computer can monitor the sensor output and modify the mission program according to a knowledge base, neural network or other adaptive sampling system/algorithm; and
- sensor health and/or status can be monitored over the network, thus avoiding a costly and time-consuming 'empty deployment'.

There are two common methods for handling data gathered by an AUV sensor. Data can be transmitted to the main AUV computer for storage, or data are stored locally in the AUV sensor itself (or both). (A less common but successful method involves using an acoustic or optical modem to transmit data to the researcher in real-time.) In either case, a high-speed method for off-loading data is desirable. AUV sensors can acquire significant quantities of data during a deployment. Research requirements often mandate rapid turn-around and redeployment; data off-load must be completed quickly. Some options include:

- high-speed (100 Mbps) Ethernet connection;
- internal interchangeable data storage media (hard disk, ram disk, tape, etc.);
- RF modem link; or
- optical IRDA or fibreoptic link.

The method of choice is certainly dependent on the sensor and research scenario; however, the highest possible data transfer speed is always desired.

A final rather significant aspect to data communication is sensor operational status. If a sensor is not integrated into the AUV's internal network for monitoring by the AUV's main computer, some method should be established for determining the status of the sensor before deployment. A sound, illumination sequence (LEDs, system lasers, etc.), unique pump motor actuation, or user communication port that can communicate sensor status should be included.

15.2.6   Optimizing the Design

There are several commonly used optimization and project management techniques for ensuring an optimum design (Onwubiko, 2000). An adequate design is one that meets functional requirements while keeping undesirable effects within acceptable limits. An optimum design is the alternative among the set of adequate designs that best meets the original objectives. Design optimization generally involves describing the design's multi-goal objective mathematically and then locating minima in the function using non-linear methods. The complexity of the problem is related to the number of variables involved. Solving optimization problems with many variables can be time-consuming and sometimes counterproductive. This is especially true if the design variables have not been carefully defined.

Details of optimization methods are beyond the scope of this chapter, but mention should be made of how to identify the design variables. The identification process can often provide significant insight towards an optimum design. The first step is to consider the design variables, such as volume, electrical current consumption, weight, sample rate, etc. Design variables are generally one of two types: continuous or discrete. Continuous variables are those that are free to assume a range of values, whereas discrete variables have a fixed value. The designer should define the variables for the sensor and group them into the two categories while taking care to keep the variables as independent of each other as possible. For example, defining length and volume as design variables could create conflicts in determining the optimum, but using the independent variables of length and width can lead to a more apparent solution. Care should also be taken to distinguish between variables and parameters. A variable might be the thickness of a pressure vessel end cap whereas a parameter might be the end cap material. Parameters can become variables and vice versa depending on compromises the designer might make. In the example of the pressure vessel and cap the type of material may not be restricted to aluminium, but the thickness might be held to a narrow range.

Next, the designer should consider constraints on the variables. Like variables, constraints are also categorized into two groups, external or internal, and they generally assume a numerical value. External constraints are those that are uncontrollable by the designer. An example might be the inability to use a hydrogen fuel cell as an energy source for the sensor because of a regulation, rule, or safety issue. Internal constraints are those imposed by the designer. These usually result from a fundamental natural law or other restraining factor; for example, the sensor has to operate using less than 50 W so to avoid overloading the host AUV's power supply system.

If the design objective has been considered appropriately, the variables, parameters and constraints together will define the design space. Once the design space is defined, a significant portion of the optimization problem has been completed. Intuitively, choosing a set of feasible design points within the design space can be helpful in gaining insight to an optimum design. A carefully selected set of feasible alternatives enables the designer to choose an empirical approach to the optimization problem. Periodic design review meetings should be held to discuss feasible alternatives and gain valuable insight. The objective opinions of technically experienced individuals who are not close to the design project can be of significant value in arriving at a 'more than adequate' design.

15.3   ADAPTING COMMERCIALLY AVAILABLE SENSORS

The most common AUV sensors are based on acoustics. Used since the 1920s, acoustic systems are perhaps the most mature marine sensor technology. Acoustic systems for bathymetry, sub-bottom profiling, positioning, velocity logging, imaging, current profiling, communication, and precision navigation are commercially available. Most acoustic sensors in AUVs are used for navigation and control. However, the inherent stability and low self noise of an AUV make it an ideal platform for acoustic imaging systems such as side-scan sonar and bottom or sub-bottom profiling. Although acoustics dominates the set of AUV-deployed commercial sensors, other commercially available sensors that use optical, wet chemistry, or electrical measurement methods have also been deployed.

Adapting commercially available sensors for use on AUVs means considering factors similar to those involved in AUV sensor design. The functional specifications are the most important requirement in choosing a sensor for experimental research. However, cost, form factor, energy requirements, etc., are all very important in the decision process.

Some instruments, such as a CTD, are very easy to integrate into an AUV. Almost any commercially available CTD can be readily adapted to the common AUV. Most CTDs are lightweight and have cylindrical form factors. Some CTDs can accommodate a range of voltages and most communicate acquired data through a low speed serial communication port that can be easily integrated into the AUV's internal network. Instruments such as chlorophyll fluorometers can be purchased in many configurations. Some are designed to simply sample the environment in which they are immersed; while others sample a ported flow in which the sample is routed from an inlet on the skin or exterior of the AUV.

There are a large number of commercial sensors available for a broad range of scientific applications. Research requirements will generally narrow the set of applicable sensors. Once a scientifically suitable set of sensors has been identified several additional factors must be considered before integration of the sensor into an AUV is attempted:

- Energy requirement: Choose the sensor that consumes the least amount of energy when it is operating in a normal measurement mode (check the specifications closely to ensure the published energy requirement does not refer to a standby or 'sleep' mode).
- Operational configurability: Choose a sensor that can be configured through simple communication port commands. This is important not only for initial set up but also for adaptive sampling scenarios that involves dynamically changing sample parameters.
- Communications configurablility: The internal AUV network must be made compatible either through a data conversion interface (node or router) or direct connection. Configuration options such as EIA RS-485 and RS-232, serial peripheral interface (SPI), or Ethernet allow the AUV designers and technicians to interface the instrument to the AUV more easily.
- Sampling method: The preferred sampling method for an AUV is ported flow, during which water is channelled to the instrument through piping or tubing. This configuration allows a 'series' arrangement that is optimal when several instruments are used on an AUV simultaneously because it avoids fine-scale spatial aliasing.
- Data storage: In some cases, it may be undesirable or impractical to interface the instrument to the AUV's network, for example, non-standard communication protocol, excessive bandwidth requirement, no available connector, etc. Therefore, the internal data storage capability of the instrument becomes an important factor. The instrument should be able to store data internally. Its data storage capacity should exceed that 'expected to be used' for the deployment duration. AUV instruments configured as 'dumb' payloads are often powered and acquiring data long before the AUV is deployed and running a mission. A good guide is to have storage capability for at least one and a half times the intended deployment duration.

- Operational status verification: This feature is not necessarily something that can be chosen between instrument manufacturers. However, choosing an instrument that can provide operational status information (if inquired) through a communication port, or some indication that is seen or heard can make the difference between an expensive, unproductive deployment and a successful one.

15.4   NOVEL SENSORS FOR AUVs

Significant advances have been made in the development of oceanographic sensors specifically designed for use on AUVs. AUVs are a useful research tool because of their ability to quickly perform large area surveys; their stable noise-free characteristics, and their lack of a restrictive tether or tow cable. Sensors used in studying marine biology, chemistry, small and large-scale physics, optical properties, bottom mapping, and 3-D imaging have been successfully deployed. There follows a brief summary of selected successfully deployed AUV sensors that includes a description of design, operational specifics, and data/results from representative field operations where available.

15.4.1   Biological Sensors

Biological processes in aquatic environments are typically coupled with ambient chemical and physical processes, as well as being affected by local geology. They can involve temporal variability ranging from hours to years and can occur over size scales from a few centimetres to thousands of kilometres. Sensors designed to sample biological parameters such as chlorophyll fluorescence, plankton abundance, and photosythentically active radiation, must measure these desired variables with sufficient frequency (sample period) and dynamic range to allow researchers to characterize them. AUVs provide the distinct advantage of *in situ* measurement capability, with high manoeuvrability and considerable depth range. They can measure a variety of biological parameters in relatively large volumes of water with minimal spatial or temporal aliasing and minimal disturbance to the environment.

15.4.1.1   Dual Light Sheet plankton counter (DLS)

Zooplankton constitute a major constituent of water column particulate matter. Studies on the types and variety of zooplankton and their relative density in the water column are important to the understanding of complex physical, biological, and chemical processes. Zooplankton influence light and sound transmission through the water and hence affect data gathered by sensors based on optics and acoustics. One method used to characterize zooplankton in the water column uses a collimated light sheet and detector (Herman, 1992). This method provides size distribution information derived from the amplitude of light attenuance caused by each plankton passing through the light sheet.

Researchers at the University of South Florida have developed an AUV-compatible instrument that uses the light sheet method (Langebrake *et al.*, 1998) and is appropriately called the DLS. The sensor uses a symmetrical sample intake port and two light sheets with a fixed separation. In theory, the symmetrical sample system lessens the probability that some zooplankton avoid being sampled. The 10 cm by 10 cm opening used on the DLS instrument minimizes the avoidance bias by reducing the opportunity for zooplankton to 'side step' the sample port. Other instruments use non-symmetric, high aspect ratio sample ports that have dimensions well within the avoidance range of mid to larger sized zooplankton.

The optical system and detection electronics for each light sheet are identical. The instrument's second light sheet measures each plankton in the orthogonal axis, so each plankton is measured (sized) twice. This allows size averaging, which reduces the error in measuring significantly non-spherical specimens. A very precise time-stamp is generated as plankton pass through each sheet. Time stamp information is used to determine time-of-flight of the particle through the instrument, which in turn is used to determine flow rate. The advantage of this method is that accurate water volume measurements are made non-intrusively, that is, plankton are not disturbed or destroyed.

The DLS has been deployed on the *Ocean Explorer* AUV in the Gulf of Mexico and in the Western Atlantic Ocean. In the deployed configuration, the DLS sample port is connected to an extender that protrudes approximately 25 cm in front of the payload nose. This sample port configuration ensures that undisturbed water is introduced to the instrument. The DLS generates raw plankton size (equivalent spherical diameter) data and post-processing software creates histograms, water flow rate, and sampled water volume data. Field tests have shown strong correlation between sampled water flow rate and AUV velocity.

The optical method used by the DLS instrument can provide general information about zooplankton distribution, however, there are several drawbacks to be considered. First, the instrument does not differentiate between living zooplankton and non-living marine snow. Second, the instrument produces histogram errors when marine particle and plankton density is great enough to result in more than one particle in the measurement light sheet at once. And third, non-symmetrical particles can be sized inaccurately since the amount of light they attenuate in the measurement light sheet depends on their orientation in the sheet (which is generally random).

The DLS instrument communicates all data via a network connection or through a serial port in real-time. It also stores the captured data internally using flash memory. The instrument is controlled by a Motorola® 68HC16 microcontroller that runs a compiled C program. The controller precision time-stamps each particle data point, handles data synchronization, storage and offload, and handles communication with externally connected devices.

15.4.1.2 Shadowed Image Particle Profiling Evaluation Recorder (SIPPER)

Another automated particle analysis system developed at the University of South Florida for use on AUVs is the Shadowed Image Particle Profiling and Evaluation Recorder or SIPPER (Samson, 1999). The SIPPER instrument uses two high-speed line scan cameras and two collimated laser generated light sheets to image transient particles. Shadow images of particles in the water are captured in two axes simultaneously. Data are post processed to permit two-view characterization and 3-D reconstruction of sampled particles.

The SIPPER instrument consists of two orthogonal optical imaging systems and a data storage system. Both optical systems are mechanically identical and are housed in separate pressure vessels. The SIPPER instrument has a total of five pressure vessels. Two 13 cm diameter cylindrical pressure vessels (imaging systems) and two similar 10 cm diameter vessels (lasers) are fastened together with external clamps and rods. The four vessels are joined together using a specially designed mitred optical interface. The fifth and largest vessel, 20 cm diameter, houses SIPPER's data handling system. The data handling system comprises of: (1) data synchronization electronics; (2) a high-speed digital data storage system; and (3) a UNIX-based processing and image-offload single board computer. The packaged instrument occupies approximately $0.3\,\mathrm{m}^3$ and weighs approximately 14 kg in water.

Each imaging system uses a 3 mW, red (635 nm) diode laser having line-generating optics as the light source. The line-generating optical element spreads the laser beam to a 10 cm width. Then a 28 cm focal length lens psuedo-collimates the beam into a 1 mm by 10 cm sheet. This light sheet passes through a sealed window in the side of the pressure vessel; it then goes through the water sampling tube where particles in the light sheet block the light transmitted from the laser. Non-particle-interrupted light is collected through a window on the opposite side of the sample tube. This light slowly converges until an achromatic imaging lens (46 mm diameter, 146 mm focal length) focuses it onto the active area of a digital line scan camera. Mirrors near the sampling tube windows and inside the camera pressure vessel fold the light path and reduce the physical size of the system.

The back-illumination method described above provides two advantages over diffuse illumination techniques. First, the depth of field of this system is greater than that of conventional imaging schemes. Second, less light is needed to illuminate the sample area. The optics design is a balanced compromise of optical path length, depth of field, and percent coverage of the sampling tube.

Each of the digital line scan cameras (one for each optical path) outputs its data to a thresholding system that differentiates between dark and illuminated pixels. Threshold data are then stored internally on a large computer hard disk (50 Gb). The threshold image data directly represents the shadow of any particle passing through the light sheet, as shown in Figure 15.3. This direct digital methodology significantly reduces the post-processing required as compared to 2-D video or photographic imaging systems. Custom software easily calculates particle volume and particle size distribution. Limited image recognition is also possible through application of commercially available knowledge-based profile recognition software.

Travelling through the water at $1\,\mathrm{m\,s^{-1}}$, the SIPPER instrument has an image resolution of approximately $50\,\mu\mathrm{m}\times100\,\mu\mathrm{m}$ across the full 10 cm sample tube. An AUV-independent 24 V lead acid gel type battery supplies the instrument's 70 W power requirement to operate under normal conditions.

15.4.2 Chemical Sensors

Similar to biological sensors, chemical sensors must have a measurement resolution and dynamic range commensurate with the characteristics of the sampled analyte.

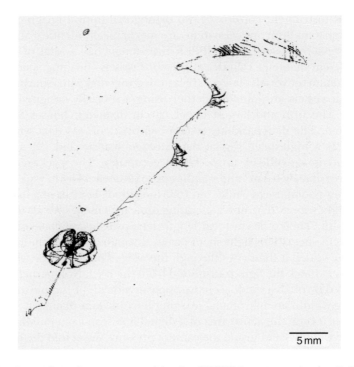

Figure 15.3   Ctenophore image captured by the SIPPER instrument in the Gulf of Mexico, 1999.

Researchers often want to measure constituents found in small concentrations in water. As such, several highly sensitive, high dynamic range sensors have been developed for AUVs to fulfil this need.

15.4.2.1   Ultraviolet spectrophotometer for measurement of nitrate

Studies that determine nitrate concentrations using ultraviolet spectroscopy have been well documented. In general, nitrate and nitrite absorb in the ultraviolet (UV). In ocean waters where there are relatively low concentrations of dissolved organic compounds, nitrate/nitrite are often the only significantly absorbing species. Halides do exist in high concentrations in seawater, but their molar extinction coefficients are a small fraction of those for nitrate and nitrite. Therefore, this method can be used to effectively construct a simple, low-cost, and low-power instrument for measuring nitrogen species concentration. Although a multispectral instrument has been built and deployed on an AUV, the prototype preceding that instrument is presented here for clarity of operating principles. Researchers at the Southampton Oceanography Centre, University of Washington, St Louis and University of Southampton have developed and deployed a lower power AUV nitrate sensor based on absorption spectroscopy (Finch *et al.*, 1998; Clayson, 2000).

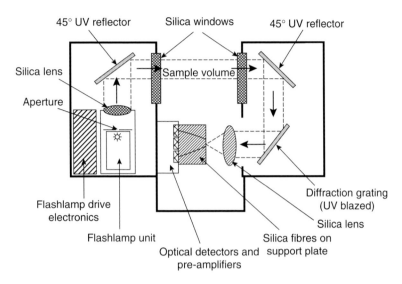

Figure 15.4   Optical arrangement of the UV spectrometer.

The instrument is designed with an exposed sample volume that comprises a 15 cm absorption path, as shown in Figure 15.4. Source light is generated using a xenon flashlamp operating at 16 Hz with a flash energy of approximately 100 mJ per pulse. Non-absorbed light is collected at the far end of the sample volume and directed through a series of wavelength resolving optical elements. The collected light is wavelength selectively reflected by a mirror and grating to reject unwanted source light and resolve three detected wavelengths: 220, 240, and 305 nm. At 220 nm, absorption due to both halides and nitrate is reasonably significant. At 240 nm, the absorption by halides is considerably less, however, nitrate absorption is smaller to a lesser degree. At 305 nm, absorption is not significantly affected by nitrogen species or halides, but is affected by particle absorption, scattering, etc. Nitrate concentration is determined by normalizing the collected data, that is, the 305 nm signal for particle absorption and the signal for halide absorption are removed to yield a relative nitrate concentration. Curve fitting to pre-measured standards data completes the calculation.

Measurements taken in the field correlate well with those taken by the described prototype ($r^2 = 0.99$). A second version of the instrument having multispectral detection capability has been developed and is actively being deployed (Griffiths *et al.*, 2000). As shown in Figure 15.4, the sensor is well suited to AUV deployment due to its open sampling scheme (no porting is required), small physical size and low energy requirement (approximately 4 W). Perhaps the sensor's most significant advantage is its ability to sample at rates as high as 1 Hz, thus providing high temporal and spatial resolution. This is an advantage compared to wet chemistry systems for measuring nutrients. Those systems can have sample periods as long as several minutes. This point is especially important to note as AUVs can dive and ascend at rates of up to a metre per second.

15.4.2.2   Underwater Mass Spectrometer (UMS)

For decades mass spectrometry has been the ultimate tool for analysing elemental and molecular constituents of solutions. The ability to apply this tool in the aquatic environment is very useful in marine research. The University of South Florida has developed an *in situ* mass spectrometer measurement system for deployment on AUVs (Fries *et al.*, 2001; Short *et al.*, 2001). The mass spectrometer is an assembly of commercially available components coupled to a novel membrane sample introduction system.

The heart of the UMS is a *Leybold Inficon Transpector* quadrupole mass filter. This unit uses an electron multiplier detector to achieve high sensitivity (approximately 1 ppb detection limit) over a mass range of 1–100 amu. The sample is introduced through a coupled membrane introduction probe via forced pumping from a peristaltic pump. The membrane consists of a tube made of a gas permeable polydimethyl siloxane (PDMS). A sample 'slug' is introduced on one side of the membrane while a vacuum on the other side allows dissolved gasses and volatile organic compounds to be introduced to the mass filter. Electron impact ionization is used to generate ions that are accelerated into the mass filter. After the mass measurement is made, the sampling system operates valves that cause a deionized 'clean' water sample to enter the membrane. A background measurement is then made for comparison with the sample measurment.

The membrane sampling system is capable of passing compounds having masses as high as 300 amu. This allows the instrument to easily measure compounds such as dimethylsulfide, chloroform, and toluene as well as low molecular weight gasses such as carbon dioxide, oxygen, and nitrogen. A typical measurement cycle-time for measuring such compounds is 5–10 min. The instrument can also be operated in a continuous measurement mode. However, in this mode there is a loss of reference data and use is limited to measurements made relative to an initial baseline.

The mass spectrometer is housed in a three-vessel tandem arrangement. The first pressure vessel serves as the sample collection system and contains a three-channel peristaltic pump and flow-injection system. The flow-injection system allows the seawater sample or deionized water to be switched (valved) to the membrane introduction system. This facilitates comparison of selected mass ion intensities against mass spectrometer background levels. A second pressure vessel houses a vacuum chamber and mass analyser, turbo vacuum pump, membrane introduction system, control electronics, and power supplies. The entire instrument is controlled through a single board Pentium® minicomputer, and can be interfaced to the host AUV through a serial or network connection. A third pressure vessel contains diaphragm pumps that provide a rough vacuum to the turbo pump. Isolating the pumps in a separate pressure vessel allows for heat management and vibration isolation. The pressure vessel also serves as a vacuum exhaust dump for the roughing pumps; this eliminates the need to pump against potentially high ambient pressures.

Once assembled the pressure vessels of the mass spectrometer are 19 cm in diameter and approximately 1 m in length. The total power required when operating at 24 V is approximately 100 W. Duration in the Florida Atlantic University *Ocean Explorer* is 4 h when used with two 220 Wh lead acid battery packs. Very low mass flow allows the vacuum system to operate for up to 10 days without servicing.

15.4.2.3 Spectrophotometric Elemental Analysis System (SEAS)

Measuring the chemical constituents of marine environments often requires high sensitivity and selectivity. To achieve this, researchers at the University of South Florida have developed an AUV sensor based on optical absorption spectroscopy. This sensor, known as the Spectrophotometric Elemental Analysis System (SEAS), uses standard reagent-based colorimetric measurement techniques coupled with a unique long pathlength optical cell to provide requisite selectivity and sensitivity (Byrne *et al.*, 2000).

The long pathlength optical cell comprises a $400\,\mu m$ inside diameter tube made of Teflon AF2400. A sample consisting of a seawater and reagent mixture is pumped through the Teflon tube. The Teflon material is unique in that it has an index of refraction less than that of seawater; the Teflon material has an index $n = 1.29$ whereas seawater has an index of approximately $n = 1.34$. Light propagates inside the Teflon tube as it would in a conventional glass optical fibre. This characteristic is what attributes to the name given to the long pathlength optical cell, the Liquid Core Waveguide or LCW. The Teflon tube is sometimes coiled to allow very long optical path-length in a relatively small volume.

A specially designed coupler allows both liquid and light to enter or exit the LCW. In the SEAS instrument, light is supplied by a broadband incandescent lamp. The lamp is fibre-optically linked to the LCW through the light/liquid coupler. Filtered sample water from a tube port is pumped to a series-connected mixing segment of tubing using a peristaltic pump. Simultaneously, a selective reagent (e.g. ferrozine to detect iron species) is pumped through a second metering peristaltic pump for mixing in the series-connected tubing segment. The reagent-mixed solution enters the LCW and quickly fills it, that is, it displaces a water-only blank. The time required to fill the LCW with mixed solution depends on the length of the LCW and the flow rate generated by the main peristaltic pump; it is typically less than 1 min. The absorbance spectrum is measured using a compact optical spectrometer with 1 nm resolution over an optical bandwidth of 280–880 nm.

It is important that the reagent-analyte colorimetric reaction provide a spectrum that contains a band in which absorption does not occur. This allows for normalization of the measurement with respect to absorption due to non-reaction related constituents, including particles, dissolved organics, etc. The concentration calculation is based on Beer's law:

$$A = \varepsilon CL, \tag{15.1}$$

where A is absorption, ε is the molar absorptivity, C is the chemical concentration and L is the pathlength. Absorption sensitivity, and thus concentration sensitivity, is linearly related to pathlength. Increasing the pathlength by an order of magnitude will generally increase the measurement sensitivity by an order of magnitude. While sensitivity increases with longer pathlengths, a wide dynamic range is achieved by choosing a suitable combination of path length, reagent concentration, and integration time applied to the spectrometer's CCD detector.

The spectrometer, sample pump, reagent metering, spectrum analysis, data storage, and communications are all handled by an internal Motorola® 68HC16 micro-controller. The microcontroller provides 8 Mb of data storage (enough for 10 h of normal operation), serial RS-232 and networked data communication, and control of all sensor functions. The completed instrument consumes 6 W at supply voltages ranging from 8 to 24 V, is 10 cm in diameter, 50 cm long and weighs 3.5 kg in air. The maximum operating design depth is 500 m using the standard anodized aluminium housing.

The SEAS instrument has been successfully deployed on the *Ocean Explorer* AUV and has been configured, through appropriate choice of reagent, to measure pH, nitrate, nitrite, ammonia, iron, and chromate. An alternate configuration that uses a UV light source within an LCW that is wrapped on a quartz mandrel has been designed for fluorescence spectroscopy. The fluorescence-configured instrument is intended for detection methods using fluorophores, fluorescent antibodies, and natural fluorescence. Figure 15.5(a) shows a typical absorbance configured SEAS instrument in a plastic pressure vessel. Figure 15.5(b) shows data from an experiment in the Gulf of Mexico during spring 2000.

15.4.2.4   Reagent-based, UV-induced fluorescence nutrient sensor

Researchers at the University of South Florida have developed a successful method for determining concentrations of nutrients that involves both fluorescence and reagent chemistry (Masserini and Fanning, 1999). This method has been integrated into an AUV sensor designed to measure nitrate, nitrite and ammonia concentrations in seawater.

The sensor is a combination of three subsystems that measure nutrients using a three-channel pulse excited flow cell fluorometer. The nitrate and nitrite subsystems use a method that involves enhancing fluorescence by creating an easily fluoresced molecular ion. In general, nitrite is acidified using a weak solution of hydrochloric acid to form nitrosium that is subsequently combined with aniline, a primary aromatic amine. The resultant ion, benzenediazonium, can be measured using UV-induced fluorescence, thus yielding a measurement of nitrite concentration. Nitrate is measured after reduction through a cadmium column; that is, nitrate is reduced to nitrite, which is measured via the previously described method in a second fluorometer channel. Nitrate concentration is determined by subtracting the measured nitrite-only concentration from the total (nitrite plus reduced nitrate) measured concentration.

The instrument uses a reverse flow injection analysis (rFIA) technique to ensure acceptable signal to noise ratio and measurement repeatability. In contrast to FIA, rFIA uses the measured sample as a carrier. Lengthening residence time, using longer fluidic pathlengths, helps to improve dispersion of constituents within the sensor's tubing. This can increase the fluorescence signal. However, its primary disadvantage is that it reduces the measurement sample rate of the instrument. The described sensor is configured to allow a reasonably detectable signal while maintaining a sample rate of approximately 20 samples per hour.

The instrument's fluidic subsystems are driven by a 16-channel peristaltic pump. Water is sampled discretely through a filtered port external to, and in front of, the

(a)

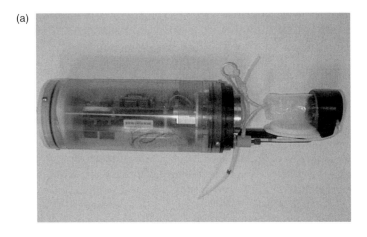

(b)

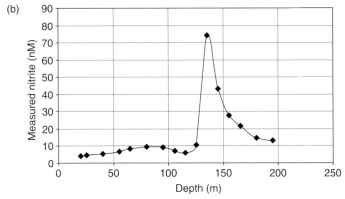

Figure 15.5 (a) Photograph of SEAS instrument; (b) Seas nitrite data, Gulf of Mexico, Spring 2000.

AUV. Automatic valves operating under the control of a preprogrammed microcontroller manage the pumping of the discrete seawater sample and reaction stages of the measurement cycle. Heaters, which speed up the reaction, are adjusted to reduce the likelihood of gas bubbles that might affect the optical fluorescence measurement.

Fluorescence measurements are made using a three-channel fluorometer for nitrite; nitrite plus reduced nitrate and ammonia. Each fluorometer channel is essentially the same, although the ammonia channel is configured differently (excitation wavelength, emission wavelength, and cell volume). A xenon flashlamp operating at a flash rate of 100 Hz (1 μs per pulse) provides light that is subsequently bandpass-filtered at 220 nm. This source of light is used for excitation. Fluorescent emission is measured at 295 nm using a photomultiplier tube and lowpass filtering circuitry. The final analogue signal is digitized by the instrument's central microcontroller and stored in local memory for post-deployment retrieval and analysis.

As is the case with the UV fluorescence-only instrument described earlier, the natural fluorescence of dissolved organics can affect the measurement. Aniline itself can also contribute to the fluorescence signal. Therefore, a background fluorescence measurement is used that accounts for dissolved organic matter (DOM), aniline and other contaminants. The final nutrient measurement value is calculated by subtracting the background level from the peak measured when the fluoresced reactant (benzenediazonium) resides in the flow cell.

Ammonia is measured using a modified Jones method (Jones, 1991). Ammonia dissolved in a pH-adjusted seawater sample is permitted to cross a Goretex-type membrane into a carrier solution containing o-phthaldialdehyde (OPA). The OPA reacts with the ammonia to form a fluorescent product, which is quantified by the third fluorometer channel. The technique requires heaters and accurate thermal regulation. As a modification to the published method, solutions are carried through a length of polytetrafluoroethelene (PTFE) tubing fitted to a specially designed heat exchanger. This allows for degassing if any dissolved gases are released during heating; it also allows for reasonably efficient heat transfer within the heaters.

The instrument has a measurement detection limit on the order of tens of nM for nitrite and nitrate and less than 10 nM for ammonia. Approximately 100 W is used when measuring three channels simultaneously. This instrument represents a unique adaptation in AUV sensor design as an extremely complex, wet chemistry sensor that has been successfully integrated and deployed *in situ*.

15.4.3 Physical Sensors

Sensors that measure aquatic physical parameters such as optical properties, turbulence, and ambient noise are also used on AUVs. Some work typically performed using ROVs has been successfully completed on AUVs, thus allowing improvements in survey completion such as rapid transects, precision constant altitude/depth, and low noise measurements.

15.4.3.1 Bottom Classification Albedo Package (BCAP)

University of South Florida researchers have packaged an ensemble of optical sensors into an instrument used to calibrate algorithms and validate satellite ocean colour data in coastal regions (Costello *et al.*, 1995). This sensor system, known as the BCAP, has been deployed aboard the Florida Atlantic University *Ocean Voyager II* and other AUVs.

The principal BCAP system components include hyperspectral (512 channel) upwelling radiance/reflectance and downwelling irradiance meters, a dual-laser range finder/chlorophyll probe, and a six-wavelength, image-intensified, CCD camera for bottom classification and object identification. The bottom reflectance and downwelling irradiance meters (developed at the University of South Florida) have a nominal resolution of 3 nm over a range of 350–900 nm. The two diode lasers function as range finders providing the high-resolution altitude determination necessary for light propagation modeling.

The six-channel, IMC-301 imager utilizes a micro-channel plate image intensifier with gallium arsenide phosphide photocathode. This gives a dynamic range of over ten orders of magnitude. It can detect fluorescence in macrophytes, seagrasses, and benthic diatoms and can record reflection (albedo) imagery and/or fluorescence

Figure 15.6 Photograph of BCAP package aboard the *Ocean Voyager II* AUV; note the black coating in the region of the instrument package to eliminate backscatter effects from light potentially reflecting off of the AUV's surface.

variations of the bottom. This system can also record pure fluorescence imagery of the bottom. The IMC-301 imager is, however, limited to six channels. To provide clear full-spectrum, high-resolution data, a calibrated 512-channel spectral radiometer obtains spectra (as a function of altitude) from a known segment of the IMC-301 image, thus providing full-spectrum measurement as well as calibration constants for the imager. Another 512-channel spectral radiometer, equipped with a cosine collector, obtains the spectra of the downwelling light field so that true albedo constants can be calculated.

Ancillary instrumentation focuses on the inherent optical properties (IOPs), physical properties and dissolved and particulate matter concentrations in the water. These sensors include a 685 nm fluorometer, a backscattering (OBS) meter (Seatech), and a CTD (Falmouth Scientific). Ancillary instrumentation that can be hosted by the BCAP sensor system include a nine-channel absorption/attenuation meter (WETLabs ac-9); 488 and 660 nm transmissometers (WETLabs), and a multi-channel back-scattering meter (HOBI Labs).

Ocean colour measurements require a solar zenith angle that allows adequate light penetration into the water column. Typically, the angle is optimum during a six-hour sampling window from 1000 to 1600 h local solar time. An ideal mission scenario would utilize multiple AUVs to maximize coverage during the sample window. Even a single AUV can provide coverage not available with a surface vessel. Assuming an AUV cruise speed of $2 \, \mathrm{m \, s^{-1}}$, a vertical ascent/descent rate of $0.5 \, \mathrm{m \, s^{-1}}$, water depth of 25 m, and vertical vehicle undulations (vertical profiling stations) at 1 km intervals, the AUV-supported BCAP could complete 43 sample stations, 86 vertical profiles (performed while underway), and map nearly 90% of the bottom along the transect. In contrast, a surface vessel with an experienced crew is able to perform a vertical optical profile, using a lowered package, in 25 min. Allowing an additional 5 min to steam to the next station, the crew could only complete 12 sample stations and would have albedo measurements of a negligible percentage of the bottom.

15.4.3.2 Ambient Noise Sonar system (ANS)

A unique sensor developed for research using an AUV is the Florida Atlantic University Ambient Noise Sonar (ANS) (Oliveri and Glegg, 1999). This system consists of a six-transducer array that can produce data for creating images of broadband ambient noise sources. This is accomplished by measuring the directivity of the acoustic field in different frequency bands. The system is useful in identifying and mapping sources of noise or imaging passive objects in the ambient noise field.

The ANS array contains six ITC 6050-C low noise hydrophones (ITC Underwater Technology). When the ANS is taking measurements, five of the six transducers are situated in a coincident plane 1.1 m behind a 'front' reference receiver. The transducers are held in place by five 1.6 m arms that are each separated by an angle of 36° (180° total). In the extended configuration, the transducers form a semicircular array with each transducer facing forward, as shown in Figure 15.7(a). Measurements are taken when the AUV is stopped and stationed on the seafloor facing the intended measurement direction. In transit, the transducer arms fold down against the AUV's outer hull, as shown in Figure 15.7(b).

Magnetic sensors that are interfaced to the AUV's internal network are mounted on the array arms so that their positions can be monitored. Arm position is generally monitored in three cases: before the AUV begins transit, before a measurement sequence begins, and while the AUV is en route.

Once the AUV has reached the desired site, the AUV's main computer sends a command to the ANS to begin operation. The AUV simultaneously activates a 'quiet' mode, so all acoustic systems, motors, etc., are powered down. The ANS computer initiates a measurement sequence, which begins by raising the arms of the array. The array arms are actuated using a single stepper motor and ball screw/nut mechanism. Each arm is attached to the mechanism by a three-pivot linkage, allowing all five arms to be raised simultaneously. Once measurements are complete, the ANS computer folds the transducer array and notifies the AUV's computer that it is ready for recovery or another measurement waypoint.

Bottom stationing is accomplished via a controllable buoyancy system. The system is designed to provide a buoyancy change of approximately 18.2 kg. A unique aspect of the design is its bias toward negative buoyancy. The concept involves providing only very slight positive buoyancy for transit while providing significantly negative buoyancy for stationed measurements. Approximately 18 kg is needed to keep the system stable on the seafloor in moderate currents.

The *Ocean Explorer* AUV is modular in design; that is, the vehicle when typically configured consists of two sections: an aft section is the completely self-contained propulsion and control system, and the forward section is the scientific payload. The controllable buoyancy system is designed to reside in the payload section yet be controlled (optionally) by the AUV's computer. This is accomplished through a networked connection within the vehicle. In the configuration, both the AUV's computer and sensor's computer can communicate with the buoyancy system.

(a)

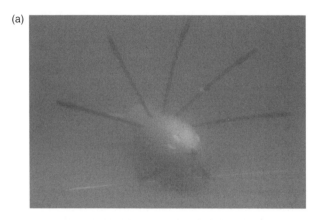

(b)

Figure 15.7 (a) ANS system shown in deployed position, AUV is bottom stationed; (b) ANS system shown in retracted position, AUV in transit.

The ANS buoyancy system uses a bladder similar to those commonly used by divers. This air-based system provides a fixed volume of positive buoyancy when the bladder is completely inflated. Three valves are used to control and regulate bladder air pressure: an inflate valve, a deflate valve, and a pressure check valve that is set to bleed pressures over 70 kPa.

The inflate and deflate valves are of the latching type so they do not require power while active. Air is supplied from a two-tank system within the payload section. One tank provides air at 21 MPa to a second regulated source tank that provides a controlled output of 1.035 MPa. This regulated pressurized air is sourced to a mechanical (second) regulator that measures the ambient hydrostatic pressure and then adjusts its output to 48 kPa over the ambient. An advantage to this arrangement is that a 48 kPa over ambient pressure is available regardless of the AUV depth as long as the AUV is at a depth less than approximately 100 m.

The sensor's main pressure vessel contains data acquisition and processing hardware for A/D conversion of the transducer's signals. A digital signal processor computes the averaged reference auto-spectra and five cross-spectra across the array.

Aliasing of the received signal is avoided by using integrated 20 kHz low pass filters and signal sample rates of up to 51.2 kHz. A built-in Pentium® PC controls the data acquisition, stores data, and generates images.

Images are generated from spectra stored during deployment. Each image requires 35.2 kb of disk memory, so 28,400 images can be stored on the ANS' single 1 Gb disk drive. At a measurement rate of 0.5 Hz, the 1 Gb disk can hold 17 h of spectra. Continuous operation over this period is achieved through an AUV-independent 1,060 Wh nickel cadmium battery system. Under normal operating conditions, the current consumption of the ANS is less than 1.3 A at 48 V.

15.4.3.3 Real-time Ocean Bottom Topography system (ROBOT)

Range scanning, illustrated in Figure 15.8, can provide real-time range information along with shape, orientation, texture, and volume of a target. Researchers at the University of South Florida have developed a range scanning system for use on AUVs. The sensor's name, ROBOT, comes from its ability to image in Real-time, Ocean Bottom Topography. Laser light, generated in a fan pattern, is used to illuminate the sea bottom surface as the host platform progresses. A 50 frame per second, 256 by 256 pixel CCD camera views the illuminated scene, captures and analyzes 'line' data in real-time and transmits it to a controlling computer.

Although high-speed and useful in mildly turbid environments, the ROBOT sometimes obtains a slightly distorted image. This is an artifact of the triangulation method used; it results from 'camera perceived' extension of the across track laser line as the scanned object's upper surfaces approach the camera's view limit. The problem is minimized via custom software and by using an AUV as the supporting deployment platform. AUVs typically have sophisticated control systems that allow for optimisation

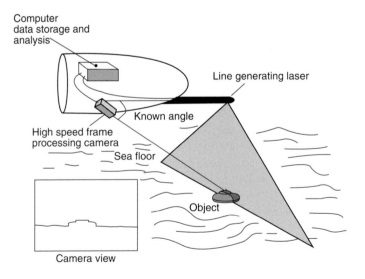

Figure 15.8 Principal of range scan imaging.

of mission parameters. Parameters such as uniform mean altitude, constant velocity (either with respect to ground or water), and accurate positioning make AUVs ideal for the ROBOT system.

ROBOT consists of a 65 mW 532 nm CW laser, high-speed CCD video camera, Pentium® based image processing computer, and nickel cadmium battery pack. The computer and camera are housed in a joined cylinder-box pressure vessel. A 30 cm by 25 cm by 12 cm aluminium rectangular housing is welded to a 25 cm by 13 cm diameter cylindrical housing. A single 2.5 cm port between the housings provides for the necessary electrical connections. The flat geometry of the computer housing and large-diameter quartz windows in the camera's cylindrical housing end cap limit the overall depth capability to 60 m. The battery pack is contained in a 6.5 cm by 11 cm diameter cylindrical pressure vessel. The laser housing is a long tube made of high-viscosity Delrin® having a dimension of 9 cm by 1 m. The laser is fixed in place with locking rings that ensure proper alignment with a radius-matched convex polycarbonate window that is glued into a fitted opening on the side of the Delrin® tube. A mirror on the end of the laser housing bends fan beam-generated laser light 90° to the axis of the laser. The tube is sealed on the interior AUV end with a flat endcap to accommodate electrical connection. The front of the tube is sealed with a conically shaped nose that provides a hydrodynamic leading surface. The entire laser housing tube fits within another tube situated within the AUV. Thumbscrews allow laser-to-camera distance to be easily adjusted and enable the laser to be retracted with minimal effort. Retracting the laser fully protects it during shipboard transit.

Data are stored in line scan form, 256 bytes at a time in the host computer. The initial processing of 'brightest pixel' or 'line' information uses a threshold process, or optionally, a successive approximation process built into the imaging camera itself. Stored line scan data are assembled into a 3-D image using MATLAB® software. Figure 15.9 shows a scanned image of a man-made structure at approximately 20 m water depth.

The across-track resolution is dependent on the flight altitude and camera field of view. The camera can record 256 pixels for its field of view. For a typical AUV flight, the altitude over bottom is expected to be approximately 3 m. For the designed view angle of 24° the pixel resolution is 5 mm. The camera is typically operated at a frame rate of 50 frames per second. At a forward speed of 1 m s^{-1} the per-line resolution is 2 cm.

15.4.3.4 Turbulence measurement

Turbulence in the water column is of interest because of its affect on sediment transport, mixing, biota, etc. The parameters impacted by water column turbulence and flow includes biological, chemical, physical, and geological; that is, accurate measurement is key to understanding the linked processes. Researchers at Florida Atlantic University have developed an AUV based system for measuring turbulence in the water column (Dhanak, 1998). This system has been deployed on the *Ocean Explorer* and *Autosub* AUVs.

The measurement system is a combination of several sensors integrated into a sting mounted package designed to sample water forward of the host vehicle. Two shear probes (Lueck, 1987) measure y and z cross-stream velocity components. A dynamic Pitot tube measures the axial (x) flow and a 3.75 cm diameter electromagnetic velocity

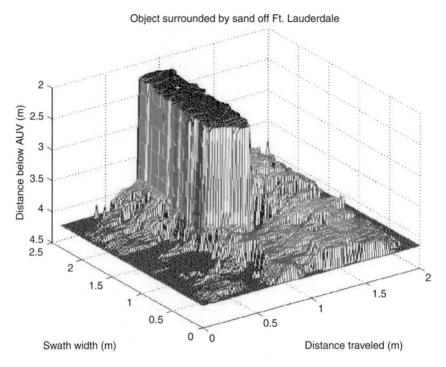

Figure 15.9   Underwater 'wall' captured at a depth of approximately 20 m by ROBOT aboard an *Ocean Explorer* AUV off of the coast of Ft. Lauderdale, Florida, 1998.

meter measures reference motion. Additional sensors include a CTD, a microstruc-ture temperature probe and three-axis accelerometers. The instrument's data storage system also captures information from networked navigational sensors on the host AUV.

The instrument is designed to measure turbulence within the length scale range of 1 cm to 1 mm and minimum spatial frequency scales of 1–100 cycles per metre. AUVs of the size of both the *Ocean Explorer* and *Autosub* provide stable platforms for such measurements. The instrument's data acquisition system includes a differential analogue front-end and a six-pole low pass filter to eliminate aliasing by a 16 bit A/D converter that is used to digitize the raw signals. Post processing is used to eliminate narrow band noise introduced by the AUV's propulsor and control motors and from the data storage system's hard disk drive. Accelerometer data provide the frequency information needed to filter vibration related noise during post processing.

The output of the instrument is shear spectra, for the *y* and *z* axes, that show the transfer of turbulent (kinetic) energy to thermal energy vs. spatial frequency. Dimen-sionally the spectra indicate energy transfer in watts per kilogram (of seawater) and spatial frequency in cycles per metre. Data taken in 1996 near the Ft. Lauderdale, Florida coast using an *Ocean Explorer* AUV show a close correlation of measured turbulence compared to Nasmyth (theoretically predicted) spectra. This indicates that the instrument performs well measuring expected turbulence in the natural environment. Spectra are generated from time-series data for selected time segments of predetermined duration, for example, 160 s. The area under a spectral curve provides the rate of kinetic to thermal energy transfer or 'dissipation rate'. Measurement of shear, density related strata, turbulence layers, thermoclines, and haloclines are some of the uses for the package. Physical oceanographers can use such data in computer models to predict small to medium scale ($1–10 \, \text{km}^2$) processes.

The turbulence measurement instrument is capable of measuring dissipation rates as low as $10^{-9} \, \text{W} \, \text{kg}^{-1}$. The instrument's probes are mounted forward of the host vehicle about 0.3 m ahead of the bow and are guarded by a 210 mm diameter guard ring. A specially engineered vibration isolation assembly helps to reduce unwanted noise generated by the AUV's mechanical systems. A separate 10 cm diameter by 44 cm long pressure vessel houses the data acquisition and storage electronics and a separate 420 Wh battery provides electrical energy for up to 48 h of operation.

15.5 MICROELECTROMECHANICAL SYSTEMS (MEMS)

Technology is developing at an ever-increasing rate. So rapid is the advancement of new technologies that it is projected that by the year 2020 the sum of human knowledge will double at a rate of once in every 73 days (Edmund, 1994). A technology that is presently revolutionizing the research world is microelectromechanical systems or MEMS (Elwenspoek and Wiegerink, 2001). MEMS is significantly impacting the way in which research is performed; the rapid completion of the mapping of the human genome, for example, is a direct result of MEMS technology.

MEMS refers to a broad group of micro-machined devices and components. MEMS devices and components are micro-machined in the sense that they are fabricated using processes very closely related to those used to make microelectronic circuits or 'chips' out of silicon. MEMS components are physically functional, that is, they themselves move or they physically act on something. Examples are microscopic gears, hinges, levers, fluid pumps, and micro-motors. MEMS components typically have dimensions about equal to the width of a human hair ($50–100 \, \mu\text{m}$). Individual micro-machined components can be integrated together to make a system or MEMS device. An integrated MEMS device is different from 'electronics only' micro devices and represents a major advance over current integrated circuit technology. While integrated circuits are limited to providing strictly electronic functions, MEMS devices provide mechanical, fluidic, optical and electronic functions all on the same chip (Kovacs, 1998; Rai-Choudhury, 2000).

Since these microscopic systems are fabricated using techniques normally applied to transistors and integrated circuits, silicon is the material most commonly used for MEMS devices. However, the fabrication of MEMS devices is not restricted to silicon. Micro-machining methods can also be applied to a large number of non-silicon materials including: polymers, plastics, metals, ceramics, etc.; this greatly increases the versatility and extends the applicability of these systems.

Some common, present-day, applications of MEMS devices include:

- genetic mapping;
- automobile air bag deployment sensors;
- read/write heads in computer disk drives;

- micro fluidic pumps;
- motion sensors;
- micro telemetry devices; and
- weapon systems.

In a real sense, the applications of MEMS technology are limitless.

What impact will technologies like MEMS have on the future of automated marine research? For those capable of keeping up with the blinding pace of new technologies, entirely new research will be enabled. The challenge will lie in keeping up; the research community will soon be faced with the not so unpleasant task of simply trying to decide which advanced system to use.

15.5.1   MEMS and AUV Sensor Applications

In general, MEMS is considered a mature technology. MEMS research has been pursued for more than two decades. Initial efforts focused on making useful structures in bulk silicon. Advancements in several MEMS fabrication techniques have produced truly useful devices that can be used as subsystems in realizable sensors. MEMS devices such as pumps, accelerometers, spectrometers, acoustic systems, and others have been fabricated using not only silicon but several other materials as well. Advancements in fabrication processes have contributed to a significant reduction in the cost of MEMS devices and since most fabrication processes use photolithographic techniques, mass production is generally straightforward. MEMS technology enjoys several of the attributes of its parent, the semiconductor industry. High capability–density and low cost make the technology very attractive for environmental sensors.

The following sections give a brief summary of representative elements that might be used to construct a basic MEMS sensor for use on an AUV. Countless devices, components, configurations and techniques have been published. Those presented are described in general terms to enable the reader to understand the basic concepts.

15.5.1.1   Micropumps

A realistic MEMS-based AUV sensor must be capable of sampling fluids. One method for generating fluid flow in a MEMS system is to take advantage of the relatively high ambient pressure of the surrounding water. An internal pre-evacuated reservoir can provide the potential energy necessary to move the requisite sample volume. Although possible and very low in energy consumption, the method has the disadvantage of moving only limited sample volume and flow rates are dependent on pressure differences. A more practical approach is to use a pump of some type. MEMS pumps utilizing piezoelectrics (van Lintel *et al.*, 1988), magnetohydrodynamics (Lemoff *et al.*, 1999), thermoneumatics (Grosjean and Tai, 1999), and several other methodologies have been demonstrated.

Consider the thermoneumatic pump referenced above. This pump uses a phased sequence of vertical heat-driven actuators to peristaltically force liquid through a constrained channel. Under each chamber is a heat section that produces vertical movement through expansion. Small piston-like segments on top of each heat section mate with a shaped surface directly above. Each section is driven 120° out of phase with adjacent sections. The result is a 'moving pocket' that captures a parcel of sample and propagates it through the pump. Operation at 4 Hz produces liquid flow rates as high as $6.3\,\mu L\ min^{-1}$.

This pump has several advantages:

- it is fabricated in silicon so it is unaffected by most chemicals and in general will not interact with the sample;
- flow rate can be precisely controlled and corresponds linearly with respect to pulse rate; and
- energy consumption is quite low (nominally 300 mW at $6\,\mu L/min^{-1}$). Note that although the energy consumption is quite low, the pumping efficiency as compared to macro-scale pumps is also low. This is due to several physical effects relating to charge distribution in the pump walls and pumped liquid, volume to surface ratios, etc.

Of course this example pump suffers the same pulsed-flow characteristic as its macro-scale counterpart.

Integrating any pump into a MEMS sensor system elicits unique challenges that lie beyond the scope of this writing. However, in general, micro-fluidic channels are etched into quartz or some other rigid substrate much like the patterns on an electronic circuit board (e.g. Schabmueller *et al.*, 1999). The pump is precisely located and bonded to this etched substrate, making fluidic contact. Appropriate fluid and electrical connections are made in the final stages of the overall sensor's fabrication process during final assembly and before packaging.

15.5.1.2 Detection

MEMS sensors, detectors and transducers have been used to detect hundreds of phenomena using thousands of methods. Using MEMS sensors to detect molecules or atoms can be challenging; the sample volume is limited by the very size constraints that define a MEMS device. For example, a typical MEMS fluidic system might have a sample measurement volume of $10\,\mu m$ by $100\,\mu m$ by $100\,\mu m$, that is, 100 pL. In a water sample containing 100 parts of the analyte in 10^{12} parts of solvent, one might expect to find only a few thousand analyte molecules.

Increasing the signal to noise ratio is imperative if MEMS sensors are to be practical. Published literature on successful MEMS accelerometers, chemical detectors, DNA analysers, flow sensors, pressure sensors, gas analysers, humidity sensors, optical sensors, acoustic sensors, and mass sensors has shown efficacy of the technology. One detection method that has several desirable characteristics, including excellent sensitivity and small size, is electrochemical immunoassay. Researchers at the University of Cincinnati (Heineman *et al.*, 1999) have developed an electrochemical immunoassay MEMS sensor, and have successfully tested it using a common research immunoglobulin antibody (mouse IgA).

After a sample is pumped into the analyser it is passed over a bed of specially treated micron-sized magnetic beads. The beads are coated with a specific antibody that is designed to recept only the analyte from a multi-component sample. The magnetic beads are immobilized in the flow stream by an externally induced electromagnetic field. Once exposed to the sample for a sufficient period of time, the beads are washed with a specially designed enzyme. The enzyme links with the analyte in much the same the way the antibody does. The antibody-coated beads, complete with attached analyte and enzyme, are released by removing the external magnetic field. The beads are then transported, through electro-osmotic flow, to an electrochemical detector.

The detector consists of a set of interdigitated finger electrodes. The enzyme, attached in the previous 'electromagnet' section, progresses through a reduction–oxidation reaction at the electrodes, inducing a current that is proportional to the quantity of analyte present. The sensitivity of this technique actually improves as the spacing in the electrode detector decreases. The frequency at which electron exchange can occur increases the closer the electrodes are, thus generating more electron transfers, or a higher current, per unit time. In this particular case, the diminishing dimension of a MEMS device improves sensitivity.

15.5.1.3   MEMS communication

Once the sample has been taken and the desired measurement has been made, information must be communicated (or stored and later communicated) to the user. This requires a connectorless method; the micro-device would most likely be destroyed or damaged by the macro-scale forces associated with a hand-engaged connector.

Wireless methods such as RF, acoustics, or optics are well suited to MEMS instruments. They also eliminate the need for a physical data connection. As with non-interfaced conventional sensors, MEMS sensors can be built with fully capable on-board data storage. The wireless data link can be used to query the sensor in real-time, send commands, or configuration information to the sensor, or used to offload stored data that have been collected during a deployment.

RF MEMS is an engineering discipline in itself. Those familiar with electrical systems theory have studied the direct analogy of mass–damper–spring mechanical systems to inductor–resistor–capacitor electrical systems. Using mechanical mass and spring structures in a MEMS device, as in Figure 15.10, enables one to create complex resonant filters, passive components, and switches. Frequencies in the 10 s of GHz have been demonstrated (Nguyen, 1999) and relatively long distance data transmissions (a few km in air) have been achieved.

Acoustic MEMS structures can also be constructed using piezoelectric and other materials. Acoustic bi-directional communications can be achieved using methodologies similar to those used in conventional acoustic modems; the primary difference is the transducer design. MEMS fabrication methodologies lend themselves to large-scale reproduction of a single specific pattern. This feature allows designers to easily fabricate acoustic transducer arrays that use spread spectrum transmission/reception, focused beam projection, phased array detectors, etc. Although limited to lower data transfer rates than RF MEMS, acoustic systems are certainly better suited for data communication in an aqueous medium.

Optical communication methods in MEMS range from very simple to extremely complex. The simplest communication system uses asynchronous opto-electronic communication. This type of free-space optical communication system uses an LED-detector pair to transmit and receive frequency shift keyed (FSK) amplitude

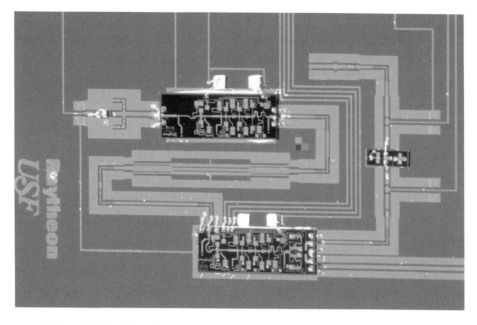

Figure 15.10   MEMS-based high frequency resonant mechanical structures.

modulated light. That is, light is amplitude modulated and subsequently frequency modulated with digital data. This method can transfer data through air over short distances of about one metre, or through water for shorter distances depending on optical attenuation. Data transfer rates as high as $4\,\mathrm{Mb\,s^{-1}}$ have been demonstrated. This simple method can be modified so that light is coupled to an optical fibre. Further, multiple wavelengths can be used over the same fibre to increase the bandwidth.

The optical communication method described is an active system. It outputs light energy sourced from the MEMS device's energy supply. For long-range high-speed optical communication, transmitted power requirements can become a significant factor in the device's energy budget. An alternative is to use a reflective element, such as a corner cube or mirror, that is physically moved to modulate reflected light. This methodology can be extremely energy-efficient. A modulated laser sends a command to a detector adjacent to the reflective element. The reflective element is then modulated (moved to sequentially reflect or not reflect light) to digitally transmit information. A detector, such as a photo multiplier tube or avalanche photodiode, is used to detect modulated reflected light. In this way several distributed sensors can be polled from a central (optically sightable) location.

Network configurations are possible in most MEMS communication scenarios. Ethernet protocols or distributed intelligence networks such as LONWorks® can be integrated into a MEMS instrument. A fully integrated MEMS instrument includes micro-electronic components for control and data storage/communication. The same hardware used in conventional sensors is integrated into a MEMS instrument using die level components rather than chip level or surface mount devices.

15.5.1.4 MEMS energy supplies

One of the distinct advantages of MEMS is extremely low power consumption. A distinct disadvantage, however, is the need for an energy source on a size scale commensurate with the device. Several approaches to this problem have been successful. One approach involves wireless transmission of energy to the device (Sasaya, 1999). Although authors in the cited reference used microwave energy, lower frequency electromagnetic fields, light energy, acoustic energy, and even heat energy (thermal gradients across the device) can be used.

The alternative is to include an energy supply in the device itself such as conventional electrochemical cell technology. The major difference between a MEMS electrochemical cell and conventional commercially available devices is mostly the form factor of the cell. In general, a cell such as a standard zinc–carbon type, is built using a laminated or layered technique. These so-called 'tape-cells' can achieve the energy density necessary to operate a MEMS device while occupying minimal volume. Of course, other electrochemical processes can be used and even fuel cell technologies have been experimentally tested.

As with MEMS communication, a hard-wired, multi-use, electrical connection is often impractical; the self-contained energy supply is preferable. With the exception of acoustics, most wireless energy transmission methods are more difficult in aqueous media. Here again, the designer must consider the goals and constraints of the planned device.

15.5.1.5 MEMS packaging

Packaging represents one of the most formidable challenges facing MEMS sensor designers. Harsh environments, such as the ocean, impose constraints related to material physical properties and chemical compatibility. MEMS devices can suffer energy density problems, that is, too much heat dissipation in too small an area. They can also be very sensitive to temperature variations, pressure, etc., and unlike conventional electronic devices where only electrons need enter or exit a device, MEMS require interfaces for physical connections.

For underwater MEMS sensors, the significant physical connection is water. However, when the water sample contains potentially corrosive elements, extreme care must be taken to choose the appropriate materials and assembly methodologies. MEMS fabrication and packaging involves etching materials such as silicon or ceramics, bonding using organic compounds, sealing using glass to metal interfaces, electroplating, chemical vapour deposition, etc. Each process can be potentially incompatible with a physical parameter of the final application. The reader is directed to the references for the finer details of MEMS packaging and related technologies.

As with conventional sensors, design optimization is extremely important. A completed sensor with fluid pump, detection system, electronics control, onboard communication, energy source, and finished package is certainly a reality using available existing technologies. Large companies such as Philips, Hewlett Packard, and Texas Instruments have invested heavily in MEMS and in fact manufacture commercial MEMS products. Academic institutions such as University of Twente (The Netherlands), Cornell University (USA), and the Universty of Tokyo (Japan) – and many more – have extensive research programs dedicated to micro and nano technologies. Sensors used to sniff amino acids, peptides, pheromones, etc. that are being conceived today will be real MEMS sensors tomorrow. The low cost and high-capability density of these devices will enable the researcher to gather data less expensively, study new or yet to be investigated phenomena, and improve the quality of gathered data.

15.6 CONCLUSIONS

AUVs can perform as excellent research tools if fitted with the appropriate sensors. Automated systems that couple vehicle technology, sensors technology, and remote intelligence will continue to mature thus providing researchers with an ever-expanding set of tools and capabilities with which to study the oceans. As computer technology advances, software and data analysis capabilities will improve. With improvements in computers and software will come advancement in autonomous vehicle control, for example, intelligent navigation, adaptive sampling, etc. The real value to the researcher will not only come in these improvements but will also come in the ability to more *accurately* measure the marine environment. Continued advancement of sensors through MEMS and nano-technologies as well as improved materials and faster electronics will ultimately provide researchers with inexpensive, reliable and highly accurate tools. Through application of these tools the future holds the promise of truly understanding the vast unknowns of the underwater world.

Acknowledgements

The author would like to recognize the contributions of Dr Scott Samson for his input on the SIPPER system, David Fries for MEMS, Dr R. Timothy Short for the UMS system, Eric Kaltenbacher, Dr Eric Steimle and Dr Robert Byrne for the SEAS instrument, David Costello and Dr Kendall Carder for the BCAP and Dr Ken Holappa on the Turbulence Measurement Package. Dr Phillip McGillivary of the US Coast Guard provided valuable background information, and Dr Shekhar Bhansali provided suggestions and information on MEMS microfluidics. Special appreciation is given to Kristen Kusek for her suggestions and input on the original manuscript and to Dr Thomas Hopkins and Dr Peter Betzer for their encouragement.

16. LOGISTICS, RISKS AND PROCEDURES CONCERNING AUTONOMOUS UNDERWATER VEHICLES

GWYN GRIFFITHS[a], NICHOLAS W. MILLARD[a] AND ROLAND ROGERS[b]

[a] Southampton Oceanography Centre, Empress Dock, Southampton SO14 3ZH, UK and
[b] QinetiQ, Winfrith Newburgh, Dorchester DT2 8XJ, UK

16.1 INTRODUCTION

Operating Autonomous Underwater Vehicles (AUVs) is not an activity that is free from risk. When AUVs were experimental platforms developed by the military and civilian research communities the major risks were mainly technical. Engineers and project managers were very conversant with these risks, and applied procedures to mitigate technical risk during the specification, design and construction of the vehicles. While AUVs share many areas of technical risk with other underwater systems, their autonomous nature means that several aspects require particular attention. These include, among others: robustness of the self-contained navigation system; reliability of the mission management software; integrity of the mission programme; reliability of the control surfaces and actuators and robustness of the collision avoidance, emergency abort and vehicle recovery systems.

While technical risk remains an important issue for AUVs, the increasing use of the technology has led to operational risk coming to the fore. In part, this is because of the success of AUV developers in achieving a high degree of reliability in the hardware and software. Operational risk in the context of this chapter is taken to include the impact of technical risk; the absolute and relative risks of different operating environments, and the risks in various deployment scenarios such as ship launched, shore launched, air launched, escorted, unescorted, supervised or unsupervised. An assessment of operational risk should also cover the questions concerning the legal regime in which the AUV is to operate.

The range of tasks tackled by AUVs continues to grow. Major projects in several countries have demonstrated missions of real utility, for example: the oceanographic survey of coastal fronts in Hero Strait, British Columbia by the Massachusetts Institute of Technology's *Odyssey II* vehicle in June 1996 (Nadis, 1997); the magnetic and physical observations made by Woods Hole Oceanographic Institution's ABE vehicle on the Juan de Fuca ridge in 1995 and 1996 (Bradley *et al.*, 1996; Tivey, 1996); a swath bathymetry survey of Oslo Fjord by *Hugin I*, a Norwegian AUV (Storkersen and Indrecide, 1997); and turbulence measurements off Florida by Florida Atlantic University's Ocean Explorer vehicle (Dhanak, 1996). All of these missions involved the use of a support ship for deployment and retrieval.

In contrast, the under sea ice cable-laying mission by International Submarine Engineering Research's *Theseus* AUV in April 1996 (McFarlane, 1997; and Chapter 13) was carried out from a camp on the sea ice north of Ellesmere Island, Canada.

The vehicle, all support equipment and personnel having been air-freighted to the camp.

This chapter draws heavily upon experience with the Autosub AUV (Millard *et al.*, 1997, 1998). The vehicle has been used on over 250 missions in coastal waters, open ocean and under Antarctic sea ice and icebergs, with deployments from ships and from shore.

Shore-based AUV missions hold many attractions for scientists and users with the promise of autonomous vehicles complementing or eventually replacing some of the work presently done from ships (Griffiths *et al.*, 1997). With an AUV deployed from a dock or harbour or from shore via a slipway or crane, only a small, inexpensive chase boat might be needed for start-up, monitoring, safety and guard duties until the AUV was safely out of confined waters. Griffiths *et al.* (1998) described a shore-based mission where the Autosub vehicle began its mission within the small, quiet harbour at Dunstaffnage Bay on the West Coast of Scotland. The vehicle travelled on the surface using DGPS navigation (Meldrum and Haddrell, 1994), proceeding out to the open sea of the Firth of Lorne through a channel some 150 m wide. A rigid inflatable chase boat and the RV *Calanus* stood by the AUV on this proof of principle mission.

If leaving and returning to harbour autonomously with an AUV can be shown to be technically feasible and acceptable to regulatory and coastal management authorities then a whole spectrum of tasks in operational and research oceanography becomes possible. AUVs launched from shore, or from near-shore subsea garages could be deployed in weather conditions that may be too hazardous for deployments from ships. A preliminary demonstration for this concept was tested during the Autonomous Vehicle Validation Experiment off Bermuda in September 1998 (Griffiths *et al.*, 2000). Such shore-based missions could study the response of the coastal ocean to hurricanes, for example.

In particular a shore-based AUV capability will be important for developing countries, which are unable to provide the finance necessary for a national survey fleet and the infrastructure required to support the vessels. The capability for these maritime nations to fully understand the economic viability of their maritime zones, to be able to monitor both third party exploitation as well as environmental contamination and to provide data relating to sustainability of the oceans under its jurisdiction could all be addressed in some measure by shore-based AUVs.

16.2   OPERATIONAL OPTIONS

AUVs, by their very nature, can be used in many different operational modes. This is in marked contrast to the case of a Remotely Operated Vehicle (ROV) where the physical connection tethers the vehicle to the surface support ship. An AUV faces no such constraint. In this section, we outline some of the modes of operation of an AUV and consider the different segments that comprise a typical mission and how choices over the mode of operation affect the risk.

16.2.1   AUV Modes of Operation

Before any campaign with an AUV takes place, it is advisable to ensure that all relevant authorities and users of the water space are well aware of the intended operations. This phase of the operational programme may well require contact with local and national authorities, setting up water space management procedures, and formulating, testing, authorising and disseminating mission plans and applying them where required. The campaign proper will begin with preparing the vehicle in a shore laboratory or workshop before the equipment is transported to the intended work area. The planned area of activity may be reached by ship, aircraft or in the case of some coastal waters by road transportation. The three forms of operational missions considered here are ship-based, shore-based and completely autonomous.

16.2.1.1 Ship-based missions

This form of mission has the AUV operating from a vessel where the command, control and communication (C^3) functions are exercised from the ship. This type of mission is the most common form to date and the one where the community has the best understanding of both the logistic requirements and risks.

16.2.1.2 Shore-based missions

This form of mission has the AUV operating from a range of shore-based facilities such as harbours with existing infrastructure, to purpose-built submerged AUV garages (see Chapter 6) to remote beaches where the only support infrastructure is that which the AUV team is able to provide. In each case the C^3 functions are exercised directly from the land based launch and recovery sites. The logistical requirements and operational risks of this form of mission are still evolving.

16.2.1.3 Completely autonomous missions

This form of mission has the AUV launched from platforms such as ships or aeroplanes or from remote seabed docking stations. The C^3 functions are exercised from a centre of operations remote from the AUV operational area. An example of this is the United States Office of Naval Research (ONR) AUV system concept – the Autonomous Undersea Sampling Network (AOSN) – described by Curtin and Bellingham (2001). The AUVs dock with seabed or moored C^3 nodes, where data collected is downloaded for analysis at the operations centre. The nodes will also allow the AUV to receive new mission instructions and possibly to recharge the batteries. This form of truly autonomous operation is the newest of the mission forms and as such, a full understanding of the logistics and the risks has yet to be developed.

16.2.2 AUV Mission Segments

These three types of AUV operational missions can be considered as being composed of a sequence of five mission segments:

- launch;
- outward transit;
- the work task;
- inbound transit;
- recovery.

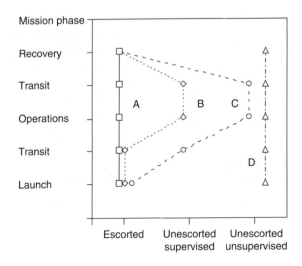

Figure 16.1   Diagrammatic representation of the degree of supervision through mission phases.

The degree of supervision during these different mission segments may vary depending on the mode of operation. A diagrammatic representation of the degree of supervision during some example missions is shown in Figure 16.1.

Now consider each of these mission segments in turn. First, launch, which may be, for example, from a surface ship, from a shore facility or from a moored docking station. If a launch is to be from ship or from shore then it will invariably be supervised. But, the level of supervision may be quite different in each case. Launch from a ship would be under the ultimate supervision of a certified mariner, probably the support vessel's master. The master has clear responsibilities in national and international maritime law for the safe conduct of all operations concerning the ship. These responsibilities would include those under the International Convention on the Safety of Life at Sea 1974 (SOLAS) and the International Management Code for the Safe Operation of Ships and for Pollution Prevention (ISM Code) 1994 (see Chapter 17).

The launch procedure from a ship is very dependent on the size of the AUV and, to some extent, to the size and capabilities of the ship. For a large AUV, such as Autosub, the safest and surest way to effect a deployment is to use a purpose-built launch and recovery system. Launch is generally easier than recovery. Figure 16.2(a) shows the moment of deployment of Autosub from its gantry when operating from the FRV *Scotia*. The gantry has a telescopic arm that takes the vehicle, on a carrier, away from the side or stern of the ship. The carrier rotates 90° to bring the vehicle in line with the side or stern. Two winches on the carrier proceed to lower the vehicle into the water; when it reaches the surface, two pins are pulled out of sockets, freeing the AUV from its lifting lines. For a small AUV, such as the REMUS, the launch procedure might involve two operators lifting the vehicle over the side of a rigid inflatable boat.

A launch from shore is more likely to be under the supervision of the technical team, probably working without the immediate advice of a certified mariner.

(a)    (b)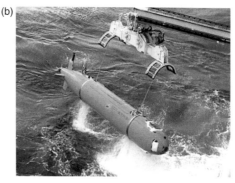

Figure 16.2   (a) Launch and (b) recovery of the Autosub vehicle from its purpose-built gantry as installed on the FRV *Scotia* in July 1999. *See* Color Plate 10.

In Figure 16.1 while missions A, B and C comprise an escorted launch; mission D's launch is unescorted. This would be the case for a launch from a moored or seabed docking station. There may or may not be supervision of a launch. Supervision could take place via a direct acoustic link or via an acoustic link to the surface and subsequent two-way telemetry by radio or satellite.

Second, consider the outward transit from the launch site to the work area. The transit might be

- escorted (Figure 16.1A and B);
- unescorted but supervised (Figure 16.1C), for example by monitoring the vehicle's progress remotely over an acoustic or radio communication link;
- or both unescorted and unsupervised (Figure 16.1D).

Third, consider the work task itself. The task might be escorted, unescorted and supervised or unescorted and unsupervised as requirements demand. In some circumstances, it might not be possible for the support ship (if there is one) to venture into the area of the work task. This might be the case when operating under sea ice, and would certainly be the case if the vehicle were to operate under polar ice shelves. Fourth, transit from the work area to the recovery position might replicate the procedure on the outward leg, or it might differ. Finally, recovery to a ship or to a shore base would usually be supervised, while recovery to a docking station would, of necessity, be automatic.

Recovery of an AUV to a ship is probably the most hazardous operation. As for launch, a purpose-built gantry system will reduce the risk. Figure 16.2(b) shows the final stage of recovering Autosub onto the FRV *Scotia*. Here, recovery began with radio communication established between the surfaced AUV and the support team on the ship. After ensuring that the mission had ended, the command to release the recovery float was sent to the vehicle. The ship was brought alongside the AUV, the recovery float grappled, the recovery line hauled inboard, pulling two lifting lines from their storage containers fore and aft within the vehicle. These lifting lines were attached to corresponding lifting lines on the two winches on the carrier. The carrier could be rotated if necessary to take out twists in the lines. With the lines clear, the vehicle was winched up to the cradle. Finally, the cradle was rotated to line up with the gantry and the carrier arm telescoped inboard. However, keeping a ship of several thousand tonnes in position next to an AUV with quite different dynamic characteristics is far from straightforward.

16.2.3 Key Issues during Mission Segments

16.2.3.1 Mission planning

Planning for the safe return of an AUV begins long before launch. What the user wants may be at variance with what the more experienced engineer/operator is happy to allow. The mission plan, jointly formulated by the user and operator, must try to satisfy the user requirement but must also be within the bounds of a strategy to avoid hazards and establish a safe, properly considered rendezvous for recovery. Hazards may include excessively rough topography during terrain following missions, strong currents, shipping and fixed structures. The mission script must be checked by a second person before downloading to the vehicle.

16.2.3.2 Pre-launch

As the scheduled launch time approaches a weather eye must be kept, remembering that launching in rough weather is generally much easier than recovery. Launching on a deteriorating weather forecast could be foolhardy, although conditions for every mission need to be considered individually. Shortly before launch, it is common procedure for a predefined checklist to be completed with no concessions made to non-compliance without very careful consideration of possible consequences. A blank checklist as used on the June 2000 Autosub missions is shown in Table 16.1.

16.2.3.3 Escort and supervision

As we have seen in Figure 16.1, the pattern of escort and supervision of an AUV might well vary during a mission. Whereas the escorted missions pose relatively well understood problems, unescorted AUV missions whether supervised or unsupervised require more detailed evaluation.

Supervision can take a number of forms. The supervisor need not be at sea with the vehicle. In this form, the supervision need not necessarily involve the constant attention of a human evaluator and decision-maker. Rather, if events warrant intervention, the operator can be alerted and action taken. Recent developments in communications technology have made it perfectly feasible to follow the progress of an AUV mission from anywhere in the world via a satellite and cell phone link or via the internet. This method of monitoring has been in routine use by the Autosub team since early 2000.

An alternative approach to AUV supervision has been taken by the REMUS group (Stokey *et al.*, 1999). Using a Portable Acoustic/Radio Geo-Referenced system the regular acoustic navigation pulses served to obtain range and bearing to the vehicles and served as 'keep alive' signals. If the vehicle failed to detect these pulses it would shut down and abort the mission.

Table 16.1 Checklist for an Autosub deployment.

GENERAL

Mission Number: # Location:
UTC/Date:

ESSENTIAL CONFIGURATION
MISSION CONTROL

- ❏ Mission History File? ...
- ❏ Mission Name ....

EMERGENCY ABORT

- ❏ Mission Time: *hr min.*
- ❏ Dive Time: *hr min.*
- ❏ Max Depth: m

DEPTH CONTROL

- ❏ Max Depth   m
- ❏ Min Altitude   m
- ❏ Min/Max Pitch -...,...... degrees

NAVIGATION

- ❏ Min Fix Type   3 GPS Datum WGS-84

ACOUSTIC TELEMETRY AND COMMAND
(1: ABORT 2: STOP 3: GO 4: SETZERO 5: SURFACE)

- ❏ Depth 0–999 m. Altitude 0–999 m.
- **ADCP down & up***(set using Nodebuilder)*
- ❏ Bottom Pings: 1, 0.   Water pings: 2,2
- ❏ Ping Period: 2.0,3.0 (sync). Delay: 1200
- ❏ Number of cells   48, 48   Cell size: 2 m
- ❏ Max BT Depth 220 m ... Salinity: 38

PRE-LAUNCH

Pre-Power Up
- ❏ LineDeploy Armed
- ❏ ARGOS 26579, 26580 on + check?
- ❏ Flashing Lights Batteries Check + ON?
- ❏ 10 kHz Acoustic Beacon Charged + ON?
- ❏ HF Dumb charged and On?
- ❏ PC Time Sync
- ❏ Mission Script Checked? By ...,...
- ❏ Mission Compiled?
- ❏ **POWER UP on Batteries.**
- ❏ Go Button ON?
- ❏ **ORBCOMM ON?**
- ❏ Data logger sync + running? at ......
- ❏ Mission Nanny Run ?
- ❏ Test mission run?
- ❏ Test Mission Stopped + set Zero ?
- ❏ **Mission Downloaded**
- ❏ Mission Check Spreadsheet run?
- ❏ EAS weight reading? ...(nom. 655)
- ❏ Leak Sensor reading? .... (nom 1.45)
- ❏ Battery Voltage? ......
- ❏ Depth sensor? 1: ... 2: ...
- ❏ Attitude Sensor? ......
- ❏ ADCP 1 and ADCP 2 working?
- ❏ ADCP1 Mbyte? .... ADFCP2 Mbyte ...
- ❏ Echo sounder Range? ......
- ❏ GPS fixes? ORBCOMM Messages?

IN WATER PRE-MISSION CHECKS

- ❏ GPS Initialised
- ❏ Mission Nanny Run?
- ❏ Mission Check Spreadsheet run?
- ❏ EAS weight reading? ...(nom. 655)
- ❏ Leak Sensor reading? ... (nom. 0.94)
- ❏ Battery Voltage? ......
- ❏ Battery Temperature? ......
- ❏ Depth sensor? 1: ...... 2: ...
- ❏ Attitude Sensor? ......
- ❏ ADCP, ADCP2 altimeter? ......
- ❏ Echo sounder Range? ......
- ❏ Acoustic telemetry?   HF ... LF ...
- ❏ Acoustic tracking? HF... LF ...
- ❏ ARGOS 26579, 26580?
- ❏ ORBCOMM Messages?
- ❏ GPS fixes?
- ❏ Seabird On?
- ❏ Turbulence Probe On ?

MISSION INFORMATION

- ❏ Mission Started: UTC.
- ❏ At: N   W.
- ❏ ETA   UTC
  At: N   W.

POST MISSION CHECKS

- ❏ Mission Ended: UTC.
- ❏ At: N   W.
- ❏ Mission **Stopped** by command?
- ❏ **GPS Fix** At end of Mission?
- ❏ Line Deploy Fired?
- ❏ EAS weight reading? ......
- ❏ Main Leak Sensor reading? ......
- ❏ Battery Voltage? ......
- ❏ Battery Temperature? ......

POST RECOVERY CHECKS

- ❏ Seabird Flushed?
- ❏ Data download?
- ❏ **Turn Vehicle Off? CAREFUL weight!**
- ❏ **ORBCOMM Switched OFF?**
- ❏ Flashing Light OFF?
- ❏ LF and HF Beacon OFF And CHARGE?

The Hugin vehicle (see Chapter 11) employs an 'acoustic umbilical' to a tracking vessel to supervise the vehicle, to provide continuous precise positioning (<3 m) and to provide quality control of the data collection.

16.2.3.4   Surface navigation by AUVs

The first rule of the Convention on the International Regulations for Preventing Collisions at Sea 1972 (COLREG) states that the rules shall apply to 'all vessels upon the high seas'. As indicated by the word 'upon', the COLREG only apply to surface navigation, there is no equivalent set of rules for subsurface navigation. If an AUV were taken to be a vessel, which at present is far from a foregone conclusion, then when it operates on the surface it would be required to comply with the COLREG rules. After a thorough review of requirements and with a working knowledge of the capabilities of current vehicles, Brown and Gaskell (2000) suggest that, by their nature, AUVs would find it very difficult to comply with all of the requirements in the COLREG. Nevertheless, because of the potential hazard of an AUV operating on the surface, courts and regulatory authorities 'will be very reluctant to exclude AUVs from the ordinary application of the Rules' (Brown and Gaskell, 2000: p. 114).

How, then, might operational risk be minimised? If the AUV is launched from a support vessel and if the vehicle is being escorted, monitoring the AUV visually and communicating by radio or by an acoustic link might arguably contribute to fulfilling the need to 'maintain a proper lookout' and to determine if a risk of collision existed. An unescorted or unsupervised AUV, when on the surface, would clearly be in breach of these rules. Brown and Gaskell (2000) conclude that, in such circumstances 'if AUVs are vessels covered by the Rules then their autonomous "navigation" or operation on the surface could be proscribed'.

Operational and legal risks may be minimised by mission strategies that do not entail the AUV surfacing away from the vicinity of a support vessel. When this procedure cannot be employed, for example because the vehicle was launched from shore, or when the vehicle is on unescorted missions, a degree of operational risk is inevitable. In practice, the risk can be partly mitigated by adopting suitable mission plans. For example, when Autosub completed 13 unescorted missions in the North Sea in the summer of 1999 a specific rendezvous procedure was adopted that minimised the time the AUV spent on the surface in the absence of the ship. When the vehicle reached its final waypoint, as estimated by its own dead-reckoning, it surfaced to obtain a GPS fix, then it would dive and transit to the correct final waypoint where it would circle at a safe depth awaiting the rendezvous with the support ship (Griffiths *et al.*, 2000). Autosub would then be summoned to the surface using an acoustic transponder activated from the support ship.

16.3   TECHNICAL RISKS WITH AN AUV

At the current stage of development, AUVs are operated most commonly by the teams that have been involved intimately in their design and build. This brings a number of advantages, including reducing risk during operations. Stokey *et al.* (1999) and Griffiths *et al.* (2002) identify human error as the dominant source of faults encountered with two quite different AUVs, namely REMUS and Autosub. One factor in

keeping human error to a minimum during operations is the knowledge and experience of the deployment team. If the deployment team has a deep understanding
of the details and track record of the vehicle, it can be of enormous benefit to the
operation at sea.

The other main sources of technical risk with AUVs are common to most items of
underwater equipment that use electromechanical systems, embedded software for
real-time control, sensors to obtain information on the environment and depend on
underwater acoustic links.

Using the fault record of the Autosub vehicle spanning its use during development
trials through engineering evaluation to use on science research campaigns, Griffiths
et al. (2002) built a statistical model of the vehicle's reliability. Each mission undertaken by the vehicle contributed to the fault database, either because one or more
faults occurred or because a fault did not occur. The latter formed the large majority
of missions. These were treated as censored observations in the statistical analysis.
Taking censored and uncensored data a Weibull distribution was fitted to the observations. The resulting probability distribution function of underway faults over 117
missions from March 1998 to June 2000 is shown in Figure 16.3. The distance travelled on a mission ranged from less than 1 to 263 km. In all, these missions covered
2,519 km. They took place in UK coastal waters, the North Sea, off Bermuda and in
the Mediterranean Sea, on familiar and unfamiliar ships.

Weibull distributions are characterised by two parameters: α, the characteristic life
and β, the shape parameter. For the distribution fitted to the data in Figure 16.3,
α was 483.9 and β was 0.773. When $\beta < 1$, the data imply that the reliability of the
vehicle did not degrade with distance travelled. This is, the vehicle did not age within
a long mission. Instead, reliability, as measured by the hazard rate (λ), here as faults
per km, improved with distance travelled.

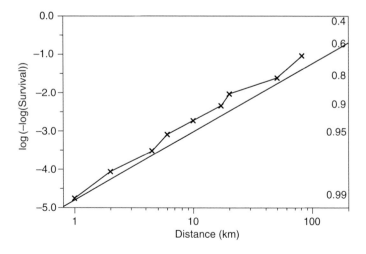

Figure 16.3 Survivability against distance with a Weibull distribution fitted to the data for
underway faults observed on Autosub missions 124–240.

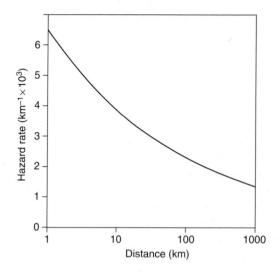

Figure 16.4 Hazard rate per km as a function of distance travelled obtained from the parameters of the Weibull distribution of faults underway shown in Figure 3.

For a Weibull distribution:

$$\lambda(x) = \frac{\beta}{\alpha}\left(\frac{x}{\alpha}\right)^{\beta-1},$$

where x is the distance to travel in a mission. The hazard rate for the distribution of Figure 16.3 is shown in Figure 16.4.

The higher hazard rate at shorter distances travelled suggest that faults occur sooner rather than later during a mission. This is not surprising. Faults such as ballasting errors and acoustic communication problems manifest themselves almost immediately the vehicle is put in the water. This type of fault cannot often be caught on deck, before the vehicle is deployed. This feature of the hazard rate of historical Autosub missions reinforces the advisability of the team's practice of tracking the vehicle at the start of all missions, even those that become unescorted through most of their length.

16.4   RISKS TO PERSONNEL

The health and safety legislation of many nations require that risk assessments should be performed, and safe systems of work put in place, if there is any significant degree of risk involved in an activity. Operating an AUV from a ship or other platform can potentially involve serious risks to personnel. The onus will be on the team leader to prepare an assessment that minimises or mitigates the risks to acceptable levels. The risks will vary with the size of vehicle, area of operation, type of deployment and recovery platform, sea state, experience and training of the team. Many of the risks will be generic to work at sea, but others may be specific to the nature and requirements of an AUV. While this chapter cannot give detailed guidance on producing risk assessments for all types of AUVs, the following are the two main hazards that have been identified and recorded in recent Autosub campaigns:

Working on an AUV close to the edge of the ship, with the safety barrier removed   The potential risks are from hypothermia if working in cold regions; from falling overboard and from manual handling and fitting in cold and wet conditions. An evaluation of the risks, if no mitigating factors were to be put in place, has shown that minor injury (small cuts, grazes) could be highly probable; injury requiring medical attention could be possible, but that major injury or long-term illness would only be a remote possibility.

To reduce these risks down to the levels of only possible minor injury the following control measures might be adopted: fit temporary barrier to the ship's stern rail; make full use of the inboard reach of the recovery/deployment system to allow adequate work area away from the edge of the ship; use mandatory safety clothing and equipment (e.g. restraining harnesses); work as a team, looking out for the welfare of each other; take breaks commensurate with the conditions and comply with prohibitions on working on deck in rough weather.

Changing the AUV's spent batteries   The potential risks include injury from crane manoeuvres, manual handling of battery packs (ca. 60 packs of 10.5 kg for Autosub), electric shock during the final assembly of fresh batteries (terminal voltage of 113 V) and when fitting the new batteries into the vehicle. An evaluation of the risks, before applying mitigating factors, has shown that it is very possible that an injury requiring medical attention occur, but that a fatal injury was a remote possibility.

To reduce these risks down to the same low level as in the first example requires that full use is made of mandatory safety clothing; that personnel have received training in manual handling; that suitable technical measures are put in place (e.g. the battery packs for Autosub are double insulated with shrouded plug-in connectors); and that the electrical work only be undertaken by experienced personnel.

16.5   OPERATIONAL RISK

This section considers the procedural and legal implications of different modes of AUV operations. For clarity and to aid the consideration of this problem we have taken as a starting point the eleven assumptions in Randall (1990). These are that the AUV

- has been classified;[1]
- has been registered;

---

1 The policy and legal issues of AUV classification, registration and insurance have been well researched and documented (PREST, 1989; Randall, 1990; Brown and Gaskell, 2000). However, these issues have yet to be fully resolved at United Kingdom (UK), European Union (EU) and international levels of legislation. These topics are also the subject of a current study by a Society for Underwater Technology (SUT) Working Group that was formed from the Society's Underwater Robotics Group. The Working Group has produced a policy document on 'The Guidelines for the Safe use of AUVs' (SUT, 2000), which it is hoped will form the UK's policy baseline for endorsement at national level. This chapter's authors were among the authors of the SUT Working Group report.

- has been insured;[2]
- is of substantial size and weight;
- will be operated in autonomous mode as the norm;[3]
- will be 'long range' in time and distance;
- will operate at slow speeds $<2.5\,\mathrm{m\,s^{-1}}$;
- will have an energy source that will not be nuclear in type;
- will have high intrinsic value;
- will be on a scientific mission rather than on a commercial or military one;[4]
- will have safety markings and signals, for example, visual, acoustic, radar and communications (positioning and recognition).[5]

In light of these assumptions the problem can be broken down into two distinct areas for consideration: procedural risk and legal access. Procedural risk is covered below; the issues concerning legal access are covered by Brown in Chapter 17. An accessible text, in question and answer format, on the issues of legal access and matters of private law, of practical value to team leaders of AUV operation has been written by Brown and Gaskell (2001).

16.5.1 Procedural Risk

It is not possible, within the scope of this chapter, to undertake a full International Maritime Organisation (IMO) Marine Safety Committee (MSC) (IMO/MSC, 1993 and 1997) Formal Safety Assessment (FSA)[6] on different AUV mission scenarios. The procedures detailed in the references have not been formulated with AUVs in mind and are still a long way from being fully developed and tested. However, they are mature enough to allow for an initial risk assessment of the AUV issues discussed here. This initial assessment follows the first two stages in the published FSA methodology.

Table 16.2 identifies hazards and gives a simple assessment of the relative risk associated with shore-based and shipbourne AUV operational deployment scenarios. The relative risk assessment shows that the shore-based AUV mission profile has a reduced risk in the areas of AUV launch and recovery. The ship borne AUV mission profile

---

2 The UK Natural Environment Research Council's AUTOSUB has been insured by the University of Southampton with an underwriter, without being classified.

3 This assumption has been made to emphasise that the AUV will operate remote from direct human intervention or control, for example, with only limited reliance on an acoustic link.

4 This assumption has been made because military and commercial missions bring their own unique policy/legal problems in relation to access to/from a coastal state's Maritime zone, for example, territorial sea. Therefore, for example, each AUV mission that falls within these two categories may have to be looked at on a case by case basis.

5 The regime of markings and signals could be based on those suggested for Ocean Data Acquisition Systems (ODAS) (UNESCO, 1969) which have been reviewed in (PREST, 1989; Randall, 1990; Brown and Gaskell, 2000).

6 The methodology as discussed in (INM/MSC, 1993) identifies five processes in the FSA. These are: identification of hazards; risk assessment; risk control options; cost benefit assessment and recommendations for decision making.

Table 16.2 Hazard identification and relative risk assessment for the proposed AUV mission profiles.

Hazard Identification	Relative Risk assessment	
	Shore-based AUV Mission	Shipborne AUV Mission
1. Risk of damage during launch of AUV	Reduced risk as the launch will be from a stable platform	Increased risk as the launch will be from a moving platform
2. Transit to site of operations		
a. Risk of collision with ship or vessel	Increased risk as AUV will, in general, have to travel longer distances to get to the site of operation	Reduced risk as AUV will, in general, have to travel shorter distances to get to the site of operation
b. Risk of loss of life or injury	Increased risk for similar reasons to 2(a). Also the safety control of the AUV will be exercised more remotely	Reduced risk for similar reasons to 2(a). Also the safety control of the AUV will be exercised less remotely
c. Risk of loss of AUV	Increased risk for similar reasons to 2(a)	Reduced risk for similar reasons to 2(a)
d. risk of theft of AUV	Increased risk for similar reasons to 2(a)	Reduced risk for similar reasons to 2(a)
3. Operational integrity of AUV compromised	Increased risk as safety and operational control of AUV will be exercised more remotely	Reduced risk as safety and operational control of AUV will, in general, be exercised less remotely
4. Risk of not recovering AUV or damage during recovery	Reduced risk as the recovery will be from a stable platform	Increased risk as the recovery will be from a moving platform

has a reduced risk in the areas of transit to AUV operational area and during the AUV operations.

It would be possible to mitigate for the increased risk assessed for the shore-based AUV mission in the transit phase by providing a small chase boat. The practicality of this mitigation may be reduced where the operational area is situated at a long distance from the shore launch site. The mitigation required to reduce the risk associated with the exercising of either operational or safety control over shore-based AUV missions is more difficult to formulate.

The reduced risk in the operational phase attributed to ship borne AUV missions is based on the safety attributes required of and associated with the host vessel and its master. A shore-based AUV mission would in practice be under the control of either the chief scientist or a senior engineer, neither of whom may not hold, and indeed are currently not required to hold, formal maritime safety or navigational qualifications. The suggested mitigation in this case would be for a qualified master mariner to be made the 'operations manager' for the scientific mission. He or she, in turn, would be required to authorise and release the planned AUV track and safety procedures to the relevant maritime authorities.

16.5.2 Liability

Although most of the potential liabilities arising from operating AUVs are common to all marine operations, two deserve particular mention. First, as vehicles controlled by either by a pre-programmed script of commands or by an adaptive path-planning controller AUVs are very dependent on the skills and experience of mission and software programmers. Human errors formed the category with most arisings in a Pareto diagram encompassing 13 categories in Griffiths *et al.* (2002). Mission plans formulated with full knowledge of the limitations of the AUV and its control software, for example, its limited endurance, navigational accuracy, and communications, should minimise the risks of liability arising out of damage to property, pollution, death or personal injury. When there is a problem, many AUVs include an emergency abort system, but the sophistication of the fault detection systems often contrast with the usually crude response – most commonly a weight drop to reach the surface. This response might be triggered irrespective of the consequence or any damage that that action might cause (Brown and Gaskell, 2000: p. 151). For example, when Autosub became trapped on a cliff in the Strait of Sicily in June 2000 the emergency abort system triggered a weight drop that lodged the vehicle even more firmly under an overhang. Engaging reverse would have been a better response in this case.

Second, liability through salvage may be far from straightforward. While recovery by a salvor would be very welcome if an errant AUV was itself in danger or was likely to cause a danger to others, false salvage of an AUV that was merely unattended could lead to a dispute. Although markings and warning lights, as advised in Technical Annex II to the draft ODAS Convention (IMO, 1984), may reduce the incidence of false salvage the risk cannot be eliminated entirely. For a truly lost AUV recovered through salvage a fee of 50–70% of the value would not be unusual (Brown and Gaskell, 2000: p. 146).

16.5.3 Insurance

AUV operators in the UK are fortunate in being close to advice from specialists in the London marine insurance market, which holds a long valued lead in innovative insurance solutions. While insurance for AUVs remains novel, it is neither difficult to obtain or expensive – indeed AUV insurance business is being actively sought. In practice, the insurance industry's experience of AUVs to date has been "particularly benign" (Edwards, 2000). This is likely to remain so while designers and developers of vehicles remain closely associated with vehicle operations at sea. In a more commercial market for AUV services, the insurance position may change but so also may the inherent reliability of the AUVs when they become production items.

16.6 CONCLUSIONS

Operating AUVs for whatever purpose entails risk. However, the balance of this risk has changed from technical in the early days of AUV development to operational risk currently associated with a maturing technology whose use is on the increase. The utility and usage of AUVs is growing in many key areas of the maritime sector. Therefore, there is now an urgent requirement for an internationally agreed set of procedures to be developed and adopted that both identifies risks and generates procedures for managing the identified risks associated with the operation of AUVs.

The United Kingdom's Code of Practice produced by the Society of Underwater Technology is a small but significant step in the right direction, SUT (2000). However, it needs to be developed further and modified to reflect current practice in other countries that use AUVs. A global Code of Practice for the safe use of AUVs must have a suitable international champion to promote the code as well as providing a forum for subsequent revision. Perhaps the most suitable and internationally competent organisation to undertake this task is the IMO.

To assist in the provision of suitable guidance the classification and registration issues discussed in Chapter 17 need to be addressed. Once this has occurred then codification of guidance will become easier to move forward. That guidance should be based on the wealth of 'customary practice' on managing AUV operational risks both in industry and academia. Some of that practice has been has been described in this chapter.

17. THE LEGAL REGIME GOVERNING THE OPERATION OF AUVs

E.D. BROWN

Professor Emeritus of International Law, Cardiff University, Cardiff, CF1 3YE UK

17.1 INTRODUCTION

The operation of Autonomous Underwater Vehicles (AUVs) is governed by both public international law and national law. The main focus of this chapter is on the international legal framework. So far as national law is concerned, States are bound by common rules of international law and are obliged to ensure that their national law and practice comply with them. Over time, it may therefore be expected that the law of the various coastal States will develop along similar lines on such matters as deployment of foreign AUVs in offshore zones. However, as is shown by the brief review of national law and State practice and procedure presented below, there is still a long way to go with this assimilation process.

For the most part, this chapter is not concerned with such private and maritime law questions as the definition in law of a 'ship', property law questions, liability and compensation, and insurance. The reader will find a full treatment of such questions in a report by Brown and Gaskell (2000).[1]

The remainder of this chapter comprises three main sections dealing with the International Legal Framework, National Law and State Practice and Procedure and AUV operations in the Southern Ocean and Antarctica. The regime governing AUV operations in Antarctica, like the general regime, comprises rules both of international law and national law. However, reflecting the unique environment of this region, this regime is a highly specialised one and deserves separate treatment.

17.2 INTERNATIONAL LEGAL FRAMEWORK

Following a brief introduction to the international law of the sea and the various maritime zones from internal waters to the high seas, this section explains the rules governing Marine Scientific Research (MSR) in those zones. This is followed by a consideration of the rights enjoyed by land-locked and geographically disadvantaged States in relation to MSR conducted in the offshore waters of neighbouring coastal States. Next, a review is provided of the rules of the United Nations Convention

---

1 This chapter is largely based on Part I of the report, which may be consulted for a more detailed, documented treatment. Part II, prepared by Professor Gaskell, deals with private and maritime law questions.

on the Law of the Sea, 1982 (UNCLOS) on the deployment of scientific research installations or equipment in the marine environment and of the attempts made to draft a convention on ocean data acquisition systems (ODAS). Finally, an account is given of the provision made in international law for the settlement of disputes.

17.2.1 Brief Introduction to International Law of the Sea and Maritime Zones

The international law of the sea is a specialised branch of public international law, the body of legal rules governing the relations of international persons, that is, entities such as sovereign States and international institutions endowed with the capacity to have rights and duties in international law. It follows that large sovereign States like the United States, small sovereign States like Vanuatu, and international organisations like the United Nations or the International Maritime Organisation (IMO) are international persons or subjects of international law, whereas natural persons and other juristic persons (e.g., individual citizens, companies or scientific research institutes) are not. A number of consequences flow from this. First, the rules of international law are binding as such only on States (or other international persons). To become binding upon citizens, companies or research institutes, they must be incorporated in national law in accordance with the constitutional process of the State concerned. Thus, in the United Kingdom, for example, treaties concluded with other States, creating rules of international law binding upon the United Kingdom as an international person, must be incorporated in legislation in order to make their rules binding as a matter of national law upon British citizens, companies or research institutes.

A second consequence is that only the international person may enter into agreements (treaties, conventions) under international law.

Finally, it should be noted that, in the event of a breach of a rule of international law, the right denied and the capacity to claim a remedy for the breach will be those of the international person, even though the impact of the breach may have been felt principally by, for example, a shipowner or research institute. For the same reason, any damages recovered would be awarded to the State in reparation for the breach of international law suffered in the person of one of its nationals.

The two principal sources of international law are treaties (conventions, agreements, protocols) and international customary law. Potentially, AUVs are governed by the general body of the international law of the sea to be found in both treaties and international customary law. The bulk of the rules are now embodied in UNCLOS. Of particular interest to AUV operations are Part XIII of the Convention, on Marine Scientific Research, and Part XII, on Protection and Preservation of the Marine Environment. The operation of AUVs is also governed by more specialised treaties, including the International Convention for the Safety of Life at Sea, the Convention on the International Regulations for Preventing Collisions at Sea, and the various instruments of the Antarctic Treaty System. AUVs may also be affected by regional agreements on the environment when operating in the regions covered by such instruments.

The precise rules of the law of the sea governing AUV operations at any particular time depend upon the location of these operations in one or more of the various maritime zones into which the seas are divided. Horizontally, the seas are divided into eight jurisdictional zones. In some cases different jurisdictional regimes apply in the three vertical subdivisions of these zones, the water column, the seabed and the subsoil. Proceeding from land, the eight zones are internal waters, archipelagic waters, the territorial sea, the contiguous zone, the Exclusive fishing zone, the Exclusive Economic Zone (EEZ), the continental shelf and the high seas. It is important to note that the regime applying in the water column of the continental shelf differs from that applicable to the seabed and subsoil and, similarly, the regime of the water column of the highs seas differs from that of the 'Area', that is, the seabed and subsoil of the seabed area lying seaward of the outer limit of the continental shelf, where seabed mining of polymetallic nodules may take place.

17.2.2 Jurisdiction Over MSR in Maritime Zones

UNCLOS has established a regime governing the conduct of MSR in the various maritime zones, whether carried out by research vessels or AUVs. As will be seen, the degree of control which the coastal State may exercise over MSR diminishes as one moves from land to high seas. Thus, at one extreme, the coastal State has full sovereignty over MSR conducted by foreign vessels in its internal waters but, at the other extreme, has no jurisdiction over MSR conducted by foreign vessels on the high seas. In the middle, in the EEZ and on the continental shelf, a 'qualified consent' regime applies and the coastal State has only limited jurisdiction over such research. The implications for AUVs may be summarised as follows:

17.2.2.1   Internal waters

The coastal State has full sovereignty over internal waters, that is, all waters landward of the baselines of the territorial sea. It follows that no MSR may be conducted by AUVs in the internal waters of a foreign State without its consent.

17.2.2.2   Archipelagic waters

Under Part IV of UNCLOS an 'archipelagic State' (a State constituted wholly or mainly by one or more archipelagos) may draw 'archipelagic straight baselines' around its archipelago/s and the enclosed waters have the status of 'archipelagic waters'. The archipelagic State has sovereignty over its archipelagic waters and foreign ships or AUVs may not carry out research in these waters without the prior authorisation of the archipelagic State.

17.2.2.3   Territorial sea

The coastal State has sovereignty over the territorial sea, extending to a maximum breadth of 12 n mile. Accordingly, Article 245 of UNCLOS provides that:

> Coastal States, in the exercise of their sovereignty, have the exclusive right to regulate, authorise and conduct marine scientific research in their territorial sea. Marine scientific research therein shall be conducted only with the express consent of and under the conditions set forth by the coastal State.

Thus, AUVs may conduct MSR in the territorial sea of a foreign State only with its express consent.

17.2.2.4 Contiguous zone

Under Article 33 of UNCLOS, the coastal State may exercise certain controls in a contiguous zone adjacent to the territorial sea and extending to a maximum breadth of 24 n mile from territorial sea baselines. However, these controls do not relate to MSR and, for purposes of jurisdiction over MSR, the waters of the contiguous zone retain the status of an EEZ (if the coastal State has claimed an EEZ) or of the high seas (if not).

17.2.2.5 Exclusive fishing zone

The concept of the exclusive fishing zone is not recognised in UNCLOS but its existence has been acknowledged by the International Court of Justice and, of course, many States, including the United Kingdom, claim an exclusive fishing zone and do not claim the more comprehensive exclusive economic zone. While the coastal State may doubtless claim jurisdiction over research related to the exploitation of fish in the exclusive fishing zone, such research is beyond the scope of this chapter. Since the area of the exclusive fishing zone will normally overlap the area of the legal continental shelf, the coastal State may have certain jurisdictional powers over MSR in the waters of this zone. However, these powers are properly considered below under the heading of the continental shelf.

17.2.2.6 Exclusive economic zone and continental shelf

Given the fact that UNCLOS deals with MSR in the EEZ and on the continental shelf together in Article 246, it is convenient to consider them together here too. The detailed rules governing MSR in the EEZ and on the continental shelf are in Part XIII of UNCLOS on 'Marine Scientific Research'. They may be summarised as follows:

The general principle – a qualified consent regime While it is true that the coastal State has the right to regulate and authorise MSR in its EEZ or on its continental shelf and that such MSR may be conducted only with its consent, it does not have an absolute discretion to withhold consent. The general rule is that in 'normal circumstances' consent has to be granted for MSR carried out 'for peaceful purposes and in order to increase scientific knowledge of the marine environment for the benefit of all mankind'. The coastal State is given discretion to withhold consent in specified circumstances and to suspend or terminate MSR in other cases. Nonetheless, the underlying policy reflected in these provisions is in favour of consent being granted and, indeed, the coastal State is required to establish rules and procedures to ensure that consent will not be delayed or denied unreasonably. Although 'normal circumstances' are not defined, they may exist in spite of the absence of diplomatic relations between the coastal and researching States.

How to apply for consent Applications have to be submitted through 'appropriate official channels', unless otherwise agreed by the States concerned. The application has to be received by the coastal State not less than six months before the expected start of the MSR project and has to contain a full description of:

(1) the nature and objectives of the project;
(2) the method and means to be used, including name, tonnage, type and class of vessels and a description of scientific equipment;
(3) the precise geographical areas in which the project is to be conducted;
(4) the expected date of first appearance and final departure of the research vessels, or deployment of the equipment and its removal, as appropriate;
(5) the name of the sponsoring institution, its director, and the person in charge of the project; and
(6) the extent to which it is considered that the coastal State should be able to participate or to be represented in the project.

Conditions to be complied with in conducting MSR Under Article 249(1), the researching State has to comply with the following conditions:

(1) ensure the right of the coastal State, if it so desires, to participate or be represented in the MSR project, especially on board research vessels and other craft or scientific research installations, when practicable, without payment of any remuneration to the scientists of the coastal State and without obligation to contribute towards the costs of the project;
(2) provide the coastal State, at its request, with preliminary reports, as soon as practicable, and with the final results and conclusions after the completion of the research;
(3) undertake to provide access for the coastal State, at its request, to all data and samples derived from the MSR project and likewise to furnish it with data which may be copied and samples which may be divided without detriment to their scientific value;
(4) if requested, provide the coastal State with an assessment of such data, samples and research results or provide assistance in their assessment or interpretation;
(5) ensure that the research results are made internationally available through appropriate national or international channels, as soon as practicable;
(6) inform the coastal State immediately of any major change in the research programme;
(7) unless otherwise agreed, remove the scientific research installations or equipment once the research is completed.

It is open to the coastal State to withhold consent if the project is of direct significance for the exploration and exploitation of natural resources, living or non-living. If, nonetheless, consent is granted it may be subject to the condition that the results of the research may be made internationally available only with the prior agreement of the coastal State.

Finally, MSR activities must not unjustifiably interfere with activities undertaken by the coastal State in the exercise of its sovereign rights and jurisdiction in the EEZ and on the continental shelf.

It should be emphasised that the researching State has a clear international legal duty to ensure compliance with these various conditions. Failure to do so may seriously prejudice freedom of MSR in foreign waters. There are many loopholes which can be used by the coastal State to frustrate the effective operation of this consent regime and it is essential that a climate of trust is established by ensuring that principal scientists and civil servants comply fully and in good faith with these conditions. Understandably, scientists frequently find compliance with such conditions irksome, especially in the context of relations with an inefficient or unhelpful foreign bureaucracy. Nonetheless, it should be remembered that the cost of failure to comply may well be an inability to secure consent for a later project.

Implied consent   Reflecting the Convention's positive attitude to MSR and recognising that it should not be frustrated by the failure of the coastal State to respond in a timely fashion, provision is made for implied consent in Article 252. The project may proceed six months after the date on which the application was lodged with the coastal State unless, within four months of that date, the coastal State has informed the researching State that: (1) it has withheld its consent under the provisions of Article 246 (see below under 'discretion to withhold consent'); or (2) the information given on the nature or objectives of the project does not conform to the manifestly evident facts; or (3) it requires further information on the matters covered in Articles 248 and 249, that is, the information supplied with the application and compliance with the conditions subject to which MSR may be conducted; or (4) outstanding obligations exist to comply with these conditions in relation to an earlier MSR project.

Discretion to withhold consent   Consent may be withheld under Article 246(5) if the project:

(1)   is of direct significance for the exploration and exploitation of natural resources, whether living or non-living;
(2)   involves drilling into the continental shelf, the use of explosives or the introduction of harmful substances into the marine environment;
(3)   involves the construction, operation or use of artificial islands, installations and structures referred to in articles 60 and 80;
(4)   contains information communicated pursuant to article 248 regarding the nature and objectives of the project which is inaccurate or if the researching State or competent international organisation has outstanding obligations to the coastal State from a prior research project.

Suspension or cessation of MSR project   Suspension may be required by the coastal State in two situations: (1) if research is not being conducted in accordance with the information contained in the application upon which consent was granted; or (2) non-compliance with the conditions of research (see 'Conditions to be complied with in conducting MSR') concerning the right of the coastal State with respect to the project. Such suspensions may be lifted once the researching State has complied with its obligations. If, however, those defects are not made good within a reasonable period of time, the coastal State may require a cessation of the project. A cessation may also be required if the conduct of research is so inconsistent with the information supplied with the application as to indicate a major change in the research project or research activities.

MSR in continental shelf beyond 200 n mile line   In areas where the continental margin extends beyond the 200 n mile outer limit of the EEZ, the coastal State is entitled to extend its legal continental shelf to the outer edge of the continental margin. However, the water column above that part of the continental shelf which lies seaward of this 200 n mile limit retains the legal status of high seas and it is accordingly provided in Article 257 that all States have the right to conduct MSR in this water column. As regards the seabed and subsoil of this area of the continental shelf, the Convention offers a useful concession in favour of freedom of MSR. One of the grounds on which the coastal State may normally withhold consent is that the project is of direct significance for the exploration and exploitation of natural resources of the continental shelf. Where, however, the project is to be undertaken on that part of the continental shelf which lies beyond the 200 n mile line, the coastal State may not withhold its consent on that ground unless the research is to be conducted within specific areas designated by the coastal State as areas in which exploitation or detailed exploratory operations are occurring or will occur within a reasonable period of time. This rule is of course without prejudice to the rights of coastal States over the continental shelf.

Measures to facilitate MSR and port calls   Port calls are often a useful facility and sometimes a necessary part of an MSR expedition. It is, therefore, useful to have the following 'measures to facilitate marine scientific research and assist research vessels' prescribed in Article 255:

> States shall endeavour to adopt reasonable rules, regulations and procedures to promote and facilitate marine scientific research conducted in accordance with this Convention beyond their territorial sea and, as appropriate, to facilitate, subject to the provisions of their laws and regulations, access to their harbours and promote assistance for marine scientific research vessels which comply with the relevant provisions of this Part.

These provisions do of course apply principally to research vessels but may indirectly benefit AUVs associated with them.

17.2.2.7   The high seas

Freedom of scientific research is expressly included among the six freedoms of the high seas listed in Article 87(1) of UNCLOS and would be enjoyed by AUVs. Such freedom has to be exercised with 'due regard' for the interests of other States in their exercise of the freedom of the high seas. It is added that due regard must be had too 'to the rights under this Convention with respect to activities in the Area'.

17.2.2.8   The 'Area' beyond the limits of national jurisdiction

The 'Area' referred to is the area of the seabed and ocean floor and subsoil thereof lying seaward of the outer limit of the legal continental shelf. It is the area to which the seabed mining regime contained in Part XI of UNCLOS applies. Provision is made in UNCLOS for MSR to be conducted by the International Seabed Authority and by States. Although the scope of these provisions is in doubt, it would appear that there should be no serious impediment to MSR carried out by AUVs operated by entities other than the Authority in the Area and in the superjacent water column if it is conducted in compliance with Article 147. Under paragraph 1 of that article, seabed mining activities in the Area must be carried out with reasonable regard for other activities in the marine environment (which would include MSR by AUVs). Correspondingly, paragraph 3 demands that other activities (which would include MSR by AUVs) in the marine environment must be conducted with reasonable regard for seabed mining activities in the Area.

17.2.3 Rights of Neighbouring Land-locked and Geographically Disadvantaged States

UNCLOS grants a privileged position both to land-locked States (L-L States) and geographically disadvantaged States (G-D States). The L-L State is defined simply as 'a State which has no sea-coast'. G-D States are defined, in relation to their rights in the EEZ, as 'coastal States, including States bordering enclosed or semi-enclosed seas, whose geographical situation makes them dependent upon the exploitation of the living resources of the exclusive economic zones of other States in the subregion or region for adequate supplies of fish for the nutritional purposes of their populations or part thereof, and coastal States which can claim no exclusive zones of their own'. Such States enjoy certain rights to participate in MSR projects under Article 254 of UNCLOS and to benefit from technical co-operation and assistance in relation to MSR under Article 266, as indicated below:

Right to receive notice of proposed MSR projects   States and international organisations submitting proposals to coastal States to undertake MSR projects in their EEZs or on their continental shelf exclusively for peaceful purposes and in order to increase scientific knowledge of the marine environment for the benefit of mankind, are required to give notice to the neighbouring L-L and G-D States of the proposed research project. No time is specified for giving such notice other than is suggested by the fact that the 'States ... which have submitted ... a project ... shall give notice ...'

Right to receive information on project   Once consent has been granted to the researching State, it must provide to neighbouring L-L and G-D States, at their request and when appropriate, the same information as it is required to provide to the coastal State. The information, which must be provided not less than six months in advance of the expected starting date of the MSR project, consists of a full description of the project, including the methods and means to be used and a description of scientific equipment. Any major change in the research programme has also to be notified.

Right to participate in project   L-L and G-D States have to be given the opportunity to participate in the project but the right is a highly qualified one. The opportunity will be given at their request, but only 'whenever feasible'. The opportunity is to be enjoyed 'through qualified experts appointed by them and not objected to by the coastal State'. Finally, the opportunity to participate must be 'in accordance with the conditions agreed for the project, in conformity with the provisions of' UNCLOS, between the coastal State and the researching State.

Right to assistance in interpreting research   The researching State must provide L-L and G-D States at their request with an assessment of the project's data, samples and research results or provide assistance in their assessment or interpretation. However, this obligation is qualified by reference to the provisions of Article 249(2), which enables the coastal State in certain circumstances to require its prior agreement for making internationally available the research results of a project of direct significance for the exploration and exploitation of natural resources.

Qualified nature of rights of L-L and G-D States   Summing up, although it is true that these States are the beneficiaries of a number of rights in relation to (1) notice of; (2) information on; and (3) participation in MSR projects in the waters of neighbouring States, and are further entitled to (4) assistance in interpreting the results of such research, these rights are subject to a number of conditions which may allow the researching State to avoid the assumption of heavy burdens in certain circumstances. This is particularly so in relation to the right to participate in projects.

17.2.4   AUVs as 'Scientific Research Installations or Equipment in the Marine Environment'

There is some doubt as to whether or not an AUV is in law a 'ship'. If it is not, it will fall within the category of what UNCLOS refers to as 'scientific installations or equipment in the marine environment'. Both possibilities are considered here. The better view probably is that, as the law stands at present, an AUV is not a 'ship' and the implications of this supposition will be considered first.

In UNCLOS, specific provision for scientific research installations or equipment in the marine environment is made in Articles 258–262. The rules there laid down are, however, of a very general character and clearly need to be supplemented by more specialised conventional rules. In the present context it will suffice to refer to the rules on deployment and use; non-interference with shipping routes; and identification markings and warning signals.

17.2.4.1   Deployment and use

Article 258 simply provides that:

> The development and use of any type of scientific research installations or equipment in any area of the marine environment shall be subject to the same conditions as are prescribed in this Convention for the conduct of marine scientific research in any such area.

In other words, the same consent regime as that reviewed above applies to the deployment of MSR instrumentalities and no further requirements are laid down other than those referred to below. Given the novel features of AUVs, it seems unlikely that the deployment of such vehicles was contemplated when these rules were drafted and it may well be that further thought will have to be given to the rules in the context of the draft Convention on ODAS, which is referred to below, or in relation to the preparation of another specialised instrument on the use of AUVs.

17.2.4.2   Non-interference with shipping routes

Article 261 requires that, 'The deployment and use of any type of scientific research installations or equipment shall not constitute an obstacle to established international shipping routes'. Here too, AUVs may present special problems and further regulations may be necessary on such matters as navigation and collision avoidance systems and the degree to which the vehicle's tracks are to be pre-programmed or 'intelligently' variable.

17.2.4.3   Identification markings and warning signals

Article 262 of UNCLOS provides that:

> Installations or equipment referred to in this section shall bear identification markings indicating the State of registry or the international organisation to which they belong and shall have adequate internationally agreed warning signals to ensure safety at sea and the safety of air navigation, taking into account rules and standards established by competent international organisations.

Pending the establishment of a register or registers for ODAS or AUVs, there is, strictly speaking, no 'State of registry' but it would be in keeping with the spirit of this passage to require markings identifying the vehicle with a particular home State.

Article 262 indicates that account is to be taken of standards established by competent international organisations and here it is appropriate to refer to a draft Convention on ODAS and to Technical Annexes attached to it, prepared by the Intergovernmental Oceanographic Commission (IOC) and the IMO. Although the draft Convention is still only a draft, its Article 16 points in what is probably the right direction. It requires that a submarine ODAS which, due to the depth at which it is deployed, 'constitutes a danger to shipping or navigation or to fishing', must either be escorted by a vessel capable of giving due warning to passing ships, or be provided with effective signals as set forth in Annex II. This Annex II is one of three Technical Annexes to the draft Convention on ODAS which were published separately so that, pending the conclusion of a Convention, States could use them on a voluntary basis as guidelines for national measures. An updated version of Annex II on Marking and Signals was circulated in 1985. Annex II makes provision for both identification marks and lights and signals.

As regards 'identification and marking', this model would require:

(1)   The clear display on an exterior surface where it can best be seen of an alpha numerical identification consisting of a unique identification number prefixed by the letters ODAS (arguably AUV would serve equally well) and suffixed by letters indicating the State in which it is registered (in the absence of a registry, the home State might be indicated). A replica of the flag of the State may also be painted on or applied to the exterior surface as a further optional means of identification.

(2)   If feasible, the name and address of the owner should also be displayed.

(3)   Surface-penetrating AUVs would have 'their visible portions' (in practice their entire surface) painted yellow.

Reference must be made to Paragraph 2 of Technical Annex II for details of 'Lights and Signals' but the main requirements may be summarised as follows:

(1) The lights and signals have to be positioned in places where they can best be seen or heard.
(2) A satisfactory radar response at a distance of at least two miles must be ensured for AUVs constituting a danger to shipping and safe navigation.
(3) Surface-penetrating AUVs must exhibit from sunset to sunrise a yellow flashing light visible all round the horizon, with, where technically feasible, a nominal range of at least five miles.
(4) They must also carry a sound signal where its installation is technically feasible.

If the assumption made above proves to be wrong, and the view is taken that an AUV is a 'ship' in law, then the required identification marks and warning signals would be those provided for in the relevant international conventions, including:

UNCLOS Articles 90–93 refer to the national flag and Article 94 to the duties of flag States, including the duty to take measures necessary to ensure safety at sea. These rules do not refer to particular identification marks or warning signals.

The Collision Regulations The Rules apply to 'vessels', which include every description of water craft used or capable of being used as a means of transportation on water (Rules 1(a) and 3(a)). Part C specifies 'Lights and Shapes'. Rule 23 specifies lights for power-driven vessels underway. Compliance with the principal prescription in this rule, which includes a masthead light, would appear to be impracticable but vessels of less than seven metres in length and whose maximum speed does not exceed seven knots may instead exhibit 'an all-round white light' and, 'if practicable', sidelights. Rule 27 prescribes more complex requirements for 'vessels not under command or restricted in their ability to manoeuvre' (defined in Rule 3) but here too vessels of less than seven metres in length are not required to exhibit the lights prescribed in this Rule.

17.2.5 Settlement of Disputes

The MSR consent regime established by UNCLOS reflects a compromise between the demands of the scientific community for protection of scientific inquiry and the legitimate concern of the coastal State to protect its interest in the waters falling under its jurisdiction. As is the case with most rules of international law, the effective implementation of this regime depends to a large extent on the good faith of, and reasonably efficient performance of their duties by, both parties. Needless to say, some governments are less sympathetic to the needs of scientific research and less efficient than others and, as a result, disputes do arise over the treatment of MSR applications. Unfortunately, the UN Convention's provision for the settlement of such disputes leaves a great deal to be desired.

So far as AUVs are concerned, three scenarios may be envisaged:

(1) Where the AUV is deployed from and closely attended and supervised by a mother research vessel. In this case, the rules on dispute settlement, referred to below would apply.

(2)   Where the AUV (however launched) is not closely attended and supervised by
      a mother research vessel and it is operating 'autonomously'; and it is accepted
      by both sides that it is in law a ship. In this case, the 'nationality' of the AUV
      could be determined and the AUV would then be subject to the jurisdiction,
      and entitled to the diplomatic protection, of its home State. If then, a foreign
      State acted in breach of the rules of international law in relation to the AUV (by,
      e.g., denying it access to its waters to conduct MSR), the owners/operators of
      the AUV could refer the matter to the home State for dispute settlement on the
      international plane in accordance with the rules on dispute settlement referred
      to below.
(3)   Where the AUV (however launched) is not closely attended and supervised by a
      mother research vessel and it is operating 'autonomously'; and it is not consid-
      ered to be a ship in law. In this case (and assuming that the AUV has no other
      link of nationality through, for example, registration on a future ODAS registry),
      the position is more difficult. Arguably, the owner of the AUV could seek the
      diplomatic protection of the owner's home State, in which case dispute settle-
      ment could be sought as in scenario (2). However, given the uncertain juridical
      status of the AUV and doubts as to its entitlement to conduct MSR, the prospects
      of proceeding in this way would not be promising.

17.2.5.1   The rules on dispute settlement

Assuming that scenario (1) or scenario (2) applies, what are the relevant rules on dispute settlement? In fact, the rules of international law on dispute settlement are very
complex and would be a matter for the Government of the home State rather than
the owners/operators of the AUV. A new system of dispute settlement was introduced
by Part XV of UNCLOS and offers a variety of alternative mechanisms, including the
International Court of Justice, the new International Tribunal for the Law of the Sea,
and arbitration. The mechanisms available for settlement of any particular dispute
would depend upon the scope of declarations on 'choice of procedure' made by
the States parties concerned. That is the general position. It has to be noted, however, that the coastal State may take advantage of a very significant escape clause in
Article 297(2) of UNCLOS, whereby it may exclude the application of these mechanisms to MSR disputes. Where the coastal State exercises this option, the AUV home
State will be left without a binding dispute settlement procedure, though in certain
cases it may have to resort to a process of 'conciliation', the result of which is not
binding.

17.3   NATIONAL LAW AND STATE PRACTICE AND PROCEDURE

17.3.1   The Wide Range of National Law and Practice

If freedom to conduct MSR by AUVs is to be protected and promoted in accordance
with the UNCLOS regime, coastal States have to ensure that their municipal law
and/or their State practice gives full effect to it. The most obvious place to look for
evidence of a State's compliance with the UN regime is in its legislation. However, to

obtain a true picture of a State's policy, it is important to examine also the degree to which it implements its legislation in good faith and how in practice it deals with applications from foreign scientists to undertake MSR in its waters. Such an examination of actual practice is of course all the more important when the question is not covered by legislation at all.

It has to be remembered that neither international law nor the scientist is too concerned about the manner in which the State complies with its international legal obligations so long as, in fact, it does so. United Kingdom practice is instructive in this respect. Even before it acceded to UNCLOS in 1997, the United Kingdom fully complied with the UNCLOS regime without having any detailed statutory requirements. Foreign applicants were required simply to submit a standard application form at least three months in advance and were then assured that:

> Conditions, other than the basic requirement to provide details of the results of a cruise within twelve months and to invite the participation of a British observer, are only placed on any proposed research cruise in order to take care of practical circumstances (e.g. bottom trawling equipment should not be used in the vicinity of fixed gear or within a mile of cables on the seabed). Copies of cruise reports should be sent to the Foreign and Commonwealth Office within three months of the completion of the cruise (or an explanation of the delay given within that period).[2]

United States practice is somewhat similar. Permission is required for MSR by foreign scientists in the territorial sea, 'concerning the continental shelf and undertaken there', and for fishery research in the 200-mile EEZ if it involves taking fish in commercial quantities.[3] However,

> The United States has not established specific procedures for the submission of requests to conduct marine scientific research requiring U.S. permission. The Department of State informs other interested agencies of the U.S. Government of such requests and coordinates a response to them. Responses ordinarily require no more than thirty days. The U.S. expects the results of marine scientific research to be made freely available. The opportunity for U.S. scientists to participate in such research is encouraged but is not ordinarily a requirement for the clearance of research requests . . .[4]

---

2 See further, 'Communication from the Permanent Mission of the United Kingdom of Great Britain and Northern Ireland to the United Nations Dated 20 May 1988', in United Nations Office for Ocean Affairs and the Law of the Sea, *The Law of the Sea. National Legislation, Regulations and Supplementary Documents on Marine Scientific Research in Areas under National Jurisdiction*, Sales No. E.89.V.9, 1989 (hereafter referred to as 'OALAS, *Nat. Leg. on MSR*'), at p. 270.
3 See further, 'Note to the United Nations of 13 October 1988 on United States Practice Relating to Marine Scientific Research' (US Department of State), in OALOS, *Nat. Leg. on MSR*, at pp. 276–277.
4 *Ibid.*, at p. 277.

The United States' very liberal attitude to MSR in the EEZ is set out in a Presidential Statement which accompanied the proclamation of an EEZ in 1983:

> While international law provides for a right of jurisdiction over marine scientific research within such a zone, the proclamation does not assert this right. I have elected not to do so because of the United States interest in encouraging marine scientific research and avoiding any unnecessary burdens. The United States will nevertheless recognize the right of other coastal States to exercise jurisdiction over marine scientific research within 200 nautical miles of their coasts, if that jurisdiction is exercised reasonably in a manner consistent with international law.[5]

It is still important, however, that coastal States should give a clear indication of their policy either by promulgating legislation or, at least, issuing a statement explaining their policy and procedure. At the very least, this ensures that the Governments concerned find the time to consider and deal with the question. It also gives the foreign scientists a clear text on which to base their applications.

Unfortunately, a review of 'national legislation, regulations and supplementary documents' on MSR published by the United Nations in 1989 shows that there is still a long way to go in transforming the UNCLOS regime of MSR into national law and practice.[6] The survey, which covers the period from 1971 to 1989, includes material from 103 of some 140 coastal States. However, the great majority of instruments reproduced make only passing reference to MSR and attention will be focussed here on the 31 States which are referred to as 'States with detailed regulations or procedures for marine scientific research'.[7] These States are very unevenly spread across the five UN regional groups and no very clear pattern emerges from a comparative study of the 33 instruments which they have produced.

Of the three African States, Gabon and Senegal specify very simple conditions and procedures, while Ghana requires highly detailed applications and undertakings. Reference is made in the Ghanaian document to UNCLOS, which it follows in a very broad sense only.

There is considerable variety too among the five Asian States reviewed. For example, at one extreme, the Malaysian Act follows UNCLOS closely; at the other extreme, India has only a framework statute and Oman, apparently, only an application form. Japan specifies a straightforward, though detailed application procedure and indicates that consent will be granted if specified conditions, consistent with UNCLOS, are met.

In Eastern Europe, the former German Democratic Republic and the former Soviet Union produced relatively detailed legislation and Bulgaria a more simple Decree, all broadly based on UNCLOS.

Although several of the nine Latin American States reviewed adopted their legislation in the mid-1970s while the Third United Nations Conference on the Law of the Sea (which adopted UNCLOS) was still at work, there is less of a common pattern than might be expected, with only Venezuela following UNCLOS closely in its 1980

---

5 'Statement by the President of 10 March 1983 Accompanying Proclamation No. 5030 on the Exclusive Economic Zone', OALOS, *Nat. Leg. on MSR*, at p. 277.
6 OALOS, *Nat. Leg. on MSR*.
7 *Ibid.*, at p. 4, where an alphabetical list of such States is given.

Directive. In some cases there are clear departures from the UN regime. For example, under Argentinian legislation, applicants receiving no reply to an application within 60 days 'must regard it as rejected'. Again, Chile requires its authorities 'to ensure that, as far as possible, all or a significant part of the processing and analysis of the data and samples obtained during the research is carried out at a location in Chilean territory selected by it'. Some of the legislation of the other Latin American States is highly detailed and to some extent based on the language of UNCLOS (or its earlier negotiating texts) but no common model, systematically based on the Convention, has so far emerged.

Among the eleven members of the group of Western European States and Others included in the review, there is a substantial degree of conformity with the UNCLOS regime. In some cases (Denmark, Greece, Italy and the United Kingdom) simple procedures are outlined in notes provided by the Governments concerned and an indication is given (except in the case of Greece) that consent is normally granted. In others (Iceland, Portugal and Spain), the procedures are more formally incorporated in legislation.

Reviewing the collection of legislation as a whole, it is clear that, as of 1989, many States had still not produced legislation or procedures in line with the UNCLOS regime; that there was a need for others to develop embryonic legislation further; and that, in the interests of clarity and international uniformity, it was desirable that States should endeavour to follow a common pattern in their legislation and procedures. Fortunately, developments in this direction have been facilitated by the publication in 1991 of a Guide to the implementation of the relevant provisions of UNCLOS. Drafted in the UN Office for Ocean Affairs and the Law of the Sea, the Guide benefited from the advice of a group of technical experts on MSR which met in New York in September 1989. The Guide follows the same sequence as applies in practice and offers sound practical advice, based on the experience of officers responsible for operating research vessel clearance procedures, on the preparation, execution and follow-up of MSR expeditions. Draft standard forms are provided in Annexes to the Guide.

The 1989 UN collection is now of course somewhat dated and has to be supplemented from other sources. Frequently, provisions on MSR are included in more comprehensive national legislation. For example, provision is made for 'Marine scientific research by foreign ships' in Canada's Oceans Act 1996. Similarly, China has recently made brief reference to MSR in its EEZ and Continental Shelf Act 1998.

17.3.2 UK Practice and the Need for a Climate of Trust

Details of UK practice and procedure are to be found in the *Principal Scientists' Guidance Notes 1999*, issued by NERC's Research Ship Unit. These *Guidance Notes*, together with a copy of the Notification Form submitted in support of a recent application to the Netherlands for consent for MSR by *Autosub* in the North Sea, are reproduced and commented on in a Report by Brown and Gaskell (2000).[8]

---

8 The *Guidance Notes* are reproduced in part in Section 4.2.1 of the *Report* and the Notification Form in Appendix II.

If there is one lesson which emerges from a study of the UK experience in applying for MSR consents from foreign States in accordance with these *Guidance Notes*, it is that the obstacles to gaining consent are seldom the result of restrictive legislation but arise rather from either bureaucratic inefficiency or a negative attitude resulting from a failure to create a climate of trust between the two States concerned. It is hoped that the advice offered below, based on that study, will be of assistance to practitioners responsible for submitting applications for diplomatic clearance for foreign cruises.

17.3.2.1   Creating a climate of trust

It has to be emphasised that, under the modern international law of the sea, the coastal State may exercise jurisdiction over MSR out to 200 miles from its coast. Even though UNCLOS requires the coastal State to grant consent, in normal circumstances, for MSR in its EEZ or on its continental shelf, both commonsense and experience suggest that foreign States will be less likely to frustrate the achievement of freedom of research if a climate of trust has been created between the two States. It is clear that difficulties can be avoided and a climate of trust promoted if scientists co-operate with officials to ensure that the following guidelines are followed:

Full information on the research project is made available to the foreign State   This involves (1) prompt and careful completion of the Notification Form. Too often in the past forms have been completed with insufficient care. This may well cause delay and arouse suspicion needlessly; and (2) the preparation of a brief description in layman's language of the purpose of the project and its ultimate benefit to science. Embassy staff and officials in foreign States may not be familiar with scientific terminology. A brief, clear description can help them to process applications speedily.

Contact and enlist assistance of foreign scientists   While diplomatic clearance must be sought through official channels, collaboration with foreign colleagues both at the planning and cruise stages can help to ensure understanding of the project and a positive response.

Ensure that the civil, oceanographic-science nature of expeditions is clearly understood   by embassy staff and, through them, foreign officials. Any suspicion of military interest will often lead to a refusal of consent.

Encourage participation by foreign scientists   Where a mother ship is involved, participation of a foreign observer should be encouraged and not simply tolerated. Training, education and mutual assistance to developing country personnel should be promoted in relation to the work of the AUVs.

Trust promotion through visits and/or memoranda of understanding   Where extensive or important work is to be undertaken in the waters of a particular State, it may well prove to be cost-effective to organise a pre-cruise visit to the officials or institutions of the State concerned in order to establish personal relationships and channels of communication through which misunderstandings can be removed and last minute or emergency changes of plan negotiated. Again, bilateral framework agreements may do much to expedite the processing of particular applications.

Post-cruise reporting   It is essential, if trust is to be engendered and maintained, that scientists and officials should be precisely aware of the extent of their obligations to supply post-cruise reports to the foreign State/s concerned and should discharge these obligations scrupulously, in full and in good faith.

Inter-State reciprocity The principle of reciprocity operates here just as much as in other areas of diplomatic relations. For example, when, in 1991, the UK refused to grant permission for a research visit to UK waters by the Soviet Research Vessel Academician B. Petrov, reference was made to the Soviet refusal in 1985 to allow the United Kingdom to conduct MSR in the Soviet EEZ in order to trace the route of caesium discharged from Sellafield.

Most of these considerations are a matter of common sense. However, the scientist may often feel that there are better ways to spend his time than in form-filling and diplomacy and it is essential, therefore, that there are experienced officials and well designed and periodically reviewed procedures to ensure that consents may be sought in the context of a carefully nurtured climate of trust.

17.4 AUV OPERATIONS IN THE SOUTHERN OCEAN AND ANTARCTICA

AUV operations in the Southern Ocean and Antarctica are governed by rules of international law, principally laid down in the series of related treaties known as the Antarctic Treaty System, and by rules of national law, most of which are the implementation in national law of the provisions of the Antarctic Treaty System.

17.4.1 The International Law Regime

AUV operations in Antarctica are subject to the general rules of international law governing MSR and to the more particular rules embodied in the Antarctic Treaty System.

In principle, the MSR regime introduced by Part XIII of UNCLOS applies in the Southern Ocean and Antarctica. As has been seen, under this regime, the consent of the coastal State is normally required for MSR conducted in its offshore waters, though the 'qualified consent' rules require consent to be granted for MSR in the waters of the continental shelf and EEZ if specified conditions are satisfied. However, this general MSR regime is significantly modified by the Antarctic Treaty System and in particular by the central instrument of that system, the Antarctic Treaty, 1959. This treaty established freedom of scientific investigation as one of the fundamental principles of the Antarctic regime. The freedom is not of course an absolute one and has to be exercised in accordance with the provisions of the following instruments which make up the Antarctic Treaty System: the Antarctic Treaty 1959; the Convention on the Protection of Antarctic Seals, 1972; the Convention for the Conservation of Antarctic Marine Living Resources, 1980; and the Protocol on Environmental Protection to the Antarctic Treaty, 1991.

17.4.2 The National Law Regime

In practice, the operator of AUVs in Antarctica need not concern himself too much with the international law regime since AUV operations there will be governed by the rules of national law which have been established to implement international law. The precise form of that national law will of course differ from State to State but a good example is offered by the relevant British legislation.

The Antarctic Act 1994 and the Antarctic Regulations 1995 made under it apply to 'Antarctica', defined to include the continent of Antarctica, all islands south of 60° South latitude, all areas of continental shelf adjacent to that continent or those islands and south of that latitude, and all sea and air space south of that latitude. Any person on a British expedition to Antarctica and any British vessel entering Antarctica would be subject to the national law regime established by these instruments. The principal elements of the regime which may impact upon AUV operations may be summarised as follows:

The permits system   Provision is made in several sections of the Act requiring permits to be obtained for various activities in Antarctica. The number and type of permits required depend upon the nature of the activity in question.

Conservation of Antarctic fauna and flora   In the interests of the conservation of Antarctic fauna and flora, UK nationals are forbidden to do various things in Antarctica except under a permit granted under the Act or under a written authorisation from another State Party to the 1991 Protocol on Environmental Protection to the Antarctic Treaty. The forbidden activities include 'use of a vehicle, vessel or aircraft in a manner that disturbs a concentration of native mammals or native birds' and doing 'anything that is likely to cause significant damage to the habitat of any native mammal, bird, plant or invertebrate'.

Special areas   The Act makes provision for two types of special area, areas restricted under the 1991 Protocol and places protected under a 1980 Convention for the Conservation of Antarctic Marine Living Resources (CCAMLR). Permits are required for entry into such special areas.

Environmental evaluation   In applying an Environmental Impact Assessment procedure introduced by the 1991 Protocol, the UK's 1995 Antarctic Regulations distinguish between:

(1)   activities likely to have a negligible impact on the environment, for which neither an Initial Environmental Evaluation (IEE) nor a Comprehensive Environmental Evaluation (CEE) is required;
(2)   'activities likely to have more than a negligible impact', referred to in the Protocol as a 'minor or transitory impact'. For such activities an IEE will normally have to be submitted, though the Secretary of State can require a draft CEE; and
(3)   'activities likely to have more than a minor or transitory impact'. For this level of expected impact, a draft CEE must be submitted, followed later by a final CEE which has to take account of comments and advice from various sources.

Whether the AUV in question were to be classified as a 'vehicle' or a 'vessel', it would be subject to the above regime (including whatever permits conditions as might be imposed) if operated in Antarctica by any person on a British expedition or from a British vessel. This appears to be so even for operations in sectors of Antarctica claimed by other States. A case could be made for saying that the consent of the foreign State would be required under the normal UNCLOS rules governing MSR. However, in practice, diplomatic clearance has not been applied for in such cases and, so far as is known, this has not given rise to any problems, even in times of political difficulty between the United Kingdom and Latin American claimant States. The view seems to be taken that the principle of freedom of scientific investigation established by the Antarctic Treaty overrides the consent regime created by UNCLOS.

Finally, it may be useful to refer to the legal position of AUV operations under ice in this region. Although the definition of Antarctica in the UK's Antarctic Act 1994 includes all the ice shelves of the continent and of islands south of 60° South latitude, neither the Act nor the Regulations made under the Act impose any special constraints on operations under ice shelves. However, as noted above, permits are required for activities in Antarctica and environmental evaluations have to be carried out. It is always possible, therefore, that the need might become evident to impose restrictive permit conditions following submission of Initial or Comprehensive Environmental Evaluations.

REFERENCES

Aage, C. and Smitt, L.W. (1994). Hydrodynamic manoeuvrability data of a flatfish type AUV. *Proceedings of the IEEE Conference Oceans'94*, Brest, France.

Aage, C. (1997). Manoeuvring simulations and trials of a flatfish type AUV. Department of Naval Architecture and Offshore Engineering, Technical University of Denmark, Larsen, M.B., Department of Automation, Technical University of Denmark, OMAE'97, Vol. IB, Tokyo.

Abacus Technology Corporation (1993). Technology assessment of advanced energy storage systems for electric and hybrid vehicles. Report prepared for the US Department of Energy Office of Transportation Technologies, April.

Abu Sharkh, S.M., Harris, M.R. and Crowder, R.M. (1994). Comparative studies of electric and hydraulic drive systems for the thrusters of remotely operated vehicles. *Proceedings of Oceanology International Conference*, Brighton, UK, 8–11 March.

Abu Sharkh, S.M., Harris, M.R. and Stoll, R.L. (1995a). Design and performance of an integrated thruster motor. *IEE International Conference on Electrical Machines and Drives EMD'95*, pp. 395–400, Durham, UK.

Abu Sharkh, S.M., Turnock, S.R. and Draper, G. (2001). Performance of a tip-driven electric thruster for unmanned underwater vehicles. *Proceedings of the 11th International Offshore and Polar Engineering Conference (ISPOE 2001)*, pp. 321–324, Stavanger, Norway, June.

Abu Sharkh, S.M., Harris, M.R., Crowder, R.M., Chappell, P.H., Stoll, R.L. and Sykulski, J.K. (1995a). Design considerations for electric drives for the thrusters of unmanned underwater vehicles. *Sixth European Conference on Power Electronics and Applications*, Vol. 3, pp. 799–804. Sevilla, Spain, September.

Agarwal, B.D. and Broutman, L.J. (1980). *Analysis and Performance of Fibre Composites.* John Wiley and Sons, New York.

Ageev, M.D. *et al.* (1999). Results of the evaluation and testing of the solar powered AUV and its subsystems. *Proceedings of the 11th International Symposium on Unmanned Untethered Submersible Technology*, AUSI, New Hampshire.

Agoros, C. (1994). US Navy unmanned underwater vehicle navigation. *Proceedings of the IEEE Conference Oceans'94*, (3): 372–377.

Alles, S., Swick, C., Hoffman, M. *et al.* (1994). A real-time hardware-in-the-loop vehicle simulator for traction assist. *International Journal of Vehicle Design*, 15(6): 597–625.

Altshuler, T.W., Vaneck, T.W. and Bellingham, J.G. (1995). Odyssey IIb: towards commercialization of AUVs. *Sea Technology*, 36(12): 15–19.

Andrews, B. (2000) DSTO Australia – UUV directions. *Proceedings of the International Conference on UUVs*, pp. 27–30, Newport, Rhode Island, April.

Anderson, B. (2000). DSTO Australia – UUV directions. *Proceedings of the International Conference on UUVs*, pp. 27–30, Newport, Rhode Island, April.

Andrews, J.B. and Cummings, D.E. (1972). A design procedure for large hub propellers. *Journal of Ship Research*, 16(3): 167–173.

Aoki, T. and Shimura, T. (1997). Fuel cell for long range AUV. *Sea Technology*, 38(8): 69–73.

Aoki, T. (2001). Deep-water AUV 'Urashima'. *Hydro International*, 5(4): 6–9.

AUSI (1996). Analysis of a solar powered AUV – preliminary design calculations. Autonomous Undersea Systems Institute Technical Report, 9612–02.

Ayela, A., Bjerrum, A., Bruun, S., Pascoal, A., Pereira, F-L., Petzelt, C. and Pignon, J-P. (1995). Development of a self-organizing underwater vehicle – SOUV, MAST-Days and EUROMAR Market, Sorrento, Italy.

Ayers, J., Witting, J., Wilbur, C., Zavracky, P., McGruer, N. and Massa, D. (2000). Biomimetic robots for shallow water mine countermeasures. *Proceedings of the Autonomous Vehicles in Mine Countermeasures Symposium*, US Naval Postgraduate School.

Bacon, S., Centurioni, L.R. and Gould, W.J. (2001). The evaluation of salinity measurements from PALACE floats. *Journal of Atmospheric and Oceanic Technology*, 18(7): 1258–1266.

Bahm, R.J. (1994). Annual mean daily total global horizontal solar radiation, Version 1.0. R. J. Bahm and Associates, Internal document.

Baunsgaard, J.P. (2001). AUVs: getting them out of the water III. *International Ocean Systems*, 6(1): 22–23.

Bech, M. (1983). The Racal Decca pilot adaptive autopilot. The International Association of Science and Technology for Development, IASTED, Lyngby, Denmark.

Belgin, E. and Boulos, R. (1973). Functional dependence of torque coefficient of coaxial cylinders on gap width and Reynolds numbers. *Transactions of the ASME, Journal of Fluids Engineering*, 122–126, March, 1973.

Bellingham, J.G. (1993). Economic ocean survey capability with AUVs. *Sea Technology*, April: 34(4): 12–18.

Bellingham, J.G., Streitlien, K., Overland, J., Rajan, S., Stein, P., Stannard, J., Kirkwood, W. and Yoerger, D. (2000). An Arctic basin observational capability using AUVs. *Oceanography*, 13(2): 64–70.

Bickell, M.B. and Ruiz, C. (1967). *Pressure Vessel Design and Analysis*. MacMillan and Co, London.

Bjerrum, A. (1997). Autonomous underwater vehicles for offshore surveys. *Technologies for Remote Subsea Operations – Forum 1997*, Aberdeen.

Bjerrum, A., Larsen, M.B. and Symonds, G. (1999). MARPOS Doppler-inertial positioning system. *Hydrofest, Society of Underwater Technology*, Aberdeen.

Bodholt, H., Ness, H. and Solli, H. (1989). A new echo-sounder system. *Proceedings of the Institute of Acoustics*, 11:196–204.

Bourgeois, B. and Harris, M. (1995). The US Navy's first generation ORCA. *Sea Technology*, 36(11): 25–33.

Bowen, M.F. (1998). A passive capture latch for ODYSSEY-class AUVs. Woods Hole Oceanographic Institution Technical Report, WHOI-98-12, 41 pp.

Bowen, M.F. and Peters, D.B. (1998). A deep sea docking station for ODYSSEY class autonomous underwater vehicles. Woods Hole Oceanographic Institution Technical Report, WHOI-98-11, 73 pp.

Bowler, M.E. (1997). Flywheel energy storage systems: current status and future prospects. *Magnetic Material Producers Association Joint Users Conference*, September.

Bradley, A.M., Yoerger, D.R. and Walden, B.B. (1996). An AB(L)E bodied vehicle. *Oceanus*, 38(1): 18–20.

Bradley, A.M., Duester, A.R., Liberatore, S.P. and Yoerger, D.R. (2000). Extending the endurance of an operational scientific AUV using lithium-ion batteries. *Proceedings UUVs 2000*, pp. 149–157, Spearhead Exhibitions Ltd, New Malden, UK.

Bradley, D. (2001). Long range AUVs for extended oceanographic measurements. *Proceedings Oceanology International Americas*, Miami Beach, April, Spearhead Exhibitions Ltd, UK, unpaginated.

Brierley, A.S. and Watkins, J.L. (2000). Effects of sea ice cover on the swarming behaviour of Antarctic krill, *Euphausia superba. Canadian Journal of Fisheries and Aquatic Sciences*, 57 (Suppl 3): 24–30.

Brierley, A.S., Ward, P., Watkins, J.L. and Goss, C. (1998). Acoustic discrimination of Southern Ocean zooplankton. *Deep-Sea Research II*, 45: 1155–1173.

Brierley, A.S., Watkins, J.L., Goss, C., Wilkinson, M.T. and Everson, I. (1999). Acoustic estimates of krill density at South Georgia, 1981–1998. *CCAMLR Science*, 6: 47–57.

Brierley, A.S. *et al.* (2002). Elevated krill abundance at sea-ice edge revealed by autonomous underwater vehicle survey. *Science*, 295: 1890–1892.

Brighenti, A., Egeskov, P. and Reinhardt, U. (1998). EURODOCKER – a universal docking-downloading-recharging system for AUVs. *Proceedings Oceanology International 98: The Global Ocean*, Vol. 3, pp. 99–107, Brighton, UK, 10–13 March.

Brown, E.D. and Gaskell, N.J.J. (2000). Report on the law, state practice and procedure relating to autonomous underwater vehicles. 200 pp + Appendices, Society for Underwater Technology, London.

Brown, E.D. and Gaskell, N.J.J. (2001). Questions and answers on the law, state practice and procedure relating to autonomous underwater vehicles. 82 pp, Society for Underwater Technology, London.

Brutzman, D.P., Kanayama, Y. and Zyda, M.J. (1992). Integrated simulation for rapid development of autonomous underwater vehicles. *Proceedings IEEE Symposium on Autonomous Underwater Vehicle Technology*, June 4–5, 1992, pp. 3–10, Washington, D.C.

Bussell, J. *et al.* (1999). Applications of the METS methane sensor to the *in situ* determination of methane over a range of time-scales and environments. *EOS Transactions of the AGU 80*, F510 (abstract).

Byrne, R., Yao, W., Kaltenbacher, E. and Waterbury, R. (2000). Construction of a compact spectrofluorometer/spectrophotometer system using flexible liquid core waveguide. *Talanta*, 50: 1307–1312.

Caesar, G.J. (1982). *The Conquest of Gaul: Book IV.III.* Translated by Handford, S.A., revised by Gardner, J.F., Penguin Book, London, first published 58 BC.

Cancilliere, F.M. (1994). Advanced UUV technology. *Proceedings of the IEEE Conference Oceans'94*, 1: 147–151.

Carey, J., Harma, D.A. and Karpiski, A.P. (1992). High energy density batteries for undersea applications. *Proceedings of the Oceans'92 Conference on Mastering Oceans Through Technology*, pp. 871–876, Newport, Rhode Island.

Carlin, R. (2000). Future UUV power sources. *Proceedings of the International UUV Symposium*, pp. 245–255, Newport, USA.

Chance, T.C., Kleiner, A.S. and Northcutt, J.G. (2000). A high-resolution survey AUV. *Sea Technology*, 41(12): 10–14.

Chance, T.S. *et al.* (2000). *The Autonomous Underwater Vehicle: A Cost-effective Alternative to Deep-towed Technology*. Integrated Coastal Zone Management, ICG Publishing Ltd.

Chappell, S.G., Komerska, R.J., Peng, L. and Lu, Y. (1999). Cooperative AUV development concept (CADCON): An environment for high-level multiple AUV simulation. *Proceedings of the 11th International Symposium on Unmanned Untethered Submersible Technology*, AUSI, New Hampshire.

Chapuis, D., Deltheil, C. and Leandri, D. (1996). Determination and influence of the main parameters for the launch and recovery of an unmanned underwater vehicle from a submarine. *Proceedings Underwater Defence Technology*, pp. 334–338.

Chin, C.S., Johnson, K.S. and Coale, K.H. (1992). Spectrophotometric determination of dissolved manganese in natural waters with 1-(2-pyridylazo)-2-napthol: application to analysis *in situ* in hydrothermal plumes. *Marine Chemistry*, 37: 65–82.

Chiodi, A.M. and Eriksen, C.C. (2002). Geostrophic fjord exchange flow observed with seaglider autonomous vehicles. *Journal of Physical Oceanography*, submitted.

Christensen, P., Lauridsen, K. and Madsen, H.Ø. (1997). Failure diagnosis and analysis for an autonomous underwater vehicle. *Proceedings of the European Safety and Reliability Conference ESREL*, Lisbon.

Clayson, C.H. (2000). Sensing of nitrate concentration by UV absorption spectrophotometry. In *Chemical Sensors in Oceanography*, Varney, M. (ed.), Gordon and Breach Science Publishers, Amsterdam.

Coates, D., Fox, C., Ideker, D. and Repplinger, R. (1993). Advanced battery systems for electric vehicle applications. *Proceedings of the 26th International Symposium on Automotive Technology and Automation (ISATA)*, pp. 429–436, Aachen, Germany.

Cockburn, J. (1993). Performance of a sodium-sulphur battery system in a test bed autonomous underwater vehicle. *Power Sources 14: The 18th International Power Sources Symposium*, pp. 327–353, Stratford-upon-Avon.

Corfield, S.J. (1995). Naval uses of underwater robotics. *SUT Colloquium Abstracts – Potential of Robotic Systems in the Seas and Oceans*, London, UK.

Cornu, J.B. (1998). High energy batteries for EVs. A whole solution to clean up the environment. *IEE Colloquium on Electric Vehicles*, Digest Number 1998/262, London.

Cornu, J.P. (1989). High performance storage battery on board of unmanned untethered submersible vehicles. *International Symposium on Unmanned Untethered Submersible Technology*, pp. 216–235, Durham, New Hampshire, USA.

Costello, D., Carder, K., Chen, R., Peacock, T. and Nettles, N. (1995). Multi-spectral imagery, hyper-spectral radiometry, and unmanned underwater vehicles: tools for the assessment of natural resources in coastal waters. *SPIE Visual Communications and Image Processing'95*, 2501: 407–415.

Coudeville, J.M. and Thomas, H. (1998). A primer: using GPS underwater. *Sea Technology*, 39(4): 31–34.

Curtin, T.B. and Bellingham, J.G. (2001). Autonomous ocean-sampling networks. *IEEE Journal of Oceanic Engineering*, 26(4): 421–423.

Cybernetix (2001). See www.cybernetix.fr.

Davis, R.E., Webb, D.C., Regier, L.A. and Dufour, J. (1991). The autonomous lagrangian circulation explorer (ALACE). *Journal of Atmospheric and Oceanic Technology*, 9: 264–285.

Davis, R.E., Sherman, J.T. and Dufour, J. (2001). Profiling ALACEs and other advances in autonomous subsurface floats. *Journal of Atmospheric and Oceanic Technology*, 18: 982–993.

Dell, R.M. (2001). Batteries: fifty years of materials development. *Solid State Ionics*, 134: 139–158.

Descroix, J.P. and Chagnon, G. (1994). Comparison of advanced rechargeable batteries for autonomous underwater vehicles. *Symposium on Autonomous Underwater Vehicle Technology*, pp. 194–197, Cambridge, MA.

Dhanak, M.R. and Holappa, K. (1996). Ocean flow measurement using an autonomous underwater vehicle. *Proceedings of Oceanology International'96: The Global Ocean – Towards Operational Oceanography*, pp. 377–383, Spearhead Exhibitions Ltd., Kingston upon Thames, UK.

Dhanak, M. and Holappa, K. (1999). An autonomous ocean turbulence measurement platform. *Journal of Atmospheric and Ocean Technology*, 16: 1506–1518.

DoD (1998). High level architecture – federation development and execution process model (FEDEP). US Dept of Defence, DMSO, Version 1.2, May. (http://www.dmso.mil).

Dreher, J., Duger, H.J., Haas, B., Schriver, U. and Habitzer, G. (1993). Lithium–cobalt oxide battery with organic electrolyte solution. *Power Sources 14: The 18th International Power Sources Symposium*, pp. 95–105, Stratford-upon-Avon.

Dubelaar, G.B., Gerritzen, P.L., Beeker, A.E.R., Jonker, R.R. and Tangen, K. (1999). Design and first results of the CYTOBUOY: an autonomous flow cytometer with wireless data transfer for *in situ* analysis of marine and fresh waters. *Cytometry*, 37: 247–254.

Dunn, P. (2000). Navy UUV master plan. *Proceedings of the International UUV Symposium*, pp. 82–92, Newport, USA.

Dunn, S.E., Smith, S.M. and Betzer, P. (1994). Design of autonomous underwater vehicles for coastal oceanography. In *Underwater Robotic Vehicles: Design & Control*, Yuh, J. (ed.), pp. 299–326, TSI Press.

East, D.J. and Bagg, M.T. (1991). Extending the reach with AUVs. *Proceedings Defence Oceanology 91*, Brighton, UK.

Edmund, N. (1994). *The General Pattern of the Scientific Method*, 2nd edn, Edmund Scientific, Tonawanda, New York, ISBN 0-9632866-3-3.

Edwards, S.R.P. (2000). AUVs. How do you insure them? *International Ocean Systems Design*, 4(1): 10–12.

Egeskov, P., Bech, M., Aage, C. and Bowley, R. (1995). Pipeline inspection using an autonomous underwater vehicle. *14th Offshore Mechanical and Arctic Engineering Conference, OMAE'95*, Copenhagen, Denmark.

Egeskov, P., Bjerrum, A., Aage, C., Pascoal, A., Silvestre, C. and Smitt, L.W. (1994). Design, construction and hydrodynamic testing of the AUV MARIUS. *Proceedings of the 1994 Symposium on Autonomous Underwater Vehicle Technology, AUV'94*, IEEE Oceanic Engineering Society, Cambridge, USA.

Elwenspoek, M. and Wiegerink, R. (2001). *Mechanical Microsensors*. Springer-Verlag, Berlin, ISBN 3-540-67582-5.

Emblem, C., Robinson, J. and Holliday, J. (2000). Submarine launched mine reconnaissance system. *Proceedings International UUV Symposium*, pp. 12–18, Newport, USA.

Englemore, J. and Morgan, T. (1989). *Blackboard Systems*. Addison Wesley, New York.

Eriksen, C.C., Osse, T.J., Light, R.D., Wen, T., Lehman, T.W., Sabin, P.L., Ballard, J.W. and Chiodi, A.M. (2001). Seaglider: A long range autonomous underwater vehicle for oceanographic research. *IEEE Journal of Oceanic Engineering*, 26(4): 424–436.

Eureka (1998). Capacitors make assault on rechargeable batteries. *Eureka*, February: 32–33.

Evans, J.C., Smith, J.S. and Keller, K.M. (2000). ASOP (active sonar object prediction): short range positioning for the autonomous SWIMMER vehicle. *Proceedings of the Fifth European Conference on Underwater Acoustics (ECUA 2000)*, Chevret, P. and Zakharia, M.E. (eds), Lyon, France, July.

Evans, J.C., Keller, K.M. and Smith, J.S. (2001a). Docking techniques and evaluation trials of the swimmer AUV: an autonomous deployment AUV for workclass ROVs. *Proceedings of the Oceans 2001 MTS/IEEE International Conference*, Honolulu, Hawaii, November.

Evans, J.C., Keller, K.M. and Smith, J.S. (2001b). Docking techniques and evaluation trials of the swimmer AUV. *Proceedings of IFAC Conference Control Applications in Marine Systems (CAMS 2001)*, Glasgow, Scotland, July.

Farmer, D.M., Vagle, S. and Booth, A.D. (1998). A free flooding acoustical resonator for measurement of bubble size distributions. *Journal of Atmospheric and Oceanic Technology*, 15: 1132–1146.

Feezor, M.D., Sorrell, F.Y., Blankinship, P.R. and Bellingham, J.G. (2001). Autonomous underwater vehicle homing/docking via electromagnetic guidance. *IEEE Journal of Oceanic Engineering*, 26(4): 515–521.

Ferguson, J. and Bane, G. (1995). Dolphin – An operational remote minehunting system. *Proceedings of the 9th Undersea Defence Technology Conference*, July.

Ferguson, J. and Pope, A. (1995). 'Theseus': multipurpose Canadian AUV. *Sea Technology*, 36(4): 19–26.

Ferguson, J. (1998). The Theseus autonomous underwater vehicle – two successful missions. *Proceedings of the Underwater Technology '98*. pp. 109–114, IEEE, Piscataway, New York.

Fernandes, P.G. and Brierley, A.S. (1999). Using an autonomous underwater vehicle as a platform for mesoscale acoustic sampling in the marine environment. *ICES CM 1999/M:01*, 16 pp.

Fernandes, P.G., Brierley, A.S., Simmonds, E.J., Millard, N.W., McPhail, S.D., Armstrong, F., Stevenson, P. and Squires, M. (2000a). Fish do not avoid survey vessels. *Nature*, 404: 35–36.

Fernandes, P.G., Brierley, A.S., Simmonds, E.J., Millard, N.W., McPhail, S.D., Armstrong, F., Stevenson, P. and Squires, M. (2000b). Addendum to fish do not avoid survey vessels. *Nature*, 407: 152–152.

Fiebig, V. (1998). The Navy's unmanned undersea vehicle program. *Proceedings of AUVSI 1998*, pp. 99–126, Huntsville, USA.

Finch, M., Hydes, D.J., Clayson, C.H., Weigl, B., Dakin, J. and William, P. (1998). A low power ultra violet spectrophotometer for measurement of nitrate in seawater: introduction, calibration and initial sea trials. *Analytica Chemica Acta*, 377: 167–177.

Flanagen, R.C. *et. al.* (1990). Design of a flywheel surge power unit for electric vehicle drives. *Proceedings of the 25th Intersociety Energy Conversion Engineering Conference*, Vol. 4, pp. 211–217.

Fofnoff, N.P. and Millard Jr, R.C. (1991). Algorithms for computation of fundamental properties of seawater. Unesco Technical Paper in Marine Science, Paris, France.

Frangos, C. (1990). Control system analysis of a hardware-in-the-loop simulation. *IEEE Transactions on Aerospace and Electronic Systems*, 26(4).

Fries, D.P., Short, T., Langebrake, L.C., Patten, J.T., Kerr, M.L., Kibelka, G., Burwell, D.C. and Jalbert, J.C. (2001). In-water field analytical technology: underwater mass spectrometry, mobile robots, and remote intelligence for wide and local area chemical profiling. *Field Analyt Chem Technol*, 5: 121–130.

Fujimot, R.M. (1990). Parallel discrete event simulation. *Communications ACM*, 33(10): 30–53.

Gade, K. (1997). Integrering av treghetsnavigasjon i en autonom undervannsfarkost. FFI/RAPPORT-97/03179, Norwegian Defence Research Establishment (in Norwegian).

Gallett, I.N.L. (1999). $10 Oil: is underwater robotics an answer? Society for Underwater Technology, London, ISBN 0-906940-35-4, 19 pp.

Galletti, A.R., Brighenti, A., Knepper, S. and Mattiuzzo, F. (2000). EURODOCKER – From design to construction. *Proceedings of the International Unmanned Undersea Vehicles Symposium*, pp. 121–125, Newport, Rhode Island, 24–28 April.

Galletti, A.R., Brighenti, A., Zugno, L. and Mattiuzzo, F. (1998). The EURODOCKER MAST Project – Update on progress. *Proceedings of the IEEE Conference Oceans'98*, Nice.

Gamo, T., Sakai, H., Nakayama,, E., Ishida, K. and Kimoto, H. (1994). A submersible flow-through analyzer for *in situ* colorimetric measurement down to 2000 m depth in the ocean. *Analytical Sciences*, 10: 843–848.

Godø, O.R. (1998). What can technology offer the future fisheries scientist – possibilities for obtaining better estimates of stock abundance by direct observations. *Journal of Northwest Atlantic Fisheries Science*, 23: 105–131.

Gongwer, C.A. (1984). Static testing of underwater thrusters. *Proceedings of the ROV'84*.

Gongwer, C.A. (1989). Relationships for self propelled underwater vehicles in steady motion. *Proceedings of the Intervention'89 Conference*, pp. 217–226.

Graham, D. (1995). Composite pressure hulls for deep ocean submersibles. *Composite Structures*, 32: 331–343.

Graham, D. (1996). Buckling of thick-section composite pressure hulls. *Composite Structures*, 35: 5–20.

Green, K. and Wilson, J.C. (2001). Future power sources for mobile communications. *IEE Electronics and Communication Engineering Journal*, February: 43–47.

Griffiths, G., Millard, N.W., Pebody, M. and McPhail, S.D. (1997). The end of research ships? Autosub – an autonomous underwater vehicle for ocean science. *Proceedings of Underwater*

Technology International, pp. 349–362, April, Aberdeen, Society for Underwater Technology, London, ISBN 0-906940-30-3.

Griffiths, G., McPhail, S.D., Rogers, R. and Meldrum, D.T. (1998). Leaving and returning to harbour with an autonomous underwater vehicle. *Proceedings of the Oceanology International'98*, Brighton, Vol. 3, pp. 75–87, Spearhead Exhibitions Ltd, New Malden, ISBN 0-900254-23-8.

Griffiths, G., Stevenson, P., Webb, A.T., Millard, N.W., McPhail, S.D., Pebody, M. and Perrett, J.R. (1999). Open ocean operational experience with the Autosub-1 autonomous underwater vehicle. *Proceedings 11th Unmanned Untethered Submersible Technology Symposium*, pp. 1–12, Durham, New Hampshire, 23–25 August.

Griffiths, G., Fernandes, P.G., Brierley, A.S. and Voulgaris, G. (2000). Unescorted science missions with the Autosub AUV in the North Sea. *Proceedings of the International Conference on UUVs*, pp. 2–10, Newport, Rhode Island, April.

Griffiths, G., Knap, A. and Dickey, T. (2000). The autonomous vehicle validation experiment. *Sea Technology*, 41(2): 35–43.

Griffiths, G., Enoch, P. and Millard, N.W. (2001). On the radiated noise of the Autosub autonomous underwater vehicle. *Journal of Marine Science*, 58: 1195–1200.

Griffiths, G. *et al.* (2002). On the reliability of the Autosub autonomous underwater vehicle. *Underwater Technology*, in press.

Grosholz, E. and Jalbert, J.C. (1998). Bio-fouling experiments and results related to the development of a solar powered autonomous underwater vehicle. Autonomous Undersea Systems Institute Technical Report, 9807-01.

Grosjean, C. and Tai, Y.C. (1999). A thermoneumatic peristaltic micropump. *Proceedings of Tranducers'99*, 3P5.10.

Hagen, E., Størkersen, N. and Vestgård, K. (1999). HUGIN – Use of UUV technology in marine applications. *Proceedings of the Oceans'99*, Seattle, WA, USA.

Hamnet, A. (1999). Fuel cells – fuelling the future. *Proceedings of the Royal Institution of Great Britain*, 70: 201–217.

Harbaugh, D.L. (1994). The ARA sealed bipolar lead–acid battery development update. *International Symposium on Automotive Technology and Automation (ISATA)*, Germany.

Hart-Smith, L.J. (1992). A scientific approach to composite laminate strength prediction (Douglas Paper 8467, *Proceedings of the 10th ASTM Symposium on Composite Materials: Testing and Design, ASTM STP 1120*, Grimes, G. C. (ed.)).

Hart-Smith, L.J. (1996). A re-examination of the analysis of in-plane matrix failures in fibrous composite laminates. *Composites Science and Technology*, 56: 107–121.

Harvald, S.V. (1983). *Propulsion and Resistance of Ships.* John Wiley and Sons, New York.

Hashin, Z. (1980). Failure criteria for unidirectional fibre composites. *Journal Applied Mechanics*, 47: 329.

Hasvold, O. (1993). A magnesium-seawater power source for autonomous underwater vehicles. *Power Sources 14: The 18th International Power Sources Symposium*, pp. 243–255, Stratford-upon-Avon.

Hawley, J.G. and Reader, G.T. (1992). Advanced power systems for autonomous unmanned underwater vehicles. *Journal of the Society of Underwater Technology*, 18(1): 24–33.

Hearn, E.J. (1977). *Mechanics of Materials.* Pergamon Press, New York.

Heineman, W., Halsall, H., Cousino, M., Wijayawardhana, C., Purushothama, S., Kradtap, S., Schlueter, K., Choi, J.W., Ahn, C. and Henderson, H. (1999). Electrochemical immunoassay with microfluidic systems. *Proceedings of the PITTCON'99*, # 690.

Henriksen, L. (1994). Real-time underwater object detection based on an electrically scanned high-resolution sonar. *Proceedings of the 1994 Symposium on Autonomous Underwater Vehicle Technology*, AUV'94, IEEE Oceanic Engineering Society, Cambridge, USA.

Herman, A. (1992). Design and calibration of a new optical plankton counter capable of sizing small zooplankton. *Deep-Sea Research*, 39: 395–415.

Herrmann, W. and Vance, T. (1997). The internet at sea. *Sea Technology*, 38: 43–45.

Hillenbrand, C. and Negahdaripour, S. (2000). Vision based navigation. *Proceedings of the International UUV Symposium*, pp. 178–208, Newport, USA.

Hitchcock, G.L., Olson, D.B., Cavendish, S.L. and Kanitz, E.C. (1996). A GPS-tracked surface drifter with cellular telemetry capabilities. *Marine Technology Society Journal*, 30: 44–49.

Horner, R.E. (1996). The key factors in the design and construction of advanced flywheel energy storage systems and their application to improve the telecommunication power back-up. *IEE ITELEC Conference*.

Holt, J.K. and White, D.G. (1994). High efficiency, counter-rotating thruster for underwater vehicles. *Proceedings of the Symposium on Autonomous Underwater Vehicle Technology*, pp. 337–339, Cambridge, Massachusetts, USA.

Hopkin, D., Seto, M. and Watt, G. (1999). A fully interactive dynamic simulation of a semi-submersible towing a large towfish. *Oceans'99 MTS/IEEE Conference Proceedings*, Vol. 3, pp. 1194–1204.

Huang, B. *et al.* (2001). High energy density, thin film, rechargeable lithium batteries for marine field operations. *Journal of Power Sources*, 97–98: 674–676.

Hughes, A.W., Abu Sharkh, S.M. and Turnock, S.R. (2000). Design and testing of a novel electromagnetic tip driven thruster for underwater vehicles. *Proceedings of the 10th International Offshore and Polar Engineering Conference (ISOPE 2000)*, pp. 299–303, Seattle, June.

Hughes, A.W., Turnock, S.R. and Abu Sharkh, S.M. (2000). CFD modelling of a novel electromagnetic tip driven thruster for underwater vehicles. *Proceedings of the 10th International Offshore and Polar Engineering Conference (ISOPE 2000)*, pp. 294–298, Seattle, June.

Hull, D. (1981). *An Introduction to Composite Materials*. Cambridge University Press, ISBN 0-521-23991-5.

Hunter, J. (1999). Core Simulation Engine API Version 2.1. University of Essex Department of Computer Science Report.

Huyer, S.A. and Grant, J.R. (2000). Computation of two-body hydrodynamics using a Lagrangian vorticity method. *Proceedings of the International Unmanned Undersea Vehicles Symposium*, pp. 126–135, Newport, Rhode Island, 24–28 April.

International Energy Agency (IEA) (1993). Electric vehicles: technology, performance and potential.

IMO/MSC (1984). Revised Technical Annex II. MSC/Circ. 372 of 14 June.

IMO/MSC (1993). International Maritime Organisation/Marine Safety Committee. The United Kingdom Government, Formal Safety Assessment. MSC 62/24/3.

IMO/MSC (1997). International Maritime Organisation/Marine Safety Committee. Outline of the draft interim guidelines for the application of Formal Safety Assessment (FSA) to the IMO rule making process. MSC 68/14.

Iwanowski, M.D. (1994). Surveillance unmanned underwater vehicle. *Proceedings of the IEEE Conference Oceans*, 94(1): 116–119.

Jalbert, J.C., Iraqui-Pastor, P., Miles, S., Blidberg, D.R., James, D. and Ageev, M.D. (1997). Solar AUV technology evaluation program. *Proceedings of the 10th International Symposium on Unmanned Untethered Submersible Technology*, AUSI.

Jalving, B. (1999). Depth accuracy in seabed mapping with underwater vehicles. *Proceedings of the Oceans'99*, Seattle, WA, USA.

Jalving, B., Kristensen, J. and Størkersen, N. (1998). Program philosophy and software architecture for the HUGIN seabed surveying UUV. *Proceedings of the Cams 98*, Japan, November.

Jefferson, D.R. (1985). Virtual time. *ACM Transactions on Programming Languages and Systems*, 7(3): 404–425.

Johannessen, O.M., Muench, R.D. and Overland, J.E. (eds.) (1994). The polar oceans and their role in shaping the global environment. Geophysical Monographs 85, American Geophysical Union.

Johns, B. and Dyke, P.P.G. (1972). The structure of the residual flow in an offshore tidal stream. *Journal of Physical Oceanography*, 2: 73–79.

Jones, R. (1991). An improved fluorescence method for the determination of nanomolar concentrations of ammonia in natural waters. *Limnology and Oceanography*, 36: 814–819.

Kaddour, A.S., Soden, P.D. and Hinton, M.J. (1998). Failure of ±55 degree filament wound glass/epoxy composite tubes under biaxial compression. *Journal of Composite Materials*, 32(18): 1618–1645.

Kalcic, M. and Kaminsky, E. (1993). Field tests of the Dolphin/EM100 over Norfolk Canyon. *Proceedings of the Survey and Mapping Conference*, pp. 61–66, June.

Keron, I.W., Graham, D. and Farnworth, J. (2000). Self-designing structures – a new approach to naval structural problems. *Journal of Defence Science*, 5(4): 416–423.

Kloser, R.J. (1996). Improved precision of acoustic surveys of benthopelagic fish by means of a deep-towed transducer. *Journal of Marine Sciences*, 53: 407–413.

Kort, L. (1934). *Der neue Düsenschrauben-Antrieb*. Werft-Reederei-Hafen, Jahrgang 15.

Kovacs, G.T.A. (1998). *Micromachined Transducers Sourcebook*. McGraw-Hill, ISBN 0-07-290722-3.

Kristensen, J. and Vestgård, K. (1998). HUGIN – An untethered underwater vehicle for seabed surveying. *Proceedings of the Oceans 98*, Nice, France, September.

Küchmann, D. and Weber, J. (1953). *Aerodynamic of Propulsion*, 1st Edn. McGraw Hill, London.

Kuroda, Y., Aramaki, K. and Ura, T. (1996). AUV test using real/virtual synthetic world. *Proceedings 1996 IEEE Symposium on Autonomous Underwater Vehicle Technology*, pp. 365–372.

Lane, D.M., Falconer, G.J., Randall, G. *et al.* (1998). Mixing simulations and real subsystems for subsea robot development – specification and development of the core simulation engine. *IEEE International Conference Oceans'98*, Nice France, 29 September–2 October.

Lane, D.M., Smith, J., Evans, J., Standeven, J. and Edwards, I. (2000a). Achieving hardware in the loop interoperability in distributed undersea robot simulation. *Undersea Weapon Simulation-Based Design Workshop*, US Naval Undersea Warfare Centre, Rhode Island, USA, 7–9 June.

Lane, D.M., Trucco, E., Petillot, Y., Tena Ruiz, I., Lebart, K., Lots, J.F. and Plakas, C. (2000b). Embedded sonar and video processing for AUV applications. *Proceedings of the Offshore Technology Conference*, Houston, May.

Langebrake, L.C. *et al.* (1998). Development of sensors and sensor systems for AUVs/UUVs. *Proceedings of Oceanology International Conference*, Vol. 3, pp. 129–148, Brighton, Uk, 10–13 March.

Larsen, M.B. and Bjerrum, A. (2000). Navigation of underwater vehicles by synthetic long baseline positioning. *IEEE MTS Conference Oceans 2000*, Providence, Rhode Island, USA.

Lee, A., James, B.D., Kuhn, I.F. and Baum, G.N. (1989). A comparative analysis of electrochemical power sources for the DARPA UUV program. *International Symposium on Unmanned Untethered Submersible Technology*, pp. 168–188, Durham, New Hampshire.

Leivant, J.I. and Watro, R.J. (1993). Mathematical foundations for time warp systems. *ACM Transactions on Programming Languages and Systems*, 15(5): 771–794.

Lemoff, A.V., Lee, A.P., Miles, R.R. and McConaghy, C.F. (1999). An AC magnetohydrodynamic micropump: Towards a true integrated microfluidic system. *Proceedings of Transducer '99*, 4C1.2 (115).

Lewis, E.V. (1988). *Principles of Naval Architecture*, Vol. 2. Society of Naval Architects and Marine Engineers, Jersey, NJ.

van Lintel, H., van de Pol, F. and Bouwstra, S. (1988). A piezoelectric micropump based on micromachining of silicon. *Sensors & Actuators A*, 15: 153–167.

Loeb, V., Siegel, V., Holm-Hansen, O., Hewitt, R., Fraser, W., Trivelpiece, W. and Trivelpiece, S. (1997). Effects of sea-ice extent and krill or salp dominance on the Antarctic food web. *Nature*, 387: 897–900.

Lots, J.F., Lane, D.M. and Trucco, E. (2000). Application of 2-D visual servoing to underwater vehicle station-keeping. *Proceedings of the MTS/IEEE Oceans 2000*.

Lots, J.F., Lane, D.M. Trucco, E. and Chaumette, F. (2001). A 2D visual servoing technique for underwater vehicle station keeping. *Proceedings of the 2001 IEEE Conference on Robotics and Automation*, Seoul, Korea, 21–25 May.

Lueck, R.G. (1987). Microstructure measurements in a thermohaline staircase. *Deep Sea Research*, 34: 1677–1688.

Madsen, H.Ø. (1997). Mission management system for an autonomous underwater vehicle. *4th IFAC Conference on Manoeuvring and Control of Marine Craft*, Brijuni, Croatia.

Maclay, D. (1997). Simulation gets into the loop. *IEE Review*, May.

Madsen, H.O., Bjerrum, A. and Krogh, B. (1996). MARTIN – an AUV for offshore surveys. *Underwater Systems Design*, 18(3): 21–25.

Madsen, B. and Bjerrum, A. (2000). Autonomous underwater vehicles in the MCM role. *Proceedings of the International UUV Symposium*, pp. 39–45, Newport, USA.

Magurran, A.E. (1999). Dynamics of pelagic fish distribution and behaviour: effects on fisheries and stock assessment ix–x. In *Fishing News Books*, Freon, P. and Misund, O. (ed.), Oxford.

Manhattan (2001). See http://www.mhtx.com/novars/index.htm.

Marcoux, L. and Bruce, G. (1993). Development of a 100 Ah rechargeable Li-SO$_2$ cell. *Power Sources 14: The 18th International Power Sources Symposium*, pp. 81–94, Stratford-upon-Avon.

Marks, R.L., Wang, H.H., Lee, M.J. and Rock, S.M. (1994). Automatic visual station keeping of an underwater robot. *Proceedings of the MTS/IEEE Oceans'94*, 2: 137–142.

Masserini, R.T. and Fanning, K.A. (2000). A sensor package for the simultaneous determination of nanomolar concentrations of nitrite, nitrate, and ammonia in seawater by fluorescence detection. *Marine Chemistry*, 68: 323–333.

McDermott (2001). See www.jraymcdermott.com.

McFarlane, J.R. (1997). The AUV revolution: tomorrow is today! *Proceedings of Underwater Technology International, Aberdeen*, pp. 323–336, Society for Underwater Technology, London, UK, ISBN 0-906940-30-3.

McPhail, S.D. and Pebody, M. (1998). Navigation and control of an autonomous underwater vehicle using a distributed, networked, control architecture. *Underwater Technology*, 23(1): 19–30.

McVee, J.D. (1994). The axisymmetric deformation of anisotropic cylindrical shells. *Marine Structures*, 7: 257–305.

Meldrum, D.T. and Haddrell, T. (1994). GPS in Autonomous underwater vehicles, *Proceedings of the Electronic Engineering in Oceanography*, pp. 394–402, The Institution of Electrical Engineers, London, July.

Meyer, A.P. (1993). Proton exchange membrane fuel cell power system for unmanned underwater vehicles. *Proceedings of the 20th Annual AUVS Technical Symposium*, pp. 1078–1088, Washington DC.

Millard, N.W., Stevenson, P., McPhail, S.D., Perrett, J.R., Pebody, M., Webb, A.W., Meldrum, D.T. and Griffiths, G. (1997). Autonomous ocean data collection using the Autosub-1 AUV. *Proceedings of the Oceanology International Pacific Rim*, Singapore, Spearhead Exhibitions Ltd, Kingston upon Thames.

Millard, N.W., Griffiths, G., Finnegan, G., McPhail, S.D., Meldrum, D.T., Pebody, M., Perrett, J.R., Stevenson, P. and Webb, A.T. (1998). Versatile autonomous submersibles – the realising and testing of a practical vehicle. *Underwater Technology*, 23(1): 7–17.

Mudge, T.D. and Lueck, R.G. (1994). Digital signal processing to enhance oceanographic observations. *Journal of Atmospheric and Oceanic Technology*, 11: 825–836.

Myers, J.J., Holm, C.H. and McAllister, R.F. (1969). *Handbook of Ocean and Underwater Engineering.* McGraw Hill, ISBN 07-044245-2.

Nadis, S. (1997). Real time oceanography adapts to sea changes. *Science,* 275: 1881–1882.

Negahdaripour, S., Xu, X. and Jin, L. (1999). Direct estimation of motion from sea floor images for automatic station-keeping of submersible platforms. *IEEE Journal of Oceanic Engineering,* 24(3): 370–382.

Niiler, P. (2001). The world ocean surface circulation. In *Ocean Circulation and Climate,* Siedler, G., Church, J. and Gould, J. (eds), pp. 193–204, Academic Press, San Diego, CA. 712 pp.

Nielsen, P.F. and Ishøy, A. (1997). High-speed communications for underwater platforms. UV97, Paris.

Northcutt, J.G., Kleiner, A.A. and Chance, T.S. (2000). A high-resolution survey AUV. *Proceedings of the Offshore Technology Conference,* Houston.

NREL (1995). Photovoltaic energy program review. National Renewable Energy Laboratory, DOE/GO-10095-082.

Olivieri, M. and Glegg, S. (1998). Using the ambient noise sonar (ANS) to probe the ocean environment in shallow water. *Proceedings of Oceans'98,* IEEE.

Onwubiko, C. (2000). *Introduction to Engineering Design Optimization.* Prentice Hall, ISBN 0-201-47673-8.

Oosterveld, M.W.C. (1972). Ducted propeller systems suitable for tugs and pushboats. *Journal of International Shipbuilding Progress,* 19.

Oosterveld, M.W.C. (1973). Ducted propeller characteristics. *RINA Symposium on Ducted Propellers,* Paper No. 4, pp. 35–69.

Osborn, T.R. (1974). Vertical profiling of velocity microstructure, *Journal of Physical Oceanography,* 4: 109–115.

Osborn, T.R., Farmer, D.M., Vagle, S., Thorpe, S.A. and Cure, M. (1992). Measurements of bubble plumes and turbulence from a submarine. *Atmosphere-Ocean,* 30: 419–440.

Patrice, R. *et al.* (2001). Understanding the second electron discharge plateau in MnO_2-based alkaline cells. *Journal of the Electrochemical Society,* 148: 448–455.

Petillot, Y., Tena Ruiz, I. and Lane, D.M. (2001). Underwater vehicle obstacle avoidance and path planning using a multi-Beam forward looking sonar. *IEEE Journal Oceanic Engineering,* April: 240–251.

Podlaha, E.J. and Cheh, H.Y. (1994a). Modelling of cylindrical alkaline cells: VI variable discharge conditions. *Journal of the Electrochemical Society,* 141: 28–35.

Podlaha, E.J. and Cheh, H.Y. (1994b). Modelling of cylindrical alkaline cells: V high discharge rates. *Journal of the Electrochemical Society,* 141: 15–27.

Post, R.F., Fowler, K. and Post, S.F. (1993). A high-efficiency electromechanical battery. *Proceedings of the IEEE,* 81(3): 462–474.

PREST (1989). Operational constraints for the autonomous vehicles: DOLPHIN and DOGGIE. University of Manchester, available from Griffiths, G., SOC, Empress Dock, Southampton, UK.

Proud, N.J., Kellsall, D.R. and Alexander, T.M. (1996). A drive system for the PIROUETTE energy storage system. *IEE Conference on Power Electronics and Drives.*

Radojcic, D. (1997). Tip-driven propellers and impellers – A novel propulsion concept. *SNAME Propellers and Shafting'97 Symposium,* Virginia Beach.

Rai-Choudhury, P. (2001). *MEMS & MOEMS: Technology and Applications.* SPIE, ISBN 0-8194-3716-6.

Randall, J. (1990). AUV system studies NUW 73A/3111 – Legal implications of AUVs. Marconi Underwater Systems. 215 PC 0260, available from Mr Tonge, A., Bae Systems, Waterlooville, Hampshire, UK.

Rae, G.J.S. and Smith, S.M. (1992). A fuzzy rule based docking procedure for autonomous underwater vehicles. *Proceedings of the Oceans'92,* Vol. 2, pp. 539–546, IEEE, New York.

Ramamurti, R. and Sandberg, W.C. (2000). Computation of 2-body flow and forces during UUV retreival. *Proceedings of the International Unmanned Undersea Vehicles Symposium*, pp. 110–120, Newport, Rhode Island, 24–28 April.

Rao, B.M.L., Hoge, H.W., Zakrzewski, J. and Shah, S. (1989). Aluminium-seawater battery for undersea vehicles. *Sixth International Conferene on Unmanned Untethered Submersible Technology*, pp. 100–108, Durham, New Hampshire.

Raynal, M. and Singhal, M. (1996). Logical time: Capturing causality in distributed systems. *Computer*, February: 49–56.

Reddy, J.N. (1997). *Mechanics of laminated Composite Plates – Theory and Analysis*. CRC Press.

Reineke, M. (1993). Evaluating photovoltaic applications. Sandia National Laboratories Technical Report, SAND 93-0120.

Reinhardt, U. *et al.* (1998). Advanced concepts for AUV docking downloading and recharging. *Proceedings of the Third European Marine Science and Technology Conference*, pp. 1403–1410. Lisbon, 23–27 May. Vol. IV: Advanced system.

Remus (2001). See http://adcp.whoi.edu/REMUS/

Riser, S.C. and Swift, D.D. (2002). Long-term measurements of salinity from profiling floats. *Journal of Atmospheric and Oceanic Technology*, in press.

Rives, P. and Borrelly, J. (1997). Visual Servoing techniques applied to underwater vehicle. *Proceedings IEEE International Conference on Robotics and Automation*, Vol. 3, pp. 1851–1856, Albuquerque, NM, USA, April.

Ross, C.T.F. (1990). *Pressure Vessels under External Pressure*. Statics and Dynamics, Elsevier Applied Science, ISBN 1-85166-433-5.

Ruiz, I.T., Petillot, Y., Lane, D.M. and Salson, C. (2001). Feature extraction and data association for auv concurrent mapping and localisation. *Proceedings of the 2001 IEEE Conference on Robotics and Automation*, pp. 2785–2790, 21–25 May, Seoul, Korea.

Rutledge, A.K. and Leonard, D.S. (2001). Role of multibeam sonar in oil and gas exploration and development. *Proceedings of the Offshore Technology Conference*, Houston, May.

Samson, S., Langebrake, L.C., Lembke, C., Patten, J. and Russell, D. (2000). Design and current results of high-resolution shadowed image particle profiling and evaluation recorder. *Proceedings of the Oceanology International 2000*, Spearhead Exhibitions, New Malden.

Sasaya, T., Shibata, T. and Kawahara, N. (1999). In-pipe wireless micro robot. *Proceedings of Transducers'99*, 3C3.06.

Say, M.G. (1983). *Alternating Current Machines*, 5th edn, p. 23, Pittman, London.

Scamans, G.M., Stannard, J.H. and Dubois, R. (1994a). High energy aluminium oxygen power source for a 21″ diameter unmanned underwater vehicle. *Oceanology International Conference*, Brighton, UK.

Scamans, G.M., Creber, D.K., Stannard, J. H. and Tregenza, J.E. (1994b). Aluminium fuel cell power sources for long range unmanned underwater vehicles. *Symposium on Autonomous Underwater Vehicle Technology*, pp. 179–186, Cambridge, MA.

Schabmueller, C.G.J., Koch, M., Evans, A.G.R. and Brunnschweiler, A. (1999). Design and fabrication of a microfluidic circuitboard. *J. Micromech. Microeng.*, 9: 176–179.

Scott, R. (1995). Remote minehunting swims ahead. *Jane's Navy International*, May/June: 27–30.

Sherman, J., Davis, R.E., Owens, W.B. and Valdes, J. (2001). The autonomous underwater glider 'Spray.' *IEEE Journal of Oceanic Engineering*, 26(4): 437–446.

Short, R., Fries, D., Kerr, M.L., Toler, S., Lembke, C., Byrne, R. and Wenner, P. (2001). Underwater mass-spectrometers for *in situ* chemical analysis of the hydrosphere. *Am. Soc. Mass Spectrom.*, 12: 676–682.

Simmonds, E.J., Bailey, M.C., Toresen, R., Couperus, B., Pedersen, J., Reid, D.G., Fernandes, P.G. and Hammer, C. (1997). 1996 ICES co-ordinated acoustic survey of ICES divisions IIIa, IVa, IVb and VIa. ICES CM H: 11.

Singh, H., Bowen, M., Hover, F., Lebas, P. and Yoerger, D. (1997). Intelligent docking for an autonomous ocean sampling network. *Proceedings of the Oceans '97, MTS/IEEE Conference Proceedings*, Vol. 2, pp. 1126–1131, Halixfax, NS, Canada, 6–9 October.

Singh, H., Bellingham, J.G., Hover, F., Lerner, S., Moran, B.A., von der Heydt, K. and Yoerger, D. (2001). Docking for an autonomous ocean sampling network. *IEEE Journal of Oceanic Engineering*, 26(4): 498–514.

Smith, C.M. (1994). Implications of low-cost distributed control systems in UUV design. *Proceedings of the AUVS Conference*, Detroit, May.

Smith, C.M., Leonard, J.J., Bennett, A.A. and Shaw, C. (1997). Feature-based concurrent mapping and localization for autonomous underwater vehicles. *IEEE Oceans '97*.

Smith, C.S. (1990). *Design of Marine Structures in Composite Materials*. Elsevier Applied Science.

Smith, C.S. (1991). Design of submersible pressure hulls in composite materials. *Marine Structures*, 4: 141–182.

Smith, P.H., James, S.D., Chua, D.L. and Lin, H.-P.W. (1993). High energy density lithium rechargeable batteries for underwater propulsion. *Power Sources 14: The 18th International Power Sources Symposium*, p. 257, Stratford-upon-Avon.

Smith, P.H., James, S.D. and Keller, P.B. (1996). Development efforts in rechargeable batteries for underwater vehicles. *Symposium on AUV technology*, IEEE, pp. 441–447.

Smith, R., Self, M. and Cheeseman, P. (1990). Estimating uncertain spatial relationships in robotics. In *Autonomous Robot Vehicles*, Cox, I. and Wilfong, G. (eds), Springer-Verlag.

Smith, S.M., An, E., Kronen, E.A.D., Ganesan, K., Park, J. and Dunn, S.E. (1996). The development of autonomous underwater vehicle based survey and sampling capabilities for coastal exploration. *Proceedings Oceans '96*, Fort Lauderdale. IEEE, Piscataway. Supplementary Proceedings, pp. 30–35.

Sounders, H.E. (1957). *Hydrodynamics in Ship Design*. Vol II, SNAME, New York.

Sparenberg, J.A. (1984). *Elements of Hydrodynamic Propulsion*. Martinus Nijhoff Publishers, The Hague, Netherlands.

Stachiw, J.D. and Kurkchubasche, R.R. (1993). Ceramics show promise in deep sea submergence housings. *Sea Technology*, 34(12): 35–41.

Stansfield, K., Smeed, D.A., Gasparini, G.P., McPhail, S.D., Millard, N.W., Stevenson, P., Webb, A.T., Vetrano, A. and Rabe, B. (2001). Deep-sea, high-resolution, hydrography and current measurements using an autonomous underwater vehicle: The overflow from the Strait of Sicily. *Geophysical Research Letters*, 28: 2645–2648.

Steiger, D. (1992). Unmanned undersea vehicle technology development in the US. *Proceedings Underwater Defence Technology*, pp. 51–57.

Stevenson, P. (1993). The durability of composites to the deep ocean environment. *Proceedings of the 2nd International Conference on Deformation and Fracture of Composites*, pp. P20/1–10, Manchester, UK (The Institute of Materials).

Stevenson, P., Graham, D. and Clayson, C.H. (1997). The mechanical design and implementation of an autonomous submersible. *Underwater Technology*, 23(1): 31–41.

Stevenson, P. (2001). AUVs: getting them out of the water I. *International Ocean Systems*, 6(1): 12–16.

Stokey, R., Purcell, M., Forrester, N., Austin, T., Goldsborough, R., Allen, B. and von Alt, C. (1997). A docking system for REMUS, an autonomous underwater vehicle. *Proceedings of the Oceans '97, MTS/IEEE Conference Proceedings*, Vol. 2, pp. 1132–1136, Halixfax, NS, Canada, 6–9 October.

Stokey, R., Austin, T., von Alt, C., Purcell, M., Goldsborough, R., Forrester, N. and Allen, B. (1999). AUV bloopers or why Murphy must have been an optimist. *Proceedings of the 11th International Symposium on Unmanned Untethered Submersible Technology*, pp. 32–40, New Hampshire, AUSI, 23–25 August.

Stokey, R., Allen, B., Austin, T., Goldsborough, R., Forrester, N., Purcell, M. and von Alt, C. (2001). Enabling technologies for REMUS docking: an integral component of an autonomous ocean-sampling network. *IEEE Journal of Oceanic Engineering*, 26(4): 487–497.

Stommel, H.M. (1955). Discussions on the relationship between meteorology and oceanography. *Journal of Marine Research*, 14: 504–510.

Stommel, H. (1989). The Slocum Mission. *Oceanography*, April: 22–25.

Storkersen, N. and Indreeide, A. (1997). HUGIN – an untethered underwater vehicle system for cost-effective seabed surveying in deep waters. *Proceedings Underwater Technology International*, pp. 337–348, Aberdeen. Society for Underwater Technology, London, UK, ISBN 0-906940-30-3.

SUT (2000). *A code of Practice for the Operation of Autonomous Underwater Vehicles*. Dering, J. (ed.), Society for Underwater Technology, London, 48 pp.

Taylor, L. and McDermott, J.R. (2001). SAILARS revolution in Subsea intervention. *Proceedings Underwater Intervention 2001*, pp. 1–12, Tampa.

Terray, E., Fornwal, B., Voulgaris, G. and Trowbridge, J.H. (in preparation). AUV-borne near-bed flow measurements.

Thom, H. (1998). A review of the biaxial strength of fibre reinforced plastics. *Composites*, 29A: 869–886.

Thomas, R. (1986). The ARCS and DOLPHIN untethered remotely operated vehicles. In *Advances in Underwater Technology, Ocean Science and Offshore Engineering, Vol. 6, Oceanology*, pp. 113–118, Graham and Trotman, London.

Thorleifson, J.M. *et al.* (1997). The Theseus autonomous underwater vehicle. *Proceedings of the Oceans'97*, Vol. 2, pp. 1001–1006, MTS/IEEE, Piscataway, NY.

Thornton, R. and Weaver, D. (1998). The near-term mine reconnaissance system. *Proceedings Unmanned Underwater Showcase 1998*, pp. 1–9, Southampton, UK.

Thorpe, S.A., Ulloa, M.J., Baldwin, D. and Hall, A.J. (1998). An autonomously recording inverted echo sounder, AIRES II. *Journal of Atmospheric and Oceanic Technology*, 15: 1347–1361.

Thorpe, S.A., Osborn, T.R., Jackson, J.F.E., Hall, A.J. and Lueck, R.G. (2002). Measurements of turbulence in the upper ocean mixing layer using Autosub. *Journal of Physical Oceanography*, in press.

Tivey, M.A. (1996). From the beginning: monitoring change in a young lava flow. *Oceanus* 39(1): 21.

Tonge, A.M. (2000). 'Marlin' – A programme update. *Proceedings of the UUVS 2000*, pp. 143–148, Spearhead Exhibitions Ltd, New Malden, UK.

Tonge, A.M. (2000). Marlin: the UK military UUV programme – a programme overview. *Proceedings International UUV Symposium*, pp. 31–38, Newport, USA.

Tonge, A.M. and Cockburn, J. (1993). Sodium–sulphur batteries and test-bed AUV. *Sea Technology*, 34: 39–43.

Trimble, G. (1996). Autonomous operation of the explosive ordnance disposal robotic work package using the CETUS untethered underwater vehicle. *Proceedings of the IEEE Symposium on Autonomous Underwater Vehicle Technology*, pp. 21–27.

Trucco, E., Petillot, Y., Tena Ruiz, I., Plakas, C. and Lane, D.M. (2000). Feature tracking in video and sonar subsea sequences with applications. *Computer Vision and Image Understanding*, 79(1): 92–122.

Tsai, S.W. and Wu, E.M. (1971). A general theory of strength for anisotropic materials. *Journal of Composite Materials*, 5: 58–80.

UNESCO (1969). Legal problems associated with ocean data acquisition systems (ODAS). International Oceanographic Commission, Technical Series, No. 5.

Velegrakis, A., Voulgaris, G., Paphitis, D. and Collins, M.B. (in preparation). Bedform Distribution and Dynamics on a Linear Tidal Ridge, Broken Bank, North Sea.

Veron, F. and Melville, W.K. (1996). Pulse-to-pulse coherent Doppler measurements of waves and turbulence: laboratory and field test. *Proceedings of the ONR Workshop on Microstructure Sensors*, 23–25 October, Mt. Hood Oregon, Agrawal, Y., Williams III, A.J. and Goodman L. (eds).

Vestgård, K. (2001). AUVs: getting them out of the water II. *International Ocean Systems*, 6(1): 19–20.

Vestgård, K., Hansen, R., Jalving, B. and Pedersen, O.A. (2001). The HUGIN 3000 Survey AUV. *Proceedings of the ISOPE 2001*, Stavanger, 18–21 June.

Vestgård, K., Klepaker, R.A. and Storkersen, N. (1998). High resolution cost efficient seabed mapping with the HUGIN UUV. *Proceedings of the UUVS'98*, pp. 63–71, Spearhead Exhibitions Ltd, New Malden, UK.

Vincent, C.A. and Scrosati, B. (1997). *Modern Batteries*. Arnold, London, 351 pp.

Voulgaris, G. and Trowbridge, J.H. (1998). Evaluation performance of the acoustic Doppler velocimeter (ADV) for turbulence measurements. *Journal of Ocean and Atmospheric Technology*, 15: 272–289.

Voulgaris, G. Trowbridge, J.H. and Terray, E. (2001). Spatial variability of bottom turbulence over a linear sand ridge: mooring deployment and AUTOSUB AUV survey cruise report. RRS Challenger, Cruise Number 146 Broken Bank, North Sea, UK, 17th–28th August, Woods Hole Oceanographic Institution, Technical Report WHOI-2001-09, 41 pp.

Voulgaris, G., Trowbridge, J.H. and Terray, E. (in preparation). Tidal and Subtidal Flow Dynamics over a Linear Sand Ridge.

Wadhams, P. (2000). *Ice in the Ocean*. Gordon and Breach Science Publishers, Abingdon Oxfordshire, 368 pp.

Webb, D.C. and Simonetti, P.J. (1999). The SLOCUM AUV: An environmentally propelled underwater glider. *Proceedings of the 11th International Symposium on Unmanned Untethered Submersible Technology*, pp. 75–85, Published by the Autonomous Undersea Systems Institute, 23–25 August.

Webb, D.C., Simonetti, P.J. and Jones, C.P. (2001). Slocum, an underwater glider propelled by environmental energy. *IEEE Journal of Oceanic Engineering*, 26(4): 447–452.

Welsh, R., Redi, J., Pointer, S.A. and Davies, J. (2000). Advances in efficient submersible acoustic mobile networks. *Proceedings of the International UUV Symposium*, pp. 149–162, Newport, USA.

Wernli, R.L. (1999). AUV's – The maturity of the technology. *Proceedings of the Oceans'99*. IEEE-MTS.

Wernli, R.L. (2000). AUV commercialisation - Who's leading the pack? *Proceedings of the MTS/IEEE Oceans 2000*, Providence, RI, USA.

Wessinger, J. and Maass, D. (1968). Theory of ducted propellers; a review. *The 7th Symposium on Naval Hydrodynamics*, pp. 1209–1264, Office of Naval Research (ONR), Italy.

Willcox, S. (1999). The Bluefin AUVs. *Proceedings of the Conference on AUV Technology*, EK121, IBC Global Conferences Ltd, London, 7–8 December.

Williams, C.D., Bose, N., Butt, M.-A. and Seto, M.L. (2000). Mast/Snorkel performance of the Dolphin AUV. *Proceedings of the SNAME Conference 2000*, Vancouver.

Wilson, S. (2000). Launching the Argo armada. *Oceanus*, 42: 17–19.

Yentsch, C.M. (1990). Environmental health: flow cytometric methods to assess our water world. In *Methods in Cell Biology 33: Flow Cytometry*, Darzynkiewicz, Z. and Crissman H. A. (eds), pp. 575–612, Academic Press, San Diego.

Young, W.C. (1989). *Roark's Formulas for Stress and Strain*. McGraw-Hill.

Young, H.W. and Phillips, S.J. (2000). Development of an autonomous semi-submersible for deploying sensors for ocean survey. *Proceedings of the International UUV Symposium*, pp. 58–63, Newport, RI.

Zedel, L., Hay, A.E., Cabrera, R. and Lohrmann, A. (1996). Performance of a single beam pulse-to-pulse coherent Doppler profiler. *IEEE Journal of Oceanic Engineering*, 21: 290–299.

INDEX

331